Nonlinear Systems and Complexity

Series Editor
Albert C. J. Luo
Southern Illinois University
Edwardsville, IL, USA

For further volumes:
http://www.springer.com/series/11433

To Cheri, Gwendoline and Luc;
To Christelle, Orane and Alix

Preface

When thinking about complex systems, one may naturally have the notion of disorder in mind, but in many ways this would be a misnomer. Indeed, actual disorder is often simpler to describe from a statistical point of view. In this sense, the concept of complexity lies actually more in the region where disorder mingles with order, such that complex systems result from the difficulty to understand the global dynamics, the interplay between these two opposite regimes, and the role of their interconnections, providing more or less an antagonist duality. In such a case, these systems are "hard" to describe and understand and end up being called complex.

This feature of complex organization arises naturally in nonlinear physics. Let it be by a process of successive bifurcations generating complex spatiotemporal patterns, or in the mixed phase space of chaotic Hamiltonian system, where typically in phase space one can find a mixture of a chaotic sea and regions with regular motion that can lead to statistical distributions with power law and scale-free tails. With this perspective in mind, we co-organized the *Chaos, Complexity and Transport* conference which was held in Marseilles during the spring of 2011. It is actually during this conference that the idea to create something which would be more than simply collecting proceedings in a book germinated. The idea was to unite in one book established scientific figures who would expose in their own personal terms their research and as such would cover different facets and points of view on how to handle or deal with nonlinear and complex systems. The title of this book was decided with this goal. In order to avoid the confusion of roles, we kept the spirit of the conference and contributing authors were chosen among the invited speakers of the conference, what de facto excluded the organizers. The authors had a *carte blanche* for their chapter, with the requirement that they would be the sole author. The result of these contributions ended up in various fields of nonlinear dynamics, Hamiltonian chaos and complex systems, and different perspectives from experimental one to the theoretical and mathematical one. We have organized them in a way which we thought reflected more or less the chronology evoked in the title *From Hamiltonian Chaos to Complex Systems* and grouped them in four different parts.

The first part concerns chaos and dynamical systems per se, and the first chapter starts with discussing the notion of weak chaos, and how this may generate problems dealing with infinite ergodic theory and anomalous transport. This chapter sets the tone of the book and not only introduces different notions such as continuous time random walks and fractional derivatives but also encompasses applications to biology while starting with the study of maps inspired from nonlinear fluid dynamics such as the Pomeau–Manneville one. The second chapter considers in the opposite a very specific and somewhat simple example of a Hamiltonian system with one and a half degree of freedom: a perturbed pendulum. It offers de facto a very pedagogical exercise of Hamiltonian chaos in the context of adiabatic theory. But beyond that, it displays a surprising new ratchet mechanism. Indeed, thorough analytic and numerical analysis is performed and transport properties are shown to display a ratchet effect and the generic raise of a net current, even though the time average of the acting perturbation force is zero.

In the second part, we have three chapters where we move to systems with large numbers of degree of freedom and in particular the case of hot plasmas is considered. In this context a kinetic approach is often necessary, as resonant interactions between particles or particles and fields are at play, and this kinetic theory bridges the gap between individual chaotic particle motion and collective effects. In the first chapter, the influence of nonlinear vibration of electrons is considered; most notably the collisionless dissipation of Landau damping is considered even in the strong nonlinear regime. This allows predictions on stimulated Raman scattering in the context of inertial fusion. A thorough comparison of the obtained results with a large full kinetic code is then performed, confirming the validity of the theory developed therein. The second contribution is dealing as well with hot fusion plasmas but in the context of magnetized confinement. As the author mentions himself, his contribution consists of two parts. In the first one, he presents his perspective on the *complexity* of doing research itself, such as retrieving valid information and relying on rigorous facts. Some diagnostics are made and then possible cures and improvements are proposed. In the second part, the author describes the interactions of electrons with Langmuir waves, in the context of a self-consistent Hamiltonian framework. The kinetic limit in terms of a Vlasov equation is then considered, and within the framework of quasi-linear theory, the regime leading to the destruction of the self-consistency is described. Finally, the last chapter deals with tokamak fusion plasmas. The problem of turbulence leading to troubles in confinement is discussed. Plasma models are introduced and the problem of closure to obtain fluid equations is presented. The solution to use adiabatic invariant of the microscopic motion of ions in order to derive a simplified kinetic equation, dubbed gyro-kinetics, and some of its latest results are presented. Finally, in this chapter, we get a summary of the state of the art considering transport issues related to magnetic confinement fusion plasmas. Also by introducing closure problems and fluid equations we are bridging the gap with the next part of the book.

In this third part, we deal with macroscopic nonlinear systems. In the first chapter, we continue with turbulence and heat problems. The experimental study of such phenomena in the fluid dynamics of soap bubbles allows to reveal the emergence of large vortices due to this turbulent heat convection. These vortices are studied in detail and shown to exhibit similarities with cyclones and large hurricanes which live on totally different length scales. Furthermore different behaviors from intermittent to Bogliano–Obukhov scaling depending on the imposed temperature gradients are found. Moreover, this work lays new perspectives for the analysis of transport induced by these large coherent vortices, for instance, the problematic of anomalous transport and Lévy flights. In the next chapter, we deal with solids and buckling of elastic sheets. In this situation as well, complexity arises, as energy can be focused/defocused, and creates complex patterns of singularities. Conversely to the expected behavior of having singularities only as a last resort when crumpling thin elastic sheets, it is shown that in some configurations and compression routes the rise of singularities corresponds merely to a transient necessary behavior inducing a change of topology, but that for high compressions a stress defocusing phenomena rises leading to the disappearance of the singularities which were concentrating most of the energy. The last chapter of this part of the book brings together solids and liquids; it does so in the framework of quantum mechanics. We somehow not only return to the probabilistic approach of complex systems already envisioned in the second part of the book and its kinetic descriptions but also anticipate the last part of the book for which stochasticity, noise, and statistics become prominent players. To be more precise a model of superfluidity, namely the one which is governed by the nonlinear Schrödinger equation is presented. The nucleation of quantized vortices and the possibility to describe a nonclassical behavior of the rotational inertia, giving rise to a super-solid phase are discussed.

Finally, in the last part of the book, we turn to stochastic behavior and complex systems beyond the physical realm. In the first chapter, we deal with the role of fluctuations in population dynamics. The influence of these fluctuations can alter the usual picture we get when thinking only in mean-field deterministic terms. The approach is described using the van Kampen system size expansion and applied in a pedagogical manner to autocatalytic reactions systems. Then in the last chapter of the book, we learn how statistical mechanics techniques can be used to tackle complex systems and in particular how traffic inference can be tackled using the Ising model. We start to learn about the belief propagation algorithm and then see how by using mean-field techniques and linear response theory and coupling it to machine learning techniques we can address problems related to traffic.

Before ending this preface, we are glad to follow a customarily tradition and close it with *acknowledgments*. It is simply true that we as editor only acted as instruments and glue and that this book's content belongs to its authors. It is then obvious that we would like to thank equally all the authors who agreed to contribute to this volume. We hope the result lives up to their expectation. We also hope that readers will enjoy as much as we did, the discovering of new ideas and the learning of new perspectives while reading each individual contribution. And last but not least, it is our great pleasure to thank Prof. A. Luo, who suggested the possibility to seize the opportunity of the conference to create something beyond simple proceedings.

Finally, we would like to thank the publisher for their patience and help during the assembling of this book, which resulted as a complex and chaotic intermittent process.

Marseilles, France Xavier Leoncini
Marseilles, France Marc Leonetti

Contents

Part III From Kinetics to Fluids and Solids

List of Contributors

Didier Bénisti
CEA, DAM, DIF, Arpajon, France

Dominique F. Escande
UMR 7345 CNRS-Aix-Marseille-Université, Marseille cedex 20, France

Duccio Fanelli
Dipartimento di Energetica, University of Florence, Florence, Italy

Cyril Furtlehner
INRIA Saclay – LRI, Université Paris-Sud, Orsay, France

Christophe Josserand
Institut D'Alembert, CNRS & UPMC (Université Paris 6), Paris, France

Hamid Kellay
Université Bordeaux1, LOMA UMR 5798 du CNRS, Talence, France

Rainer Klages
School of Mathematical Sciences, Queen Mary University of London, London, UK

Alain Pocheau
IRPHE, Aix-Marseille Université, Marseille Cedex 13, France

Yanick Sarazin
CEA, IRFM, Saint-Paul-Lez-Durance, France

Alexei Vasiliev
Space Research Institute, Moscow, Russia

Part I
Low Dimensional Chaos

Chapter 1
Weak Chaos, Infinite Ergodic Theory, and Anomalous Dynamics

Rainer Klages

Abstract This book chapter introduces to the concept of weak chaos, aspects of its ergodic theory description, and properties of the anomalous dynamics associated with it. In the first half of the chapter we study simple one-dimensional deterministic maps, in the second half basic stochastic models, and eventually an experiment. We start by reminding the reader of fundamental chaos quantities and their relation to each other, exemplified by the paradigmatic Bernoulli shift. Using the intermittent Pomeau–Manneville map the problem of weak chaos and infinite ergodic theory is outlined, defining a very recent mathematical field of research. Considering a spatially extended version of the Pomeau–Manneville map leads us to the phenomenon of anomalous diffusion. This problem will be discussed by applying stochastic continuous time random walk theory and by deriving a fractional diffusion equation. Another important topic within modern nonequilibrium statistical physics are fluctuation relations, which we investigate for anomalous dynamics. The chapter concludes by showing the importance of anomalous dynamics for understanding experimental results on biological cell migration.

1.1 Introduction

Deterministic dynamical systems involving only a few variables can exhibit *complexity* reminiscent of many-particle systems if the dynamics is *chaotic*, as is quantified by the existence of a positive Lyapunov exponent [1–4]. Such systems, which may be called *small* because of their small number of degrees of freedom [5], can display an intricate interplay between nonlinear microscopic dynamical properties and macroscopic statistical behavior leading to highly nontrivial fluctuations

R. Klages (✉)
School of Mathematical Sciences, Queen Mary University of London,
Mile End Road, London E1 4NS, UK
e-mail: r.klages@qmul.ac.uk

X. Leoncini and M. Leonetti (eds.), *From Hamiltonian Chaos to Complex Systems: A Nonlinear Physics Approach*, Nonlinear Systems and Complexity 5,
DOI 10.1007/978-1-4614-6962-9_1,

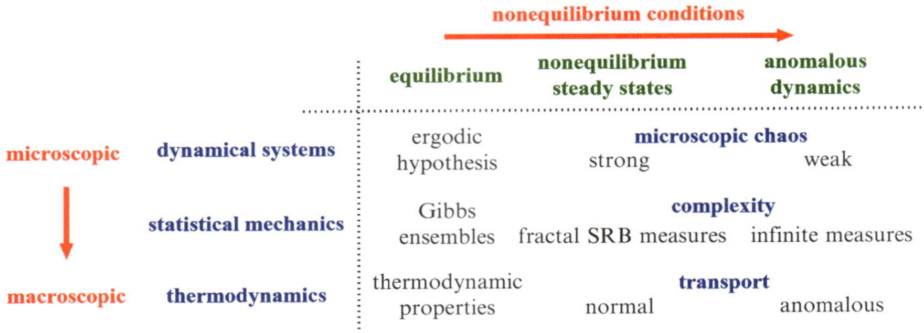

Fig. 1.1 Conceptual foundations of a theory of nonequilibrium statistical physics based on dynamical systems theory by motivating the topic of this book chapter, which is represented by the third column

of physical observables. This becomes particularly interesting in nonequilibrium situations when these systems are exposed to external gradients or fields. Despite their complexity, examples of these systems are still amenable to detailed analysis by means of dynamical systems theory in combination with stochastic theory. Hence, they provide important paradigms to construct a theory of nonequilibrium statistical physics from first principles: Based on the chaotic hypothesis, which generalizes Boltzmann's ergodic hypothesis, SRB measures were studied as nonequilibrium equivalents of the Gibbs ensembles of equilibrium statistical mechanics. This novel approach led to the discovery of fundamental relations characterizing nonequilibrium transport in terms of microscopic chaos [6–9], such as formulas expressing transport coefficients in terms of Lyapunov exponents and dynamical entropies, equations relating the nonequilibrium entropy production to the fractality of SRB measures, and fluctuation relations, which are now widely studied as a fundamental property of nonequilibrium processes [5, 8, 10, 11].

The interplay between these different levels of description in modern nonequilibrium statistical mechanics is illustrated by the second column in Fig. 1.1, in analogy to the theory of equilibrium statistical mechanics sketched in the first column. As is represented by the third column, however, more recently scientists learned that random-looking evolution in time and space also occurs under conditions that are weaker than requiring a positive Lyapunov exponent [12, 13]. It is now known that there is a wealth of systems exhibiting zero Lyapunov exponents, meaning that the separation of nearby trajectories is weaker than exponential. This class of dynamical systems is called *weakly chaotic*. Examples include maps with indifferent fixed points, polygonal particle billiards, and Hamiltonian systems with sticky islands in phase space [8, 12–14].

Weakly chaotic systems exhibit *anomalous dynamics* characterized by novel properties such as ageing, which reflects an extremely weak relaxation towards equilibrium involving more than one time scale in the decay of correlations. Other surprising properties are the existence of Lévy-type probability distributions obeying generalized central limit theorems [15, 16] and the non-equivalence of time and ensemble averages, called weak ergodicity breaking [17]. These physical phenomena were observed experimentally in a wide variety of systems, such as in the anomalous statistics of blinking quantum dots, in the anomalous diffusion of atoms in optical lattices, in plasma physics, and even in cell and animal migration [14, 17–19].

Recent work in ergodic theory, on the other hand, has led to mathematically rigorous results about some of the physically relevant phenomena mentioned above. It turned out that there is an intimate connection between the mechanism generating weakly chaotic dynamics and the existence of non-normalizable, so-called *infinite invariant measures* [12, 20, 21]. The ergodic theory of generalized random walks driven by weak chaos and of other systems exhibiting infinite invariant measures, which is called *infinite ergodic theory*, has thus the potential of providing a sound mathematical basis for some of the physical phenomena displayed by anomalous dynamics.

This book chapter gives a brief introduction to important aspects of the above topics in four sections: As a warm-up, the beginning of Sect. 1.2 briefly reminds us of the concept of deterministic chaos in simple dynamical systems as quantified by a positive Lyapunov exponent. On this basis, we will introduce to the phenomenon of weak chaos, and the idea of infinite ergodic theory will be outlined. The chapter concludes by putting different forms of deterministic chaos into perspective. Section 1.3 relates these concepts and ideas to the problem of anomalous diffusion in deterministic systems. Here we make a transition to stochastic theory by studying these systems also from a stochastic point of view. For this purpose we use a generalization of ordinary random walk theory, called continuous time random walk (CTRW) theory. In a scaling limit, this theory leads to generalized diffusion equations involving fractional derivatives. Section 1.4 introduces to the topic of fluctuation relations, which generalize the Second Law of Thermodynamics and other fundamental thermodynamic relations to small systems far away from equilibrium. After discussing transient fluctuation relations (TFRs) for a very basic type of stochastic dynamics as an example, we explore the validity of such relations for generalizations of this dynamics yielding anomalous diffusion. In Sect. 1.5 we relate this line of theoretical reasoning about anomalous dynamics to biophysical reality by studying the case of biological cell migration. After briefly introducing to the problem of cell migration, we report experimental results on fundamental statistical physical properties of migrating cells, extracted from statistical data analysis. We conclude this section with a stochastic modeling of these experimental results by using a generalized, fractional Fokker-Planck type equation. We summarize our discussion of this book chapter in the final Sect. 1.6.

The title of this review is inspired by a conference that the author had the pleasure to organize together with R. Zweimüller, E. Barkai, and H. Kantz at the Max Planck Institute for the Physics of Complex Systems, Dresden, in Summer 2011, which bears exactly the same title [22]. However, naturally this chapter reflects the author's very personal take on this topic and his own research. The subsequent second section is to some extent based on the review [23] by combining it with ideas from [8, 24]. The third section builds on [25, 26]. The fourth section incorporates material from the review [27] and from [28], the fifth one draws on [29].

1.2 Chaos and Anomalous Dynamics

In this section we focus on purely deterministic dynamics modeled by two simple but paradigmatic one-dimensional maps: the famous Bernoulli shift, as a model for strong chaos characterized by a positive Lyapunov exponent, and the Pomeau–Manneville map, as an example exhibiting weak chaos with zero Lyapunov exponent. We start by briefly reminding the reader of basic concepts of dynamical systems theory and ergodic theory such as Lyapunov exponents, ergodicity, SRB measures, and Pesin's theorem, illustrated for the Bernoulli shift. Reference [23] provides a more tutorial exposition of most of these ideas. By switching to the Pomeau–Manneville map we find that generalizations of these concepts are needed in order to describe the model's weakly chaotic dynamics. This motivates the mathematical problem of infinite ergodic theory, which is intimately related to defining suitably generalized chaos quantities assessing weak chaos, and a generalization of Pesin's theorem. In the final part of this chapter we propose a generalized hierarchy of chaos, based on the existence of different types of stretching between two nearby trajectories, which we use to characterize chaotic dynamics.

1.2.1 Deterministic Chaos in a Simple Map

The main vehicle of our approach in this and the next section are one-dimensional time-discrete maps $F: J \to J$, $J \subseteq \mathbb{R}$ obeying

$$x_{n+1} = F(x_n) \, , \, n \in \mathbb{N}_0 \, , \tag{1.1}$$

which defines the equations of motion of our deterministic dynamical systems. For a given initial condition x_0 we have $x_n = F^n(x_0)$. A particularly simple example of F are piecewise linear maps, such as the paradigmatic Bernoulli shift [1, 2, 4, 6]

$$B: [0,1) \to [0,1) \, , \, B(x) := 2x \bmod 1 \; = \begin{cases} 2x \, , \; 0 \le x < 1/2 \\ 2x - 1 \, , \; 1/2 \le x < 1 \end{cases} \tag{1.2}$$

depicted in Fig. 1.2. This simple system exhibits a very complicated dynamics governed by sensitivity to initial conditions, as can be quantified by calculating its Lyapunov exponent [2, 30]: Consider two points that are initially displaced from

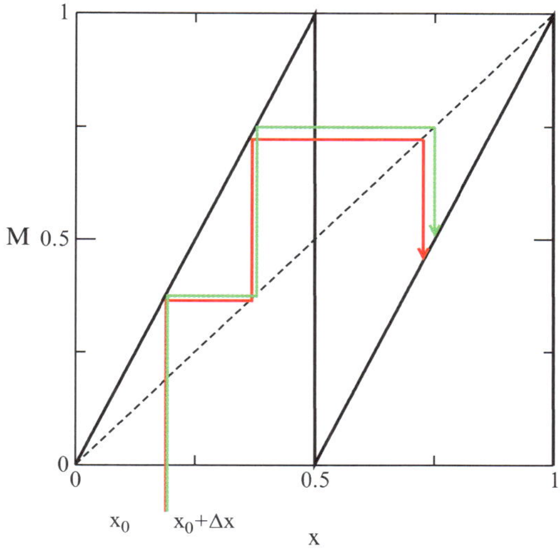

Fig. 1.2 The Bernoulli shift Eq. (1.2) and two trajectories starting from two nearby initial conditions x_0 and $x_0' = x_0 + \Delta x_0$ displaced by $\Delta x_0 \ll 1$

each other by $\Delta x_0 := |x_0' - x_0|$ with Δx_0 "infinitesimally small" such that x_0, x_0' do not hit different branches of the Bernoulli shift $B(x)$ around $x = 1/2$.[1] We then have

$$\Delta x_n := |x_n' - x_n| = 2\Delta x_{n-1} = 2^2\Delta x_{n-2} = \cdots = 2^n\Delta x_0 = \mathrm{e}^{n\ln 2}\Delta x_0. \tag{1.3}$$

We thus see that there is an exponential separation between two nearby points as we follow their trajectories, where the rate of separation $\lambda(x_0) := \ln 2$ is the (local) Lyapunov exponent of $B(x)$. Since $\lambda(x_0) > 0$, the system displays an exponential dynamical instability and is hence called chaotic (in the sense of Lyapunov) [2–4, 30].

Writing down the analogue of Eq. (1.3) for a given differentiable map F, we get

$$\Delta x_n = |x_n' - x_n| = |F^n(x_0') - F^n(x_0)| =: \mathrm{e}^{n\lambda(x_0)}\Delta x_0 \, (\Delta x_0 \to 0), \tag{1.4}$$

which we can take as the definition of the Lyapunov exponent $\lambda(x_0)$ that comes in as the exponential stretching rate on the right-hand side. Solving this equation for $\lambda(x_0)$ by using the chain rule, it is not too hard to see [4] that this simple procedure of calculating λ can be generalized in terms of the time (or Birkhoff) average

$$\lambda(x) = \lim_{n\to\infty} \frac{1}{n} \sum_{i=0}^{n-1} \ln |F'(x_i)| \tag{1.5}$$

[1]This condition could be eliminated by defining a metric on a circle [4].

with $x = x_0$. If the dynamical system defined by the map F is ergodic, the time average does not depend on the initial condition for a typical x, $\lambda = \lambda(x) = \text{const}$. It can be shown that the Bernoulli shift is ergodic [6], and indeed, following Eq. (1.5), for B we trivially find that $\lambda = \ln 2$ for all x. In particular, according to Birkhoff's theorem [6, 31–33], for ergodic systems the time average is equal to the ensemble average, which for the Lyapunov exponent of one-dimensional maps reads

$$\lambda = \langle \ln |F'(x)| \rangle_{\mu^*} := \int_J \mathrm{d}\mu^* \ln |F'(x)| \, . \tag{1.6}$$

Here μ^* is the invariant measure of the map. If the map exhibits an SRB measure [34–36], we have

$$\mathrm{d}\mu^* = \rho^*(x)\, \mathrm{d}x, \tag{1.7}$$

where $\rho^*(x)$ holds for the invariant density of the map. That is, the measure μ^* has the nice property that it can be obtained by integrating a density,

$$\mu^*(A) = \int_A \mathrm{d}x\, \rho^*(x)\, , A \subseteq J, \tag{1.8}$$

which simplifies the calculation of the ensemble average Eq. (1.6). For the Bernoulli shift it is not too difficult to see [3] that, for typical initial conditions, the invariant density is $\rho^*(x) = 1$. By combining Eqs. (1.6) and (1.7), we get

$$\lambda = \int_0^1 \mathrm{d}x \rho^*(x) \ln 2 = \ln 2. \tag{1.9}$$

This result is equal to the time average calculated above and confirms the result obtained from our handwaving argument Eq. (1.3).

Lyapunov exponents are not the only quantities assessing the chaotic character of a dynamical system. Pesin's Theorem [6, 34, 35] states that for closed C^2 Anosov [6, 35] systems the Kolmogorov-Sinai (or metric) entropy h_{KS} is equal to the sum of positive Lyapunov exponents. For one-dimensional maps that are expanding [3, 4],

$$\forall x \in J \quad |F'(x)| > 1, \tag{1.10}$$

this theorem boils down to

$$\lambda = h_{\mathrm{KS}}, \tag{1.11}$$

where [2, 6]

$$h_{\mathrm{KS}} := \lim_{n\to\infty} -\frac{1}{n} \sum_{w \in \{W_i^n\}} \mu^*(w) \ln \mu^*(w). \tag{1.12}$$

Here $\mu^*(w)$ is the SRB measure of an element w of the partition $\{W_i^n\}$, and n defines the level of refinement of the partition. Note that in Eq. (1.12) we have assumed that the partition is generating [31, 35, 37]. If $h_{\mathrm{KS}} > 0$ one sometimes speaks of measure-theoretic chaos [3]. For the Bernoulli shift it is not too hard to calculate h_{KS} from

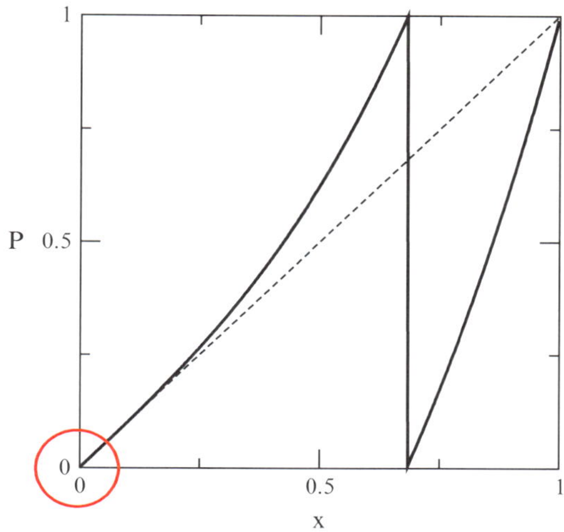

Fig. 1.3 The Pomeau–Manneville map Eq. (1.13) for $a = 1$ and $z = 3$. Note that there is a marginal fixed point at $x = 0$ leading to the intermittent behavior depicted in Fig. 1.4

first principles leading to $h_{\mathrm{KS}} = \ln 2$ [2,3], which combined with our previous result for the Lyapunov exponent is in line with Pesin's theorem. This theorem can be formulated under weaker assumptions, and it is believed to hold for a wider class of dynamical systems than stated above. We remark that typically the KS-entropy is much harder to calculate for a given dynamical system than Lyapunov exponents. Hence, Pesin's theorem is often employed in the literature for indirectly calculating the KS-entropy.

1.2.2 Weak Chaos and Infinite Ergodic Theory

Let us now consider a nonlinear generalization of our previous piecewise linear model, which is known as the *Pomeau–Manneville map* [38],

$$P_{a,z}(x) = x + ax^z \mod 1, \tag{1.13}$$

see Fig. 1.3, where following Eq. (1.1) the dynamics is defined by $x_{n+1} = P_{a,z}(x_n)$. This map has the two control parameters $a \geq 1$ and the exponent of nonlinearity $z \geq 1$. For $a = 1$ and $z = 1$ the map reduces to the Bernoulli shift Eq. (1.2) for $z > 1$ it provides a nontrivial nonlinear generalization of it. The nontriviality is due to the fact that in this case the stability of the fixed point at $x = 0$ becomes *marginal* (sometimes also called indifferent, or neutral), $P'_{a,z}(0) = 1$. This implies that the map is non-hyperbolic, because [39],

$$\nexists N > 0 \,\text{such that}\, \forall x \forall n \geq N \, |(P^n_{a,z})'(x)| \neq 1, \tag{1.14}$$

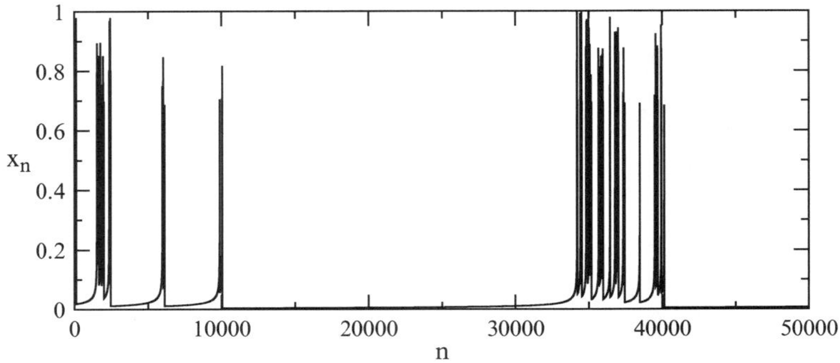

Fig. 1.4 Phenomenology of intermittency in the Pomeau–Manneville map Fig. 1.3: The plot shows the time series of position x_n versus discrete time step n for an orbit generated by the map Eq. (1.13) which starts at a typical initial condition x_0

which is related to the fact that the map is not expanding anymore according to Eq. (1.10). Since the map is smooth around $x = 0$, the dynamics resulting from the left branch of the map is determined by the stability of this fixed point, whereas the right branch is still of "Bernoulli shift type" generating ordinary chaotic dynamics. There is thus a competition in the dynamics between these two different branches as illustrated by the time series displayed in Fig. 1.4: One can observe that long periodic "laminar phases" determined by the marginal fixed point around $x = 0$ are interrupted by "chaotic bursts" reflecting the Bernoulli shift-type part of the map with slope $a > 1$ around $x = 1$. This phenomenology is the hallmark of what is called *intermittency* [1,2].

This seemingly small nonlinear modification of the Bernoulli shift has dramatic consequences for the whole dynamics of the new system. We discuss them step by step following our exposition of the Bernoulli shift dynamics in Sect. 1.2.1: The invariant density of the Pomeau–Manneville map can be calculated to [24,40–43]

$$\rho^*(x) \sim x^{1-z} \; (x \to 0). \tag{1.15}$$

This singularity reflects the stickiness of trajectories to the marginally unstable fixed point at $x = 0$. Correspondingly, the measure obtained by integrating this density is *non-normalizable* for $z \geq 2$ yielding the *infinite invariant measure*

$$\mu^*(x) = \int_x^1 \mathrm{d}y\, \rho^*(y) \to \infty \; (x \to 0). \tag{1.16}$$

The branch of ergodic theory exploring the ergodic properties of infinite measure preserving dynamical systems is thus called *infinite ergodic theory*; see [20,44,45] for introductions to this topic and [12] for an in-depth mathematical treatment. The marginal fixed point has also an impact on the dispersion of nearby trajectories, which can be calculated to [24,42,43,46]:

$$\Delta x_n \sim \exp\left(n^{\frac{1}{z-1}}\right) \Delta x_0 \; (z > 2). \tag{1.17}$$

In contrast to the Bernoulli shift, which according to Eq. (1.3) exhibits exponential sensitivity to initial conditions, here we thus have a weaker *stretched exponential sensitivity*. By repeating the calculation leading to Eq. (1.5), it is not hard to see that Eq. (1.17) yields a zero Lyapunov exponent,

$$\lambda = 0, \tag{1.18}$$

despite the fact that Fig. 1.4 displays irregular dynamics. Dynamical systems where the separation of nearby trajectories grows weaker than exponential, which implies that the corresponding Lyapunov exponents are zero, have been coined *weakly chaotic* [13, 47–49]. We remark, however, that this denotation is not used unambiguously in the literature. Most importantly, the standard definitions of Lyapunov exponents for expanding and hyperbolic systems Eqs. (1.5) and (1.6) yield no good indicators of irregular dynamics anymore, because they do not capture the sub-exponential dispersion of trajectories. It is thus desirable to come up with generalized definitions of ordinary chaos quantities, which enable us to still assess this different type of chaotic behavior by calculating quantities that yield nonzero values.

The way to achieve this goal is shown by advanced concepts of infinite ergodic theory and corresponding generalized ergodic theorems. Recall that Birkhoff's theorem implies that for observables which are Lebesgue integrable, $f \in L^1$, we have [6, 31–33]

$$\frac{1}{n}\sum_{i=0}^{n-1} f(x_i) = \langle f \rangle_{\mu^*} . \tag{1.19}$$

However, it turns out that for $z \geq 2$ the Birkhoff sum on the left-hand side does not converge anymore. Surprisingly, it becomes a random variable that depends on initial conditions, and the equation breaks down. This non-equivalence between time and ensemble averages became known as *weak ergodicity breaking* in the physics literature, see, e.g., [14, 17, 50] and further references therein. It was observed experimentally in the anomalous statistics of blinking quantum dots and plays also a crucial role for the anomalous diffusion of atoms in optical lattices [14, 17, 50]. Note that physicists typically refer to ergodicity as the equality between time and ensemble average, whereas mathematicians usually define ergodicity via indecomposability [31, 32]. Equation (1.19) then follows from this definition by using Birkhoff's theorem. This should be kept in mind when referring to a weak ergodicity breaking.

In case of $z \geq 2$ and $f \in L^1$ for our map, the nature of the breakdown of Eq. (1.19) is elucidated by the *Aaronson-Darling-Kac theorem* [21, 45, 51] :

$$\frac{1}{a_n}\sum_{i=0}^{n-1} f(x_i) \xrightarrow{d} \mathcal{M}_\alpha \langle f \rangle_{\mu^*} \; (n \to \infty), \tag{1.20}$$

where the arrow holds for convergence in distribution. Here $\mathcal{M}_\alpha$, $\alpha \in [0,1]$, denotes a nonnegative real random variable distributed according to the normalized Mittag-Leffler distribution of order α, which is characterized by its moments:

$$\langle \mathcal{M}_\alpha^r \rangle = r! \frac{(\Gamma(1+\alpha))^r}{\Gamma(1+r\alpha)}, \; r \geq 0. \tag{1.21}$$

For the Pomeau–Manneville map $P_{a,z}$ one can prove [51] that $a_n \sim n^\alpha$ with $\alpha := 1/(z-1)$. Integrating Eq. (1.21) with respect to Lebesgue measure m suggests

$$\frac{1}{n^\alpha} \sum_{i=0}^{n-1} \langle f(x_i) \rangle_m \sim \langle f \rangle_{\mu^*}. \tag{1.22}$$

Note that for $z < 2$ one has to choose $\alpha = 1$, because the map still exhibits an SRB measure, and Eq. (1.22) becomes an equality. However, for $z \geq 2$ we have an infinite invariant measure that cannot be normalized; hence here Eq. (1.22) remains a proportionality, unless we fix this constant by other constraints.

These known facts from infinite ergodic theory motivate to suitably define generalized chaos quantities, which assess weakly chaotic dynamics by yielding nonzero values. Following the left-hand side of Eq. (1.22), by choosing $f(x) = \ln |P'_{a,z}(x)|$, we define the *generalized Lyapunov exponent* as

$$\Lambda := \lim_{n\to\infty} \frac{\Gamma(1+\alpha)}{n^\alpha} \sum_{i=0}^{n-1} \langle \ln |M'(x_i)| \rangle_m. \tag{1.23}$$

The inclusion of the gamma function in the numerator is not obvious at this point; however, it turns out to be convenient when calculating Λ for the Pomeau–Manneville map [24]. Interestingly, it is precisely the same canonical choice as is made in other areas of anomalous dynamics [26]. Analogously, we amend Eq. (1.12) to define the *generalized KS entropy* as

$$H_{\mathrm{KS}} := \lim_{n\to\infty} -\frac{\Gamma(1+\alpha)}{n^\alpha} \sum_{w \in \{W_i^n\}} m(w) \ln m(w). \tag{1.24}$$

Note that here we define H_{KS} with respect to the Lebesgue measure m. Both quantities can be calculated independently for the piecewise linearization of $P_{a,z}$ proposed in [46] by applying the thermodynamic formalism [3, 52] in combination with transfer operator methods [53–55]. As a result, one obtains [24]

$$H_{\mathrm{KS}} = \Lambda, \tag{1.25}$$

which may be considered as a generalization of Pesin's formula Eq. (1.11) to anomalous dynamics. Related generalizations of chaos quantities, and other versions of a generalized Pesin formula, have been discussed in [42, 43, 56, 57]. We remark, however, that in the mathematical literature there is the well-known formula by

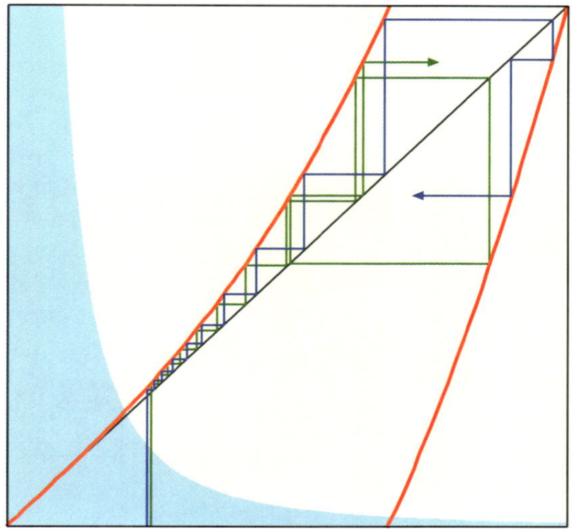

Fig. 1.5 Illustration of the interplay between weak chaos, infinite measures, and anomalous dynamics in the Pomeau–Manneville map Eq. (1.13) shown by the *thick bent lines*: Anomalous dynamics is indicated by the irregular behavior of the two single trajectories. Weak chaos is exemplified by the divergence of the two trajectories starting at nearby initial conditions. The *grey area in the lower left corner* depicts the shape of the infinite invariant density, which diverges at the marginal fixed point of the map [22]

Rokhlin [58], which for the Pomeau–Manneville map reads [40, 51]

$$h_{\mathrm{Kr}} = \langle \ln |P'_{a,z}(x)| \rangle_{\mu^*} . \tag{1.26}$$

Here the left-hand side holds for the so-called Krengel entropy. In case of finite invariant measures one can show that $h_{\mathrm{KS}} = h_{\mathrm{Kr}}$, the right hand is the Lyapunov exponent defined via the ensemble average Eq. (1.6), and Rokhlin's formula boils down to Pesin's formula Eq. (1.11). For infinite invariant measures, one can show that $h_{\mathrm{Kr}} = H_{\mathrm{KS}}$ [24]. Combining Rokhlin's formula with the integrated form of the Aaronson-Darling-Kac theorem Eq. (1.22) by using $f(x) = \ln |P'_{a,z}(x)|$, exploiting the definition Eq. (1.23) for the generalized Lyapunov exponent, and by fixing the constant of proportionality in Eq. (1.22) with respect to Lebesgue initial measure, one recovers Eq. (1.25). One may thus argue that, within this setting, Rokhlin's formula is a generalization of Pesin's formula for infinite measure-preserving transformations and that Eq. (1.25) is an illustration of it, worked out for the example of the Pomeau–Manneville map [24].

The main theoretical objects of discussion in this subsection are shown together in Fig. 1.5. This figure actually represents the logo of the conference about *Weak chaos, infinite ergodic theory, and anomalous dynamics* that was referred to in the introduction [22], from which the title of this book chapter derives.

1.2.3 A Generalized Hierarchy of Chaos

We conclude this section by embedding the previous results into the more general context of irregular deterministic dynamics [8]. There exist in fact further fundamental types of dynamics that are intermediate between strongly chaotic, in the sense of exponential sensitivity quantified by a positive Lyapunov exponent, and trivially being non-chaotic in terms of purely regular dynamics. These different types of irregular dynamics can be characterized by suggesting a classification of chaotic behavior based on the dispersion of initially infinitesimally close trajectories.

We start from the general expression for the asymptotic growth of the displacement $\Delta(t)$ of two trajectories generated by dynamics in continuous time t in the form of [46]

$$\ln \Delta(t) \sim t^{\nu_0} (\ln t)^{\nu_1}\,,\, 0 \leq \nu_0\,,\, \nu_1 \in \mathbb{R}. \tag{1.27}$$

If $\nu_0 = 1\,,\, \nu_1 = 0$ we recover the usual exponential dynamical instability of Eq. (1.3),

$$\Delta(t) \sim \exp(\lambda t), \tag{1.28}$$

representing *Lyapunov chaos* [2], whose strength is well quantified by the maximal positive Lyapunov exponent λ. As discussed before, if $\Delta(t)$ grows weaker than exponential, one speaks of *weak chaos* [13,47–49]. The regime of Eq. (1.27) with $0 < \nu_0 < 1$ or $\nu_0 = 1$ and $\nu_1 < 0$, which is typical for intermittent dynamics, was characterized as *sporadic* by Gaspard and Wang [46]; cf. Eq. (1.17) and our respective discussion, as well as further details of this dynamics as presented in the following section. Here the dynamical instability is either of stretched exponential type or exponential with logarithmic corrections,

$$\Delta(t) \sim \exp(t^{\nu_0} (\ln t)^{\nu_1}). \tag{1.29}$$

Equation (1.27) with $\nu_0 = 0$ and $\nu_1 = 1$, on the other hand, yields purely algebraic dispersion,

$$\Delta(t) \sim t^{\nu_2}\,,\, 0 < \nu_2, \tag{1.30}$$

for which Zaslavsky and Edelman [59, 60] suggested the term *pseudochaos*.[2] Note that algebraic dispersion with logarithmic corrections may also exist,

$$\Delta(t) \sim t^{\nu_2} (\ln t)^{\nu_3}\,,\, \nu_3 \in \mathbb{R}, \tag{1.31}$$

covering a slightly larger class of dynamical systems. A prominent class of dynamical systems exhibiting algebraic dispersion are polygonal billiards; two examples are depicted in Fig. 1.6. They represent the special case of pseudochaotic dynamics with $\nu_2 = 1$ for which the dispersion is strictly linear in time:

---

[2]We remark that in [59–61] one finds several slightly different definitions of pseudochaos. Here we refer to the first one stated in [59].

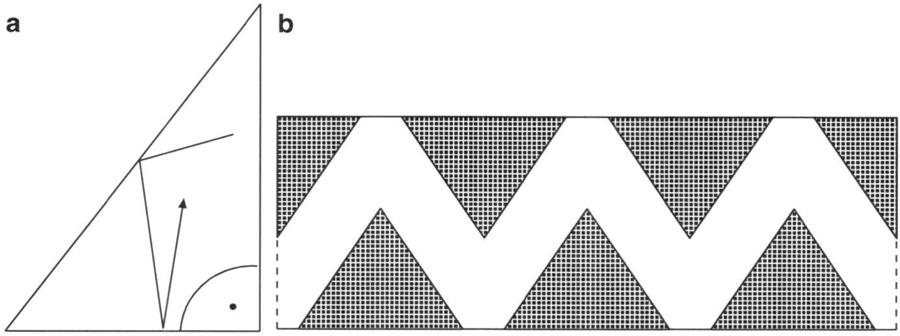

Fig. 1.6 Two simple examples of two-dimensional polygonal billiards [8]: A particle with unit velocity moves inside the depicted geometric domains by scattering elastically with their boundaries. (**a**) Shows a right triangular billiard [62] and (**b**) the triangle channel, where a unit cell with triangular scatterers is spatially continued along the line [63]

$$\Delta(t) \sim t. \tag{1.32}$$

However, in contrast to Lyapunov chaos and our weakly chaotic generalizations, here the linear dispersion does not actually capture the essential mechanism leading to dynamical randomness. For example, according to this classification both free flights and polygonal billiards of genus one, which clearly exhibit regular dynamics, are also pseudochaotic. As is discussed for the example of rational billiards, e.g., in [8], in polygonal billiards complicated topologies reflecting the existence of corners, which yield pseudohyperbolic fixed points and pseudointegrability, provide the source of nontrivial irregular dynamics. One is thus tempted to speak of *topology-induced chaos*[3] as a subclass of pseudochaos if there is linear dispersion on surfaces that are not integrable. *Pseudointegrable* rational billiards then form a subclass of topology-induced chaotic dynamical systems. Surprisingly, systems with linear dispersion generating nontrivial dynamics due to complicated topological structures may still exhibit ergodic and transport properties as they are usually associated with Lyapunov unstable chaotic dynamical systems. The trivial end point of this attempt of a generalized classification of chaotic dynamics on the basis of dispersion is simply the purely regular, or periodic, case of $\Delta(t) = \text{const}$.

1.3 Anomalous Diffusion

We now establish a cross-link between weakly chaotic dynamics as discussed in the previous section and the problem of deterministic diffusion. The main question we address is what type of diffusion arises if we suitably spatially extend a simple

[3]This should not be confused with *topological chaos* as defined in [7].

dynamical system exhibiting anomalous dynamics. We first set up our model, which can be considered as a purely deterministic, anomalous version of a random walk on the line, and introduce the concept of anomalous diffusion. We then outline CTRW theory, a standard tool in the theory of stochastic processes to study anomalous diffusion. The results obtained from this theory, worked out for our model, are compared to results from computer simulations. We conclude this chapter by deriving on the basis of this theory a generalized, fractional diffusion equation that approximately reproduces the probability density function (PDF) of our model.

1.3.1 A Simple Model Generating Anomalous Diffusion

A straightforward way to define a spatially extended dynamical system based on the Pomeau–Manneville map discussed in Sect. 1.2.2 is as follows: By using

$$P_{a,z}(x) = x + ax^z \ , \ 0 \leq x < \frac{1}{2} \tag{1.33}$$

of Eq. (1.13) without the modulus, as well as reflection symmetry,

$$P_{a,z}(-x) = -P_{a,z}(x), \tag{1.34}$$

we continue this map onto the whole real line by a *lift of degree one* [64–66]:

$$P_{a,z}(x+1) = P_{a,z}(x) + 1. \tag{1.35}$$

The resulting model [67, 68] is displayed in Fig. 1.7. Here points are not restricted anymore onto the unit interval. Due to the coupling between different unit cells by eliminating the modulus, there are now "jumps" possible from unit interval to unit interval. One may thus think of this dynamical system as a fully deterministic, anomalous version of a simple random walk on the line. A basic question is now which type of diffusion is generated by this model? As usual, the diffusive behavior is quantified by the mean square displacement (MSD) defined by

$$\langle x^2 \rangle := \int \mathrm{d}x \, x^2 \rho_n(x), \tag{1.36}$$

where $\langle x^2 \rangle$ is the second moment of the position PDF $\rho_n(x)$ at time step n. Starting from a given initial PDF $\rho_0(x)$ at time step $n = 0$, points, or point particles, will spread out over the whole real line, as quantified by $\rho_n(x)$. Surprisingly, by calculating this MSD both analytically and from computer simulations, one finds [67, 68] that for $z > 2$

$$\langle x^2 \rangle \sim n^{\alpha}, \quad \alpha < 1 \quad (n \to \infty). \tag{1.37}$$

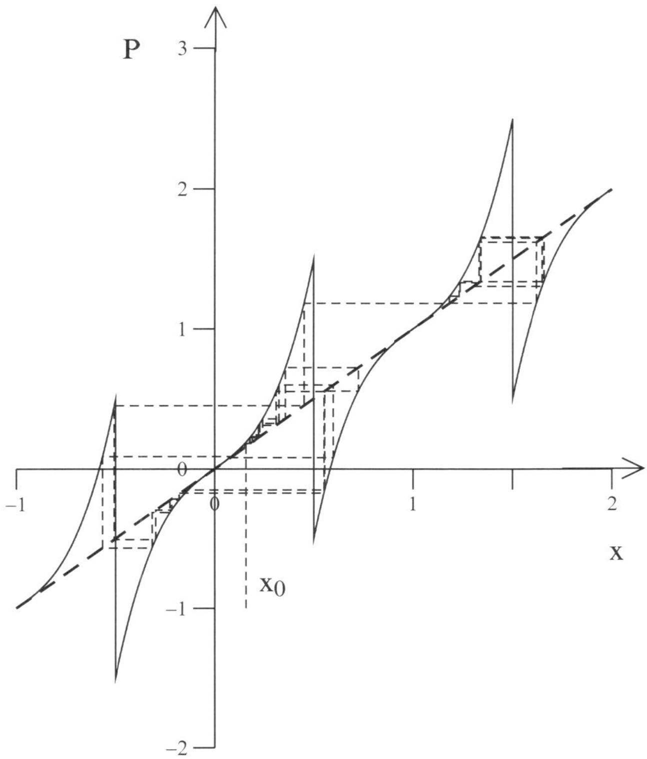

Fig. 1.7 The Pomeau–Manneville map Fig. 1.3, Eq. (1.13), lifted symmetrically onto the whole real line such that it generates subdiffusion

If one defines the diffusion coefficient of the system in the standard way by

$$D := \lim_{n\to\infty} \frac{\langle x^2 \rangle}{2n}, \tag{1.38}$$

Eq. (1.37) implies that $D = 0$, despite the fact that particles can go anywhere on the real line as illustrated in Fig. 1.7. While a process like Brownian motion leads to $D > 0$, here we thus encounter a very different type of diffusion: If the exponent α in the temporal spreading of the MSD Eq. (1.37) of an ensemble of particles is not equal to one, one speaks of *anomalous diffusion* [14, 18]. If $\alpha < 1$ one says that there is *subdiffusion*, for $\alpha > 1$ there is *superdiffusion*. In case of linear spreading with $\alpha = 1$ one refers to *normal diffusion*. The constant

$$K := \lim_{n\to\infty} \frac{< x^2 >}{n^{\alpha}}, \tag{1.39}$$

where in case of normal diffusion in one dimension $K = 2D$, is called the *generalized diffusion coefficient*.[4] For our simple map model depicted in Fig. 1.7 we will first calculate the MSD analytically by means of stochastic CTRW theory. By comparing the analytical formula with results from computer simulations, we will then focus on how K behaves as a function of z for fixed a revealing some interesting, nontrivial properties.

1.3.2 Continuous Time Random Walk Theory

Pioneered by Montroll, Weiss, and Scher [69–71], CTRW theory yields perhaps the most fundamental theoretical approach to explain anomalous diffusion [72–74]. In further groundbreaking works by Geisel et al. and Klafter et al., this method was then adapted to sub- and superdiffusive deterministic maps [67, 68, 75, 76]:

The basic assumption of this approach is that diffusion can be decomposed into two stochastic processes characterized by waiting times and jumps, respectively. Thus one has two sequences of independent identically distributed random variables, namely, a sequence of positive random waiting times $T_1, T_2, T_3, \ldots$ with PDF $w(t)$ and a sequence of random jumps $\zeta_1, \zeta_2, \zeta_3, \ldots$ with PDF $\lambda(x)$. For example, if a particle starts at point $x = 0$ at time $t_0 = 0$ and makes a jump of length ζ_n at time $t_n = T_1 + T_2 + \cdots + T_n$, its position is $x = 0$ for $0 \le t < T_1 = t_1$ and $x = \zeta_1 + \zeta_2 + \cdots + \zeta_n$ for $t_n \le t < t_{n+1}$. The probability that at least one jump is performed within the time interval $[0, t)$ is then $\int_0^t \mathrm{d}t' w(t')$ while the probability for no jump during this time interval reads $\Psi(t) = 1 - \int_0^t \mathrm{d}t' w(t')$. The master equation for the PDF $P(x,t)$ to find a particle at position x and time t is then

$$P(x,t) = \int_{-\infty}^{\infty} \mathrm{d}x' \lambda(x - x') \int_0^t \mathrm{d}t'\, w(t - t')\, P(x', t') + \Psi(t)\delta(x), \qquad (1.40)$$

which has the following probabilistic meaning: The PDF to find a particle at position x at time t is equal to the PDF to find it at point x' at some previous time t' multiplied with the transition probability to get from (x', t') to (x, t) integrated over all possible values of x' and t'. The second term accounts for the probability of remaining at the initial position $x = 0$. The most convenient representation of this equation is in Fourier-Laplace space:

$$\hat{\tilde{P}}(k,s) = \int_{-\infty}^{\infty} \mathrm{d}x\, \mathrm{e}^{ikx} \int_0^{\infty} \mathrm{d}t\, \mathrm{e}^{-st} P(x,t), \qquad (1.41)$$

where the hat stands for the Fourier transform and the tilde for the Laplace transform. Respectively, this function obeys the Fourier-Laplace transform of

[4] In detail, the definition of a generalized diffusion coefficient is a bit more subtle [26].

Eq. (1.40), which is called the Montroll-Weiss equation [69–71]:

$$\hat{\tilde{P}}(k,s) = \frac{1-\tilde{w}(s)}{s}\frac{1}{1-\hat{\lambda}(k)\tilde{w}(s)}. \tag{1.42}$$

The Laplace transform of the MSD can be obtained by differentiating the Fourier-Laplace transform of the PDF:

$$\widetilde{\langle x^2(s)\rangle} = \int_{-\infty}^{\infty} \mathrm{d}x\, x^2 \tilde{P}(x,s) = -\left.\frac{\partial^2 \hat{\tilde{P}}(k,s)}{\partial k^2}\right|_{k=0}. \tag{1.43}$$

In order to calculate the MSD within this theory, it thus suffices to know $\lambda(x)$ and $w(t)$ generating the stochastic process. For one-dimensional maps of the type of Eqs. (1.33) and (1.34), by exploiting the symmetry of the map, the waiting time distribution can be calculated from the approximation

$$x_{n+1} - x_n \simeq \frac{\mathrm{d}x_t}{\mathrm{d}t} = a x_t^z, \quad x \ll 1, \tag{1.44}$$

where we have introduced the continuous time $t \geq 0$. This equation can be solved for x_t with respect to an initial condition x_0. Now one needs to define when a particle makes a "jump," as will be discussed below. By inverting the solution for x_t, one can then calculate the time t a particle has to wait before it makes a jump as a function of the initial condition x_0. This information determines the relation between the waiting time PDF $w(t)$ and the as yet unknown PDF of injection points:

$$w(t) \simeq P_{\mathrm{in}}(x_0)\left|\frac{\mathrm{d}x_0}{\mathrm{d}t}\right|. \tag{1.45}$$

Making the assumption that the PDF of injection points is uniform, $P_{\mathrm{in}} \simeq 1$, the waiting time PDF is straightforwardly calculated from the knowledge of $t(x_0)$. The second ingredient that is needed for the CTRW approach is the jump PDF. Standard CTRW theory takes jumps between neighboring cells *only* into account leading to the ansatz [67, 68]:

$$\lambda(x) = \delta(|x| - 1). \tag{1.46}$$

It turns out that in order to qualitatively reproduce the dependence of the generalized diffusion coefficient $K = K(z,a)$ Eq. (1.39) on the map's two control parameters z and a, one needs to modify the standard theory at three points [25, 26]: Firstly, the waiting time PDF must be calculated according to the unit interval $[0,1]$, not according to $[-0.5, 0.5]$, which is an alternative but not appropriate choice [77, 78], yielding

$$w(t) = a(1 + a(z-1)t)^{-\frac{z}{z-1}}. \tag{1.47}$$

However, this PDF also accounts for *attempted* jumps to another cell, since after a step the particle may stay in the same cell with a probability of $(1-p)$. The latter quantity is roughly determined by the size of the escape region $p=(1-2x_c)$ with x_c as a solution of the equation $x_c + ax_c^z = 1$. We thus model this fact, secondly, by a jump length distribution in the form of

$$\lambda(x) = \frac{p}{2}\delta(|x|-l) + (1-p)\delta(x). \tag{1.48}$$

Thirdly, in order to capture the dependence of K on z for fixed a, we define a typical jump length as

$$l = \{|[M_{a,z}(x)]|\}, \tag{1.49}$$

where the square brackets stand for the floor function, which gives the coarse-grained displacement in units of elementary cells. The curly brackets denote both a time and ensemble average over points leaving a box. Note that for capturing the dependence of K on a for fixed z a different definition of the jump length is appropriate [25, 26]. Working out the modified CTRW approximation sketched above by taking these three details into account one obtains the result for the exponent α of the MSD, Eq. (1.37):

$$\alpha = \begin{cases} 1, & 1 \le z < 2, \\ \frac{1}{z-1}, & 2 \le z, \end{cases} \tag{1.50}$$

which matches to the standard theory [67, 68]. This result is in excellent agreement to simulations for a broad range of control parameters z and a. Building on this result, the generalized diffusion coefficient can be calculated to

$$K = pl^2 \times \begin{cases} a^\gamma \sin(\pi\gamma)/\pi\gamma^{1+\gamma}, & 0 < \gamma < 1, \\ a(1-1/\gamma), & 1 \le \gamma < \infty, \end{cases} \tag{1.51}$$

with $\gamma := 1/(z-1)$, which for $z \ge 2$ is identical to α of Eq. (1.50). In Fig. 1.8 this analytical approximation for K is compared with computer simulation results as a function of z for fixed a. There is good qualitative agreement between theory and simulations for $z > 2$, which converge asymptotically to each other for large z. For $z < 2$ there is a reasonable qualitative agreement, though quantitative deviations, for z close to 2, while the theory does not work anymore for $z \to 0$, a problem that is discussed in [26].

Remarkably, the $K(z)$ obtained from simulations does not appear to be a smooth function of the control parameter, which is at variance with the prediction of CTRW theory Eq. (1.51). This non-smooth parameter dependence is not due to numerical errors (which here are very difficult to assess, as discussed below) but a well-known phenomenon for this type of systems. It has first been discovered for the normal diffusive case of this map with $z = 1$, where the diffusion coefficient $D = K/2$ has been studied both numerically and analytically as a function of the slope a as a

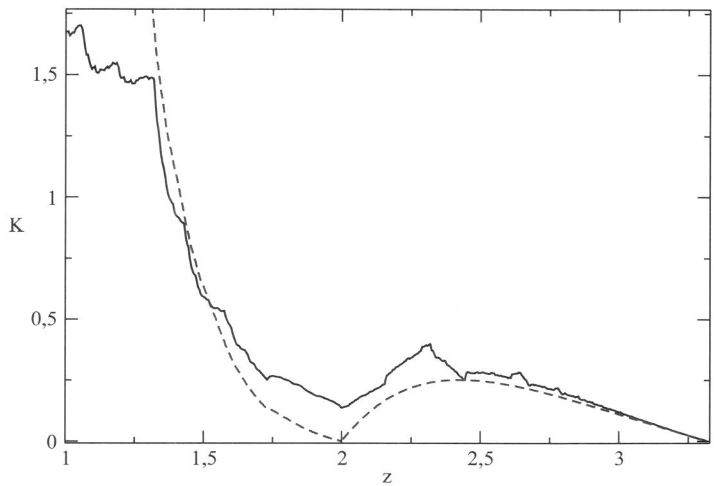

Fig. 1.8 The generalized diffusion coefficient K, Eq. (1.39), for the spatially extended Pomeau–Manneville map displayed in Fig. 1.7 as a function of z for $a = 5$. The *bold black line* depicts computer simulation results. The *dashed line* corresponds to the modified CTRW approximation Eqs. (1.49) and (1.51) [26]

control parameter [79, 80]. Note that for $z = 1$ the Pomeau–Manneville map boils down to a parameter-dependent, generalized version of the Bernoulli shift Eq. (1.2).

We do not wish to further elaborate on the fractal parameter dependencies of transport coefficients in simple deterministic dynamical systems, an interesting phenomenon that has been discussed in detail in [8, 77]. Rather, we focus on the behavior of the generalized diffusion coefficient at the point $z = 2$. According to the exponent α of the MSD given by Eq. (1.50), here the map exhibits a transition from normal to anomalous diffusion, which one may classify as a dynamical phase transition [3, 81]. As can be seen in Fig. 1.8, right around this transition point, there are significant deviations between CTRW theory and the simulation results. Most notably, at $z = 2$, the CTRW approximation forms a non-differentiable little wedge by predicting $K(2) = 0$, whereas the simulations yield $K(2) > 0$. By increasing the computation time one indeed finds very slow convergence of the simulation data towards the CTRW solution [26].

The explanation of these deviations, and of the phenomenon of a complete suppression of the generalized diffusion coefficient right at the transition point, is obtained by carrying out a refined analysis by means of CTRW theory. For a long time it was known already that at $z = 2$, the MSD behaves like $\langle x^2 \rangle \sim n/\ln n\,(n \to \infty)$ [67, 68]. Note that according to our definition of the generalized diffusion coefficient Eq. (1.39) this logarithmic dependence was incorporated into the strength of the diffusion coefficient, otherwise our analytical CTRW approximation would not have been continuous at $z = 2$. By taking into account higher-order terms when performing the CTRW theory calculations, which correspond to lower-order terms in time for the MSD, one arrives at

$$\langle x^2 \rangle \sim \begin{cases} \frac{n}{\ln n}\,, n < n_{\mathrm{cr}} \text{ and } \sim n\,, n \gg n_{\mathrm{cr}}\,, & z < 2, \\ \frac{n}{\ln n}\,, & z = 2, \\ \frac{n^{\alpha}}{\ln n}\,, n < \tilde{n}_{\mathrm{cr}} \text{ and } \sim n^{\alpha}\,, n \gg \tilde{n}_{\mathrm{cr}}\,, & z > 2. \end{cases} \tag{1.52}$$

Here n_{cr} and $\tilde{n}_{\mathrm{cr}}$ are crossover times that can be calculated exactly in terms of the map's control parameters. For $z \to 2$ both these crossover times diverge, and one arrives at the asymptotic $n/\ln n$ dependence mentioned before. The perhaps surprising fact is that around the transition point these multiplicative logarithmic corrections still survive for long but finite time, in agreement with computer simulation results. In other words, these logarithmic corrections lead to an ultraslow convergence of the simulation results thus explaining the deviations between long-time CTRW theory and simulations shown in Fig. 1.8. But more importantly, these logarithmic terms dominate the strength of the generalized diffusion coefficient around the transition point from normal to anomalous diffusion eventually yielding a full suppression of this quantity right at the transition point. It can be conjectured that the presence of such multiplicative logarithmic corrections around transition points between normal and anomalous diffusion is a typical scenario in this type of systems [26].

1.3.3 A Fractional Diffusion Equation

We now turn to the PDFs generated by the lifted map Eq. (1.13). As we will show now, CTRW theory not only predicts the power α correctly but also the form of the coarse grained PDF $P(x,t)$ of displacements. Correspondingly the anomalous diffusion process generated by our model is not described by an ordinary diffusion equation but by a generalization of it. Starting from the Montroll-Weiss equation and making use of the expressions for the jump and waiting time distributions Eqs. (1.46) and (1.47), we rewrite Eq. (1.42) in the long-time and large-space asymptotic form:

$$s^{\gamma}\hat{\hat{P}} - s^{\gamma-1} = -\frac{pl_i^2}{2cb^{\gamma}}k^2\hat{\hat{P}} \tag{1.53}$$

with $c = \Gamma(1-\gamma)$ and $b = \gamma/a$. For the initial condition $P(x,0) = \delta(x)$ of the PDF we have $\hat{P}(k,0) = 1$. Interestingly, the left-hand side of this equation corresponds to the definition of the *Caputo fractional derivative* of a function G [82, 83],

$$\frac{\partial^{\gamma} G}{\partial t^{\gamma}} := \frac{1}{\Gamma(1-\gamma)} \int_0^t \mathrm{d}t' \,(t-t')^{-\gamma} \frac{\partial G}{\partial t'}, \tag{1.54}$$

in Laplace space,

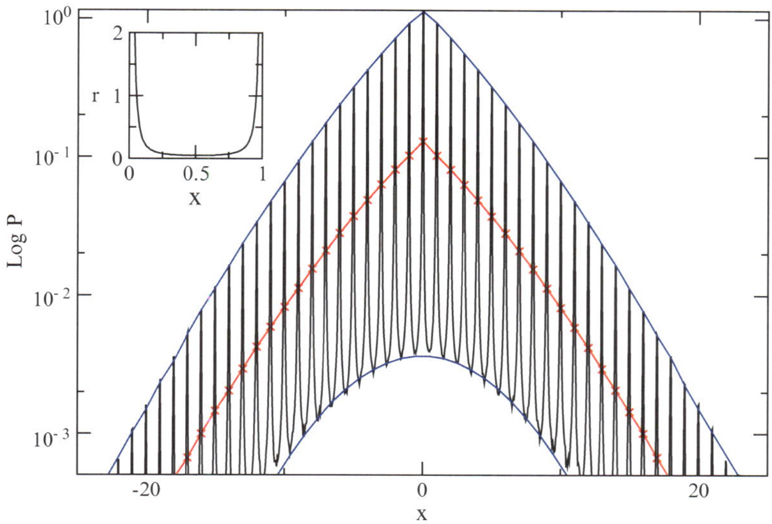

Fig. 1.9 Comparison of the probability density obtained from simulations of the lifted map Eq. (1.13) (*oscillatory structure*) with the analytical solution Eq. (1.57) of the fractional diffusion equation Eq. (1.56) (*continuous line in the middle*) for $z = 3$ and $a = 8$. The probability density was computed from 10^7 particles after $n = 10^3$ iterations. For the generalized diffusion coefficient in Eq. (1.57) the simulation result was used. The *crosses (x)* represent the numerical results coarse grained over unit intervals. The *upper* and the *lower curves* correspond to fits with a stretched exponential and a Gaussian distribution, respectively. The *inset* depicts the probability density function for the map on the unit interval with periodic boundaries

$$\int_0^\infty \mathrm{d}t\, \mathrm{e}^{-st} \frac{\partial^\gamma G}{\partial t^\gamma} = s^\gamma \tilde{G}(s) - s^{\gamma-1} G(0). \tag{1.55}$$

Thus, fractional derivatives come naturally into play as a suitable mathematical formalism whenever there are power law memory kernels in space and/or time generating anomalous dynamics; see, e.g., [18, 84] for short introductions to fractional derivatives and [82] for a detailed exposition. Turning back now to real space, we thus arrive at the time-fractional diffusion equation

$$\frac{\partial^\gamma P(x,t)}{\partial t^\gamma} = D \frac{\partial^2 P}{\partial x^2} \tag{1.56}$$

with $D = K\Gamma(1+\gamma)/2$, $0 < \gamma < 1$, which is an example of a *fractional diffusion equation* generating subdiffusion. Note the existence of the gamma function in the numerator defining D, which is analogous to the appearance of the gamma function in our generalized chaos quantities Eqs. (1.23) and (1.24). For $\gamma = 1$ we recover the ordinary diffusion equation. The solution of Eq. (1.56) can be expressed in terms of an M-function of Wright type [83] and reads

$$P(x,t) = \frac{1}{2\sqrt{D}t^{\gamma/2}} M\left(\xi, \frac{\gamma}{2}\right). \tag{1.57}$$

Figure 1.9 demonstrates an excellent agreement between the analytical solution Eq. (1.57) and the PDF obtained from simulations of the map Eq. (1.13) if it is coarse grained over unit intervals. However, it also shows that the coarse graining eliminates a periodic fine structure that is not captured by Eq. (1.57). This fine structure derives from the "microscopic" invariant density of an elementary cell (with periodic boundaries) as shown in the inset of Fig. 1.9 [77]. The singularities are due to the marginal fixed points of the map, where points are trapped for long times. Remarkably, that way the microscopic origin of the intermittent dynamics is reflected in the shape of the PDF on the whole real line: From Fig. 1.9 it is seen that the oscillations in the density are bounded by two functions, the upper curve being of a stretched exponential type while the lower is Gaussian. These two envelopes correspond to the laminar and chaotic parts of the motion, respectively.

1.4 Anomalous Fluctuation Relations

After having accomplished a transition from deterministic dynamics to stochastic modeling in the previous section, for the remainder of this chapter we fully focus on stochastic systems. First, we discuss a remarkable finding in nonequilibrium statistical mechanics that was widely investigated over the past two decades, which are fluctuation relations. After providing a brief outline of what they are and why they are important, we first study an example of them for one of the most simple types of stochastic dynamics, which is Brownian motion modeled by an ordinary Langevin equation. Along these lines, we then consider generalized versions of Langevin dynamics exhibiting anomalous diffusion. For these types of dynamics we check again for fluctuation relations and in one case obtain a different, new form of such a formula. We argue that generalized, anomalous fluctuation relations should be important to understand nonequilibrium fluctuations in glassy dynamics.

1.4.1 Fluctuation Relations

Fluctuation relations (FRs) denote a set of symmetry relations describing large-deviation properties of the PDFs of statistical physical observables far from equilibrium. First forms defining one subset of them, often referred to as *Fluctuation Theorems*, emerged from generalizing fluctuation–dissipation relations to nonlinear stochastic processes [85, 86]. They were then discovered as generalizations of the Second Law of Thermodynamics for thermostated dynamical systems, i.e., systems interacting with thermal reservoirs, in nonequilibrium steady states [87–90]. Another subset, called *work relations*, generalizes a relation between work and free energy, known from equilibrium thermodynamics to nonequilibrium situations

[91, 92]. These two fundamental classes were later on amended and generalized by a variety of other FRs from which they can partially be derived as special cases [93–96]. Research performed over the past ten years has shown that FRs hold for a great variety of systems thus featuring one of the rare statistical physical principles that is valid even very far from equilibrium; see, e.g., [5, 8, 10, 97–100] and further references therein. Many of these relations have meanwhile been verified in experiments on small systems, i.e., systems on molecular scales featuring only a limited number of relevant degrees of freedom [11, 101–105].

So far FRs have mostly been studied for dynamics exhibiting normal diffusion. This raises the question to which extent the "conventional" FRs derived for these cases are valid for anomalous dynamics [27, 28]. Theoretical results for generalized Langevin equations [106–109], Lévy flights [110, 111], and continuous-time random walk models [112] as well as computer simulations for glassy dynamics [113] showed both validity and violations of the various types of conventional FRs referred to above, depending on the specific type of anomalous dynamics considered and the nonequilibrium conditions that have been applied [28].

In this section we outline how the two different fields of FRs and anomalous dynamics can be cross-linked in order to explore to which extent conventional forms of FRs are valid for anomalous dynamics. With the term *anomalous fluctuation relations* we thus refer to deviations from conventional forms of FRs, which are due to anomalous dynamics. Here we focus on generic types of stochastic anomalous dynamics by only checking TFRs, which describe the approach from a given initial distribution towards a (non)equilibrium steady state. As a warm-up, we first derive the conventional TFR for the trivial case of Brownian motion of a particle moving under a constant external force modeled by standard Langevin dynamics. We then consider a straightforward generalization of this type of dynamics in form of long-time correlated Gaussian stochastic processes. For two fundamental, different versions of this dynamics, we check for the existence of conventional TFRs under the simple nonequilibrium condition of a constant external force.

1.4.2 Fluctuation Relations for Ordinary Langevin Dynamics

Consider a particle system evolving from some initial state at time $t = 0$ into a nonequilibrium steady state for $t \to \infty$. A famous example that has been investigated experimentally [101] is a colloidal particle immersed into water and confined by an optical harmonic trap, see Fig. 1.10. The trap is first at rest but then dragged through water with a constant velocity v^*.

The key for obtaining FRs in such systems is to compute the PDF $\rho(\xi_t)$ of suitably defined dimensionless entropy production ξ_t over trajectory segments of time length t. The goal is to quantify the asymmetry between positive and negative entropy production in $\rho(\xi_t)$ for different times t since, as we will demonstrate in a moment, this relation is intimately related to the Second Law of Thermodynamics. For a very large class of systems and under rather general conditions, it was shown

Fig. 1.10 Sketch of a colloidal particle confined within a harmonic trap that is dragged through water with a constant velocity v^*, cf. the experiment by Wang et al. [101]

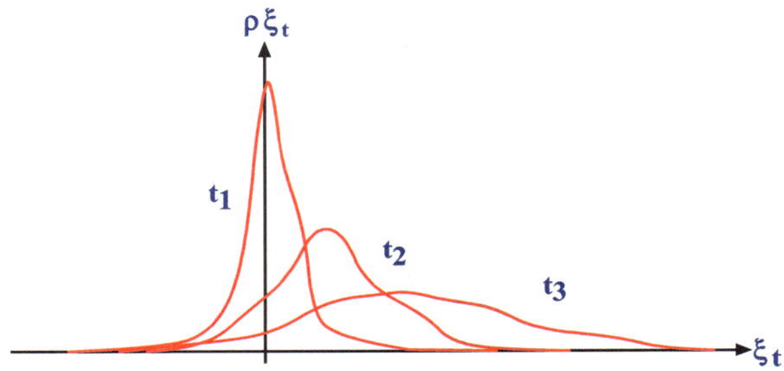

Fig. 1.11 Illustration of the dynamics of the probability density function for entropy production $\rho(\xi_t)$ for different times $t_1 < t_2 < t_3$

that the following equation holds [10, 98, 114]:

$$\ln \frac{\rho(\xi_t)}{\rho(-\xi_t)} = \xi_t . \tag{1.58}$$

Given that here we consider the transient evolution of a system from an initial into a steady state, this formula became known as the TFR. We may call the left-hand side the fluctuation ratio. Relations exhibiting this functional form have first been proposed in the seminal work by Evans et al. [87], although in the different situation of considering nonequilibrium steady states. Such a steady state relation was proved a few years later on by Gallavotti and Cohen for deterministic dynamical systems, based on the chaotic hypothesis [89, 90]. The idea to consider such relations for transient dynamics was first put forward by Evans and Searles [88].

Figure 1.11 displays the temporal evolution of the PDF for entropy production in such a situation. The asymmetry of the evolving distribution, formalized by the fluctuation relation Eq. (1.58), is in line with the Second Law of Thermodynamics. This easily follows from Eq. (1.58) by noting that

$$\rho(\xi_t) = \rho(-\xi_t)\exp(\xi_t) \geq \rho(-\xi_t), \tag{1.59}$$

where ξ_t is taken to be positive or zero. Integration from zero to infinity over both sides of this inequality after multiplication with ξ_t and defining the ensemble average over the given PDF as $\langle \ldots \rangle = \int_{-\infty}^{\infty} d\xi_t \, \rho(\xi_t) \ldots$ yields

$$\langle \xi_t \rangle \geq 0. \tag{1.60}$$

As a warm-up, we may first check the TFR for the ordinary overdamped *Langevin equation* [115]

$$\dot{x} = F + \zeta(t), \tag{1.61}$$

with a constant external force given by F and Gaussian white noise $\zeta(t)$. Note that for the sake of simplicity, here we set all the other constants that are not relevant within this specific context equal to one. For Langevin dynamics with a constant force the entropy production ξ_t defined by the heat, or equivalently the dissipative work, is simply equal to the mechanical work [116]

$$W_t = Fx(t). \tag{1.62}$$

It follows that the PDF of entropy production, which here is identical to the one for the mechanical work, is trivially related to the PDF of the position x of the Langevin particle via

$$\rho(W_t) = F^{-1} \rho(x,t). \tag{1.63}$$

This is very convenient, since it implies that all that remains to be done in order to check the TFR Eq. (1.58) is to solve the Fokker-Planck equation for the position PDF $\rho(x,t)$ for a given initial condition. Here and in the following, we choose $x(0) = 0$, i.e., in terms of position PDFs we start with a delta-distribution at $x = 0$. Note that for ordinary Langevin dynamics in a given potential, typically the equilibrium density is taken as the initial density [116, 117]. However, since in the following we will consider dynamics that may not exhibit a simple equilibrium state, without loss of generality here we make a different choice.

For the ordinary Langevin dynamics Eq. (1.61) modeling a linear Gaussian stochastic process, the position PDF is Gaussian exhibiting normal diffusion [115, 118]:

$$\rho(x,t) = \frac{1}{\sqrt{2\pi\sigma_{x,0}^2}} \exp\left(-\frac{(x - \langle x \rangle)^2}{2\sigma_{x,0}^2} \right). \tag{1.64}$$

With the subscript zero we denote ensemble averages in case of zero external field. By using the PDF-scaling Eq. (1.63) and plugging this result into the TFR Eq. (1.58), we easily derive that the TFR for the work W_t holds if

$$\langle W_t \rangle = \frac{\sigma_{W_t,0}^2}{2}, \tag{1.65}$$

which is nothing else than an example of the *fluctuation–dissipation relation of the first kind* (FDR1) [115, 119]. We thus arrive at the seemingly trivial but nevertheless important result that for this simple Gaussian stochastic process, the validity of

FDR1 Eq. (1.65) implies the validity of the work TFR Eq. (1.58). For a full analysis of FRs of ordinary Langevin dynamics we refer to van Zon and Cohen [116, 117].

Probably inspired by the experiment of [101], typically Langevin dynamics in a harmonic potential moving with a constant velocity has been studied in the literature [107–109, 120], cf. Fig. 1.10. Note that in this slightly more complicated case the (total) work is not equal to the heat [116]. While for the work one recovers the TFR in its conventional form Eq. (1.58) in analogy to the calculation above, surprisingly the TFR for heat looks different for large enough fluctuations. The origin of this phenomenon has been discussed in detail in [117]; related effects have been reported in [98, 121, 122]. However, in the following we check for another source of deviations from the conventional TFR Eq. (1.58) that is due to the existence of microscopic correlations in form of anomalous dynamics. In order to illustrate the main ideas it suffices to consider a nonequilibrium situation simply generated by a constant external force.

1.4.3 Fluctuation Relations for Anomalous Langevin Dynamics

In our presentation of this section we follow [28], which may be consulted for further details. Our goal is to check the TFR Eq. (1.58) for *Gaussian stochastic processes* generating anomalous diffusion. These processes are defined by the overdamped generalized Langevin equation

$$\int_0^t \mathrm{d}t' \dot{x}(t')\gamma(t-t') = F + \zeta(t) \tag{1.66}$$

with Gaussian noise $\zeta(t)$ and friction that is modeled with a memory kernel $\gamma(t)$. By using this equation a stochastic process can be defined that exhibits normal statistics but with anomalous memory properties in form of non-Markovian long-time correlated Gaussian noise. Equations of this type can be traced back at least to work by Mori and Kubo around 1965 (see [119] and further references therein). They form a class of standard models generating anomalous diffusion that has been widely investigated, see, e.g., [115, 123, 124]. FRs for this type of dynamics have more recently been analyzed in [106–109]. Examples of applications for this type of stochastic modeling are given by generalized elastic models [125], polymer dynamics [126], and biological cell migration [29]. We split this class into two specific cases:

1.4.3.1 Correlated Internal Gaussian Noise

We speak of *internal* Gaussian noise in the sense that we require the system to exhibit the *fluctuation–dissipation relation of the second kind* (FDR2) [115, 119]:

$$\langle\zeta(t)\zeta(t')\rangle \sim \gamma(t-t'), \tag{1.67}$$

again by neglecting all constants that are not relevant for the main point we wish to make here. We now consider the specific case that both the noise and the friction are correlated by a simple power law:

$$\gamma(t) \sim t^{-\beta}\,,\ 0<\beta<1. \tag{1.68}$$

Because of the linearity of the generalized Langevin equation (1.66) the position PDF must be the Gaussian Eq. (1.64), and by the scaling of Eq. (1.63) we have $\rho(W_t) \sim \rho(x,t)$. It thus remains to solve Eq. (1.66) for mean and variance, which can be done in Laplace space [28] yielding *subdiffusion*,

$$\sigma_{x,F}^2 \sim t^{\beta}, \tag{1.69}$$

by preserving the FDR1 Eq. (1.65). Here and in the following we denote ensemble averages in case of a nonzero external field with the subscript F. For Gaussian stochastic processes we have seen in the previous section that the conventional work TFR follows from FDR1. Hence, for the above power-law correlated internal Gaussian noise, we recover the conventional work TFR Eq. (1.58).

1.4.3.2 Correlated External Gaussian Noise

As a second case, we consider the overdamped generalized Langevin equation

$$\dot{x} = F + \zeta(t), \tag{1.70}$$

which represents a special case of Eq. (1.66) with a memory kernel modeled by a delta-function. Again we use correlated Gaussian noise defined by the power law

$$\langle\zeta(t)\zeta(t')\rangle \sim |t-t'|^{-\beta},\ 0<\beta<1, \tag{1.71}$$

which one may call *external*, because in this case we do not postulate the existence of FDR2. The position PDF is again Gaussian, and as before $\rho(W_t) \sim \rho(x,t)$. However, by solving the Langevin equation along the same lines as in the previous case, here one obtains *superdiffusion* by breaking FDR1:

$$\langle W_t\rangle \sim t, \quad \sigma_{W_t,F}^2 \sim t^{2-\beta}. \tag{1.72}$$

Calculating the fluctuation ratio, i.e., the left-hand side of Eq. (1.58), from these results yields the *anomalous work TFR*:

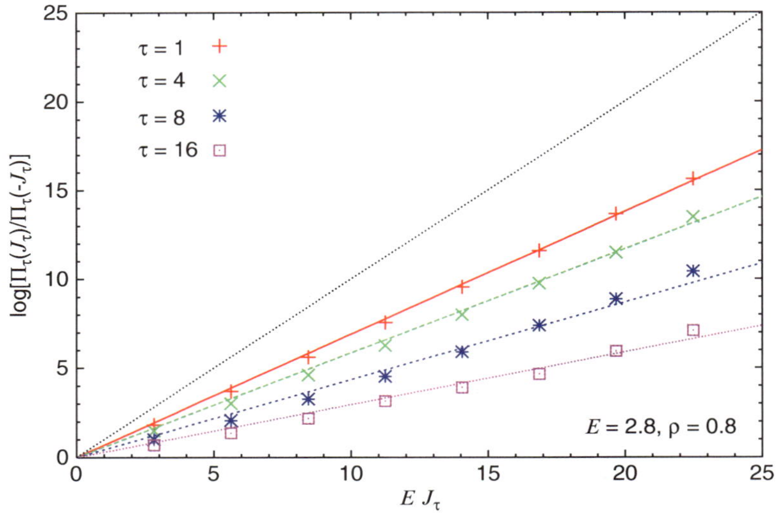

Fig. 1.12 The fluctuation ratio $\ln(\Pi_\tau(J_\tau)/\Pi_\tau(J_\tau))$ for the entropy production $W_\tau = EJ_\tau$ with particle current J_τ and field strength E for particle density ρ at different times τ. The *full line*, with slope one, displays the result of the conventional FR Eq. (1.58) in a nonequilibrium steady state. The figure is from [113]

$$\ln \frac{\rho(W_t)}{\rho(-W_t)} = C_\beta t^{\beta-1} W_t \quad 0 < \beta < 1, \tag{1.73}$$

where C_β is a constant that depends on physical parameters [28]. Comparing this equation with the conventional form of the TFR Eq. (1.58) one observes that the fluctuation ratio is still linear in W_t thus exhibiting an exponential large-deviation form [111]. However, there are two important deviations: (1) the slope of the fluctuation ratio as a function of W_t is not equal to one anymore, and in particular (2) it decreases with time. We may thus classify Eq. (1.73) as a *weak violation of the conventional TFR*.

We remark that for driven glassy systems FRs have already been obtained displaying slopes that are not equal to one. Within this context it has been suggested to capture these deviations from one by introducing the concept of an "effective temperature" [120, 127, 128]. As far as the time dependence of the coefficient is concerned, such behavior has recently been observed in computer simulations of a paradigmatic two dimensional lattice gas model generating glassy dynamics [113]. Figure 1.12 shows the fluctuation ratio as a function of the entropy production at different times τ as extracted from computer simulations of this model, where the PDF has first been relaxed into a nonequilibrium steady state. It is clearly seen that the slope decreases with time, which is in line with the prediction of the anomalous TFR Eq. (1.73). To which extent the nonequilibrium dynamics of this lattice gas model can be mapped onto the generalized Langevin equation Eq. (1.70) is an open question.

In summary, for Gaussian stochastic processes with correlated noise, the existence of FDR2 implies the existence of FDR1, and FDR1 in turn implies the existence of work TFR in conventional form. That is, the conventional work TFR holds for internal noise. However, there is a weak violation of the conventional form in case of external noise yielding a pre-factor that is not equal to one and in particular depends on time.

1.5 Anomalous Dynamics of Biological Cell Migration

In order to illustrate the importance of anomalous dynamics for realistic situations, in this final section of our book chapter we discuss experiments and theory about the migration of single biological cells crawling on surfaces as an example. Here we focus on cells in an equilibrium situation, i.e., not moving under the influence of any external gradients or fields. This case is investigated by extracting results for the MSD and for the position PDFs from experimental data. We then show how the experimental results can be understood by a mathematical model in form of a fractional Klein-Kramers equation. As far as MSD and velocity autocorrelation function are concerned, this equation bears some similarity to a generalized Langevin equation that is of the same type as the one that has been discussed in Sect. 1.4.3. Our presentation in this section is based on [29].

1.5.1 Cell Migration

Nearly all cells in the human body are mobile at a given time during their life cycle. Embryogenesis, wound-healing, immune defense, and the formation of tumor metastases are well-known phenomena that rely on cell migration [129–131]. Figure 1.13 depicts the path of a single biological cell crawling on a substrate measured in an *in vitro* experiment [29]. At first sight, the path looks like the trajectory of a Brownian particle generated, e.g., by the ordinary Langevin dynamics of Eq. (1.61). On the other hand, according to Einstein's theory of Brownian motion, a Brownian particle is *passively* driven by collisions from the surrounding fluid molecules, whereas biological cells move *actively* by themselves converting chemical into kinetic energy. This raises the question whether the random-looking paths of crawling biological cells can really be understood in terms of simple Brownian motion [132, 133] or whether more advanced concepts of dynamical modeling have to be applied [134–138].

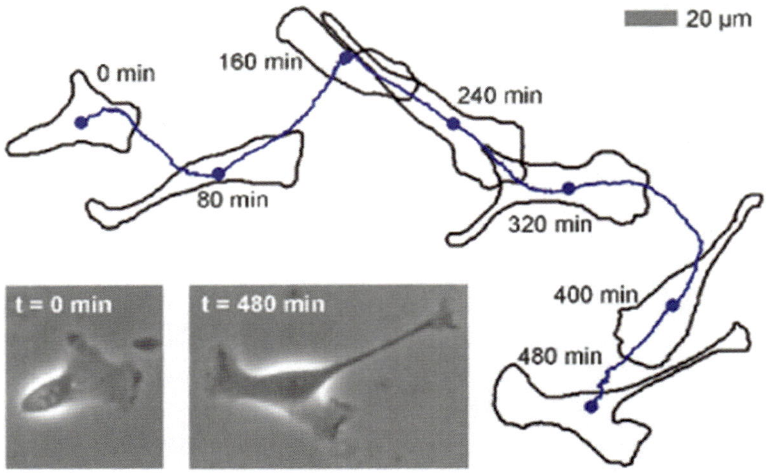

Fig. 1.13 Overlay of a biological cell migrating in vitro on a substrate. The cell frequently changes its shape and direction during migration, as is shown by several cell contours extracted during the migration process. The *inset* displays phase contrast images of the cell at the beginning and to the end of its migration process [29]

1.5.2 Experimental Results and Statistical Analysis

The cell migration experiments that we now discuss have been performed on two types of tumorlike migrating *transformed renal epithelial Madin Darby canine kidney (MDCK-F)* cell strains: wild-type (NHE^+) and NHE-deficient (NHE^-) cells. Here NHE^+ stands for a molecular sodium hydrogen exchanger that either is present or deficient. It can thus be checked whether this microscopic exchanger has an influence on cell migration, which is a typical question asked by cell physiologists. The cell diameter is about 20–50 μm and the mean velocity of the cells about 1 μm/min. Cells are driven by active protrusions of growing actin filaments (*lamellipodial dynamics*) and coordinated interactions with myosin motors and dynamically reorganizing cell-substrate contacts. The leading edge dynamics of a polarized cell proceeds at the order of seconds. Thirteen cells were observed for up to 1,000 min. Sequences of microscopic phase contrast images were taken and segmented to obtain the cell boundaries shown in Fig. 1.13; see [29] for full details of the experiments.

According to the Langevin description of Brownian motion outlined in Sect. 1.4.2, Brownian motion is characterized by a MSD $\sigma_{x,0}^2(t) \sim t \ (t \to \infty)$ designating normal diffusion. Figure 1.14 shows that both types of cells behave differently: First of all, MDCK-F NHE^- cells move less efficiently than NHE^+ cells resulting in a reduced MSD for all times. As is displayed in the upper part of this figure, the MSD of both cell types exhibits a crossover between three different dynamical regimes. These three phases can be best identified by extracting the time-

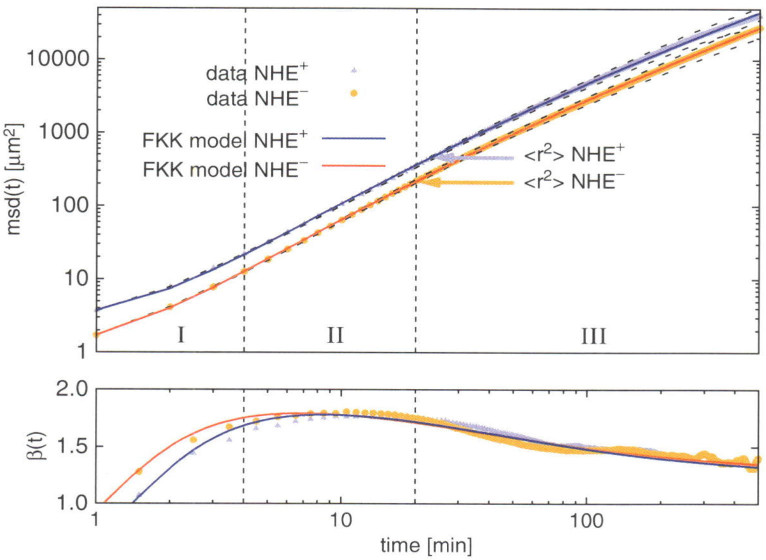

Fig. 1.14 *Upper part*: double-logarithmic plot of the mean square displacement (MSD) as a function of time. Experimental data points for both cell types are shown by *symbols*. Different time scales are marked as phases I, II, and III as discussed in the text. The *solid lines* represent fits to the MSD from the solution of our model; see Eq. (1.80). All parameter values of the model are given in [29]. The *dashed lines* indicate the uncertainties of the MSD values according to Bayes data analysis. *Lower part*: logarithmic derivative $\beta(t)$ of the MSD for both cell types as defined by Eq. (1.74)

dependent exponent β of the MSD $\sigma_{x,0}^2(t) \sim t^\beta$ from the data, which can be done by using the logarithmic derivative

$$\beta(t) = \frac{d\ln msd(t)}{d\ln t}. \tag{1.74}$$

The results are shown in the lower part of Fig. 1.14. Phase I is characterized by an exponent $\beta(t)$ roughly below 1.8. In the subsequent intermediate phase II, the MSD reaches its strongest increase with a maximum exponent β. When the cell has approximately moved beyond a square distance larger than its own mean square radius (indicated by arrows in the figure), $\beta(t)$ gradually decreases to about 1.4. Both cell types therefore do not exhibit normal diffusion, which would be characterized by $\beta(t) \to 1$ in the long-time limit, but move anomalously, where the exponent $\beta > 1$ indicates superdiffusion.

We next study the PDF of cell positions. Since no correlations between x and y positions could be found, it suffices to restrict ourselves to one dimension. Figure 1.15a, b reveals the existence of non-Gaussian distributions at different times. The transition from a peaked distribution at short times to rather broad distributions at long times suggests again the existence of distinct dynamical processes acting

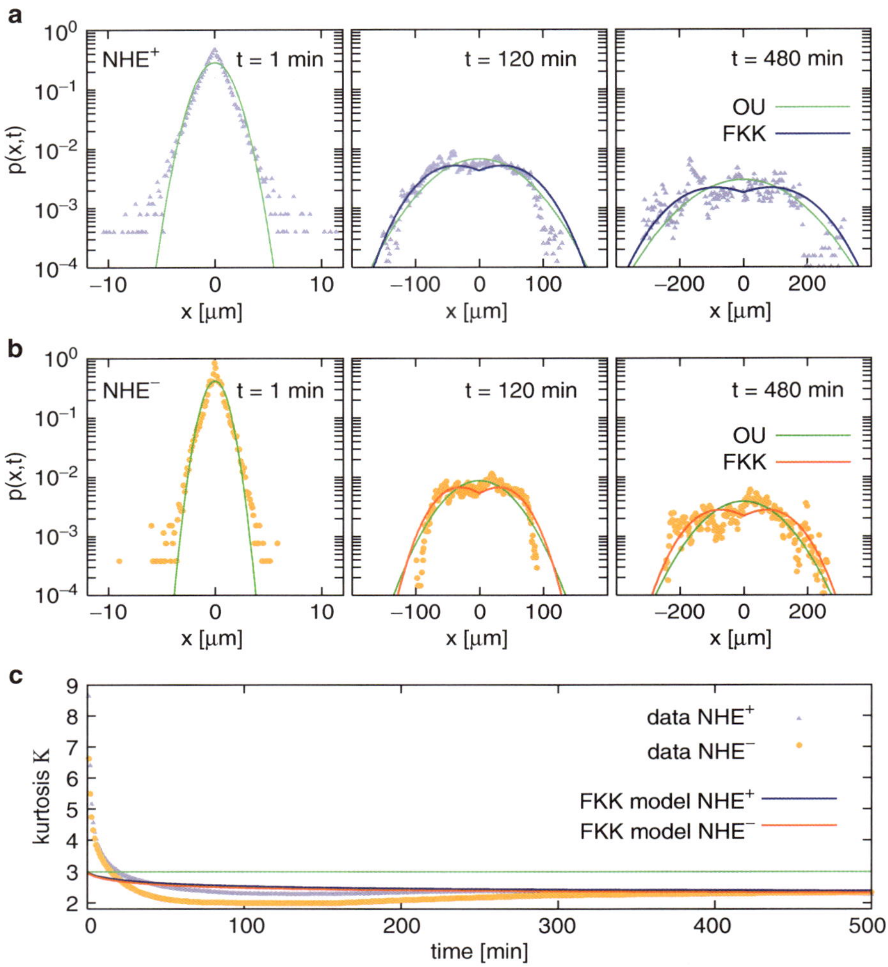

Fig. 1.15 Spatiotemporal probability distributions $P(x,t)$. (**a**), (**b**) Experimental data for both cell types at different times in semilogarithmic representation. The *dark lines*, labeled FKK, show the long-time asymptotic solutions of our model Eq. (1.76) with the same parameter set used for the MSD fit. The *light lines*, labeled OU, depict fits by the Gaussian distributions Eq. (1.64) representing Brownian motion. For $t = 1$ min both $P(x,t)$ show a peaked structure clearly deviating from a Gaussian form. (**c**) The kurtosis $\kappa(t)$ of $P(x,t)$, cf. Eq. (1.75), plotted as a function of time saturates at a value different from the one of Brownian motion (line at $\kappa = 3$). The other two lines represent $\kappa(t)$ obtained from the model Eq. (1.76) [29]

on different time scales. The shape of these distributions can be quantified by calculating the *kurtosis*:

$$\kappa(t) := \frac{\langle x^4(t)\rangle}{\langle x^2(t)\rangle^2}, \tag{1.75}$$

which is displayed as a function of time in Fig. 1.15c. For both cell types $\kappa(t)$ rapidly decays to a constant that is clearly below three in the long-time limit. A value of three would be the result for the spreading Gaussian distributions characterizing Brownian motion. These findings are another strong manifestation of the anomalous nature of cell migration.

1.5.3 Stochastic Modeling

We now present the stochastic model that we have used to reproduce the experimental data yielding the fit functions shown in the previous two figures. The model is defined by the *fractional Klein-Kramers equation* [139]:

$$\frac{\partial \rho}{\partial t} = -\frac{\partial}{\partial x}[v\rho] + \frac{\partial^{1-\alpha}}{\partial t^{1-\alpha}}\gamma_\alpha \left[\frac{\partial}{\partial v}v + v_{\text{th}}^2 \frac{\partial^2}{\partial v^2}\right]\rho \quad , 0 < \alpha < 1. \tag{1.76}$$

Here $\rho = \rho(x, v, t)$ is the PDF depending on time t, position x, and velocity v in one dimension, γ_α is a friction term; and $v_{\text{th}}^2 = k_B T/M$ stands for the thermal velocity squared of a particle of mass $M = 1$ at temperature T, where k_B is Boltzmann's constant. The last term in this equation models diffusion in velocity space. In contrast to Fokker-Planck equations, this equation features time evolution both in position and velocity space. What distinguishes this equation from an ordinary Klein-Kramers equation, the most general model of Brownian motion [118], is the presence of the Riemann-Liouville fractional derivative of order $1 - \alpha$,

$$\frac{\partial^{1-\alpha}}{\partial t^{1-\alpha}}\rho = \frac{\partial}{\partial t}\left[\frac{1}{\Gamma(\alpha)}\int_0^t \mathrm{d}t' \frac{\rho(t')}{(t-t')^{1-\alpha}}\right], \tag{1.77}$$

in front of the terms in square brackets. Note that for $\alpha = 1$ the ordinary Klein-Kramers equation is recovered. The analytical solution of this equation for the MSD has been calculated in [139] to

$$\sigma_{x,0}^2(t) = 2v_{\text{th}}^2 t^2 E_{\alpha,3}(-\gamma_\alpha t^\alpha) \quad \to \quad 2\frac{D_\alpha t^{2-\alpha}}{\Gamma(3-\alpha)} \quad (t \to \infty) \tag{1.78}$$

with $D_\alpha = v_{\text{th}}^2/\gamma_\alpha$ and the *two-parametric* or *generalized Mittag-Leffler function* (see, e.g., Chap. 4 of [14] and References [82, 140]):

$$E_{\alpha,\beta}(z) = \sum_{k=0}^{\infty} \frac{z^k}{\Gamma(\alpha k + \beta)}, \ \alpha, \beta > 0, z \in \mathbb{C}. \tag{1.79}$$

Note that $E_{1,1}(z) = \exp(z)$; hence $E_{\alpha,\beta}(z)$ is a generalized exponential function. We see that for long times Eq. (1.78) yields a power law, which reduces to the long-time Brownian motion result in case of $\alpha = 1$.

In view of the experimental data shown in Fig. 1.14, Eq. (1.78) was amended by including the impact of random perturbations acting on very short time scales for which we take Gaussian white noise of variance η^2. This leads to [141]

$$\sigma^2_{x,0;\text{noise}}(t) = \sigma^2_{x,0}(t) + 2\eta^2. \tag{1.80}$$

The second term mimics both measurement errors and fluctuations of the cell cytoskeleton. In case of the experiments with MDCK-F cells [29], the value of η can be extracted from the experimental data and is larger than the estimated measurement error. Hence, this noise must largely be of a biological nature and may be understood as being generated by microscopic fluctuations of the lamellipodia in the experiment.

The analytical solution of Eq. (1.76) for $\rho(x,v,t)$ is not known; however, for large friction γ_α this equation boils down to a fractional diffusion equation for which $\rho(x,t)$ can be calculated in terms of a Fox function [142]. The experimental data in Figs. 1.14 and 1.15 was then fitted consistently by using the above solutions with the four parameters $v^2_{\text{th}}, \alpha, \gamma$, and η^2 in Bayesian data analysis [29].

In summary, by statistical analysis of experimental data, we have shown that the equilibrium migration of the biological cells under consideration is anomalous. Related anomalies have also been observed for other types of migrating cells [134–138]. Our experimental results are coherently reproduced by a mathematical model in form of a stochastic fractional equation. We now elaborate on possible physical and biological interpretations of our findings.

First of all, we remark that the solutions of Eq. (1.76) for both the MSD and the velocity autocorrelation function match precisely to the solutions of the generalized Langevin equation [124]:

$$\dot{v} = -\int_0^t \mathrm{d}t'\, \gamma(t-t')v(t') + \xi(t). \tag{1.81}$$

Here $\xi(t)$ holds for Gaussian white noise and $\gamma(t) \sim t^{-\alpha}$ for a time-dependent friction coefficient with a power law memory kernel, which alternatively could be written by using a fractional derivative [124]. For $\gamma(t) \sim \delta(t)$ the ordinary Langevin equation is recovered. Note that the position PDF generated by this equation is Gaussian in the long-time limit and thus does not match to the one of the fractional Klein-Kramers equation Eq. (1.76). However, alternatively one could sample from a non-Gaussian $\xi(t)$ to generate a non-Gaussian position PDF. Strictly speaking, despite equivalent MSD and velocity correlations, Eqs. (1.76) and (1.81) define different classes of anomalous stochastic processes. The precise cross-links between the Langevin description and the fractional Klein-Kramers equation are subtle [143] and to some extent still unknown. The advantage of Eq. (1.81) is that it allows more straightforwardly a possible biophysical interpretation of the origin of the observed anomalous MSD and velocity correlations, at least partially, in terms of the existence of a memory-dependent friction coefficient. The latter, in turn, might be explained by anomalous rheological properties of the cell cytoskeleton, which consists of a complex biopolymer gel [144].

Secondly, what could be the possible biological significance of the observed anomalous cell migration? There is an ongoing debate about whether biological organisms such as albatrosses, marine predators, and fruit flies have managed to minimize the search time for food in a way that matches to optimizing search strategies in terms of stochastic processes; see [145, 146] and further references therein. In particular, it has been argued that Lévy flights are superior to Brownian motion in order to find sparsely, randomly distributed, replenishing food sources [145]. However, it was also shown that in other situations *intermittent dynamics* is more efficient than pure Lévy motion [145]. For our cell experiment, both the experimental data and the theoretical modeling suggest that there exists a slow diffusion on short time scales, whereas the long-time motion is much faster, which resembles intermittency as discussed in [145]. Hence, the results on anomalous cell migration presented above might be biologically relevant in view of suitably optimized foraging strategies.

1.6 Summary

This chapter highlighted some fundamental aspects of anomalous dynamics: The scene was set by Sect. 1.2, which reviewed basic ideas of weak chaos by establishing cross-links to infinite ergodic theory. This branch of ergodic theory provides a rigorous mathematical approach to study weakly chaotic dynamical systems. In particular, we proposed suitable definitions of generalized chaos quantities assessing weakly chaotic dynamics by yielding a generalized version of Pesin's theorem. We also outlined a generalized hierarchy of chaos on the basis of different functional forms of the dispersion exhibited by nearby trajectories of a deterministic dynamical system. In Sect. 1.3 we related these concepts to the problem of anomalous diffusion by spatially extending our previously discussed simple map model. Applying stochastic CTRW theory to this model in comparison to computer simulations, we learned about an intricate dynamical phase transition between normal and anomalous diffusion, governed by multiplicative logarithmic corrections in the MSD. We also derived a fractional diffusion equation that reproduced the subdiffusive diffusive dynamics of this model on coarse scales. The subsequent Sect. 1.4 elaborated on fluctuation relations, here understood as a large-deviation symmetry property of the work probability distributions generated by a given stochastic dynamics far from equilibrium. We familiarized ourselves with the conventional form of transient work fluctuation relations derived from standard Langevin dynamics before exploring anomalous generalizations of Langevin equations. One of them reproduced the conventional form of fluctuation relations, whereas the other one yielded a generalized, anomalous fluctuation relation. The precise form of the resulting fluctuation relation appeared to be intimately related to whether or not fluctuation–dissipation relations are broken. In our final main section we related our previous theoretical ideas to the experimental problem of studying biological cell migration. By extracting the MSD and the position probability distributions from experimental data, we found that the dynamics exhibited by these cells was anomalous, showing different behavior on different time scales, by eventually yielding superdiffusion for long times. On the basis of these experimental results we suggested a stochastic theoretical model of cell migration in form of a fractional Klein-Kramers equation, which coherently reproduced our experimental findings.

In summary, we traversed quite an anomalous scientific landscape of different but related topics: Starting from simple deterministic maps and their ergodic theory description we switched to basics of anomalous stochastic processes, studied both normal and anomalous stochastic fluctuations very far from equilibrium in terms of Langevin dynamics by ending up with anomalously crawling biological cells. We thus meant to illustrate the third column displayed in the very first Fig. 1.1 of the introduction, by also explaining the title of this contribution. Within a larger scientific context, one may consider our discussion as an indication that a novel theory of anomalous nonequilibrium processes is presently emerging. In contrast to standard nonequilibrium statistical mechanics, this dynamics is inherently nonstationary, due to the weak chaos by which it is generated. This mechanism leads to important physical consequences like anomalous transport, which can be tested in experiments. On the side of theoretical physics this approach asks for further generalizations of recently developed fundamental concepts, perhaps leading to a weakly chaotic hypothesis, to the identification of the physically relevant measures characterizing such systems, and to deriving experimentally measurable consequences such as generalizations of ordinary large-deviation properties and fluctuation relations. However, these questions also motivate further mathematical work in upcoming directions of infinite ergodic theory to provide a formal framework and rigorous results for parts of the physical theory.

Acknowledgment Each of the four sections reflects the collaboration with colleagues, without whom the work presented here would not have been possible. The second section benefitted very much from discussions with R. Zweimüler, whom the author thanks very much for a lot of mathematical insight into aspects of infinite ergodic theory. Particularly, the author is indebted to his former postdoc P. Howard, who did brilliant work on calculating generalized chaos quantities for the Pomeau–Manneville map. Regarding the third section, credit goes to his former Ph.D student N.Korabel for joint work that formed part of his Ph.D thesis. A.V. Chechkin significantly contributed to the same section as well as performed major research on the topic covered by the fourth one. The author is deeply indebted to him for his long-term collaboration on anomalous stochastic processes. P. Dieterich was the driving force behind the project reviewed in the fifth section. The author thanks him for much insight into the biophysical aspects of biological cell migration. Finally, he wishes to thank the editors of this book for their patience with this book chapter.

References

1. H. Schuster, *Deterministic Chaos*, 2nd edn. (VCH Verlagsgesellschaft mbH, Weinheim, 1989)

2. E. Ott, *Chaos in Dynamical Systems* (Cambridge University Press, Cambridge, 1993)
3. C. Beck, F. Schlögl, *Thermodynamics of Chaotic Systems*. Cambridge Nonlinear Science Series, vol. 4 (Cambridge University Press, Cambridge, 1993)
4. K. Alligood, T. Sauer, J. Yorke, *Chaos - An Introduction to Dynamical Systems* (Springer, New York, 1997)
5. R. Klages, W. Just, C. Jarzynski (eds.), *Nonequilibrium Statistical Physics of Small Systems: Fluctuation Relations and Beyond*. Reviews of Nonlinear Dynamics and Complexity (Wiley-VCH, Berlin, 2013)
6. J. Dorfman, *An Introduction to Chaos in Nonequilibrium Statistical Mechanics* (Cambridge University Press, Cambridge, 1999)
7. P. Gaspard, *Chaos, Scattering, and Statistical Mechanics* (Cambridge University Press, Cambridge, 1998)
8. R. Klages, *Microscopic Chaos, Fractals and Transport in Nonequilibrium Statistical Mechanics*. Advanced Series in Nonlinear Dynamics, vol. 24 (World Scientific, Singapore, 2007)
9. C. Castiglione, M. Falcioni, A. Lesne, A. Vulpiani, *Chaos and Coarse Graining in Statistical Mechanics* (Cambridge University Press, Cambridge, 2008)
10. D. Evans, D. Searles, Adv. Phys. **51**, 1529 (2002)
11. C. Bustamante, J. Liphardt, F. Ritort, Phys. Today **58**, 43 (2005)
12. J. Aaronson, *An Introduction to Infinite Ergodic Theory*. Mathematical Surveys and Monographs, vol. 50 (American Mathematical Society, Providence, 1997)
13. G. Zaslavsky, D. Usikov, *Weak Chaos and Quasi-Regular Patterns*. Cambridge Nonlinear Science Series (Cambridge University Press, Cambridge, 2001)
14. R. Klages, G. Radons, I. Sokolov (eds.), *Anomalous Transport: Foundations and Applications* (Wiley-VCH, Berlin, 2008)
15. M. Shlesinger, G. Zaslavsky, J. Klafter, Nature **363**, 31 (1993)
16. J. Klafter, M.F. Shlesinger, G. Zumofen, Phys. Today **49**, 33 (1996)
17. F. Stefani, J. Hoogenboom, E. Barkai, Phys. Today **62**, 34 (2009)
18. R. Metzler, J. Klafter, Phys. Rep. **339**, 1 (2000)
19. R. Metzler, J. Klafter, J. Phys. A: Math. Gen. **37**, R161 (2004)
20. J. Aaronson, in *Descriptive Set Theory and Dynamical Systems*, ed. by M. Foreman et al. London Mathematical Society lecture notes, vol. 277 (Cambridge University Press, Cambridge, 2000), pp. 1–29
21. M. Thaler, R. Zweimüller, Probab. Theory Relat. Fields **155**, 15 (2006)
22. R. Klages, R. Zweimüller, E. Barkai, H. Kantz, Weak chaos, infinite ergodic theory, and anomalous dynamics (2011), http://www.pks.mpg.de/~wchaos11
23. R. Klages, in *Reviews of Nonlinear Dynamics and Complexity*, vol. 3 (Wiley-VCH, Berlin, 2010), pp. 169–227
24. P. Howard, R. Klages, Entropy and stretching rates in intermittent maps (2009). Unpublished
25. N. Korabel, A. Chechkin, R. Klages, I. Sokolov, V. Gonchar, Europhys. Lett. **70**, 63 (2005)
26. N. Korabel, R. Klages, A. Chechkin, I. Sokolov, V. Gonchar, Phys. Rev. E **75**, 036213/1 (2007)
27. R. Klages, A. Chechkin, P. Dieterich, in *Nonequilibrium Statistical Physics of Small Systems: Fluctuation Relations and Beyond*. Reviews of Nonlinear Dynamics and Complexity (Wiley-VCH, Berlin, 2013), pp. 259–282
28. A. Chechkin, R. Klages, J. Stat. Mech.: Theor. Exp. **03**, L03002/1 (2009)
29. P. Dieterich, R. Klages, R. Preuss, A. Schwab, Proc. Natl. Acad. Sci. **105**, 459 (2008)
30. C. Robinson, *Dynamical Systems* (CRC Press, London, 1995)
31. A. Katok, B. Hasselblatt, *Introduction to the Modern Theory of Dynamical Systems*. Encyclopedia of Mathematics and its Applications, vol. 54 (Cambridge University Press, Cambridge, 1995)
32. V. Arnold, A. Avez, *Ergodic Problems of Classical Mechanics* (W.A. Benjamin, New York, 1968)
33. M. Toda, R. Kubo, N. Saitô, *Statistical Physics*, 2nd edn. Solid State Sciences, vol. 1 (Springer, Berlin, 1992)
34. L.S. Young, J. Stat. Phys. **108**, 733 (2002)

35. J.P. Eckmann, D. Ruelle, Rev. Mod. Phys. **57**, 617 (1985)
36. V. Baladi, *Positive Transfer Operators and Decay of Correlations*. Advanced Series in Nonlinear Dynamics, vol. 16 (World Scientific, Singapore, 2000)
37. R. Badii, A. Politi, *Complexity: Hierarchical Structures and Scaling Physics* (Cambridge University Press, Cambridge, 1997)
38. Y. Pomeau, P. Manneville, Commun. Math. Phys. **74**, 189 (1980)
39. R. Devaney, *An Introduction to Chaotic Dynamical Systems*, 2nd edn. (Addison-Wesley, Reading, 1989)
40. M. Thaler, Israel J. Math. **46**, 67 (1983)
41. R. Zweimüller, Nonlinearity **11**, 1263 (1998)
42. N. Korabel, E. Barkai, Phys. Rev. Lett. **102**, 050601/1 (2009)
43. N. Korabel, E. Barkai, Phys. Rev. E **82**, 016209/1 (2010)
44. M. Thaler, Infinite ergodic theory. Course notes from "The Dynamic Odyssey", CIRM 2001 (2001), http://www.sbg.ac.at/mat/staff/thaler/thaler_english.htm
45. R. Zweimüller, Surrey notes on infinite ergodic theory. Course notes from the "LMS Graduate school on Ergodic Theory", Surrey (2009), http://homepage.univie.ac.at/roland.zweimueller/MyPub/SurreyNotes.pdf
46. P. Gaspard, X.J. Wang, Proc. Nat. Acad. Sci. USA **85**, 4591 (1988)
47. S. Galatolo, Nonlinearity **16**, 1219 (2003)
48. H. van Beijeren, Physica D **193**, 90 (2004)
49. R. Artuso, G. Cristadoro, Chaos **15**, 015116/1 (2005)
50. A. Rebenshtok, E. Barkai, J. Stat. Phys. **133**, 565 (2008)
51. R. Zweimüller, Ergod. Theor. Dyn. Syst. **20**, 1519 (2000)
52. D. Ruelle, *Thermodynamic Formalism*. Encyclopedia of Mathematics and its Applications, vol. 5 (Addison-Wesley, Reading, 1978)
53. T. Prellberg, J. Slawny, J. Stat. Phys. **66**, 503 (1992)
54. S. Tasaki, P. Gaspard, J. Stat. Phys. **109**, 803 (2002)
55. S. Tasaki, P. Gaspard, Physica D **187**, 51 (2004)
56. T. Akimotoa, Y. Aizawa, Chaos **20**, 033110/1 (2010)
57. A. Saa, R. Venegeroles, J. Stat. Mech: Theor. Exp. **03**, P03010/1 (2012)
58. G. Keller, *Equilibrium States in Ergodic Theory*. London Mathematical Society Student Texts, vol. 42 (Cambridge University Press, Cambridge, 1998)
59. G. Zaslavsky, Phys. Rep. **371**, 461 (2002)
60. G. Zaslavsky, M. Edelman, in *Perspective and Problems in Nonlinear Science*, ed. by E. Kaplan, J. Marsden (Springer, New York, 2003), pp. 421–443
61. G. Zaslavsky, B. Carreras, V. Lynch, L. Garcia, M. Edelman, Phys. Rev. E **72**, 026227/1 (2005)
62. R. Artuso, G. Casati, I. Guarneri, Phys. Rev. E **55**, 6384 (1997)
63. B. Li, L. Wang, B. Hu, Phys. Rev. Lett. **88**, 223901/1 (2002)
64. H. Fujisaka, S. Grossmann, Z. Physik B **48**, 261 (1982)
65. T. Geisel, J. Nierwetberg, Phys. Rev. Lett. **48**, 7 (1982)
66. M. Schell, S. Fraser, R. Kapral, Phys. Rev. A **26**, 504 (1982)
67. T. Geisel, S. Thomae, Phys. Rev. Lett. **52**, 1936 (1984)
68. G. Zumofen, J. Klafter, Phys. Rev. E **47**, 851 (1993)
69. E. Montroll, G. Weiss, J. Math. Phys. **6**, 167 (1965)
70. E. Montroll, H. Scher, J. Stat. Phys. **9**, 101 (1973)
71. H. Scher, E. Montroll, Phys. Rev. B **12**, 2455 (1975)
72. J. Bouchaud, A. Georges, Phys. Rep. **195**, 127 (1990)
73. G. Weiss, *Aspects and Applications of the Random Walk* (North-Holland, Amsterdam, 1994)
74. W. Ebeling, I. Sokolov, *Statistical Thermodynamics and Stochastic Theory of Nonequilibrium Systems* (World Scientific, Singapore, 2005)
75. T. Geisel, J. Nierwetberg, A. Zacherl, Phys. Rev. Lett. **54**, 616 (1985)
76. M. Shlesinger, J. Klafter, Phys. Rev. Lett. **54**, 2551 (1985)

77. R. Klages, *Deterministic Diffusion in One-dimensional Chaotic Dynamical Systems* (Wissenschaft & Technik-Verlag, Berlin, 1996)
78. R. Klages, J. Dorfman, Phys. Rev. E **55**(2), R1247 (1997)
79. R. Klages, J. Dorfman, Phys. Rev. Lett. **74**, 387 (1995)
80. R. Klages, J. Dorfman, Phys. Rev. E **59**, 5361 (1999)
81. X. Wang, Phys. Rev. A **39**, 3214 (1989)
82. I. Podlubny, *Fractional Differential Equations* (Academic Press, New York, 1999)
83. F. Mainardi, in *Fractals and Fractional Calculus in Continuum Mechanics*, ed. by A. Carpinteri, F. Mainardi. CISM Courses and Lecture Notes, vol. 378 (Springer, Berlin, 1997), pp. 291–348
84. I. Sokolov, J. Klafter, A. Blumen, Phys. Today **55**, 48 (2002)
85. G. Bochkov, Y. Kuzovlev, Physica A **106**, 443 (1981)
86. G. Bochkov, Y. Kuzovlev, Physica A **106**, 480 (1981)
87. D. Evans, E. Cohen, G. Morriss, Phys. Rev. Lett. **71**, 2401 (1993)
88. D. Evans, D. Searles, Phys. Rev. E **50**, 1645 (1994)
89. G. Gallavotti, E. Cohen, Phys. Rev. Lett. **74**, 2694 (1995)
90. G. Gallavotti, E. Cohen, J. Stat. Phys. **80**, 931 (1995)
91. C. Jarzynski, Phys. Rev. Lett. **78**, 2690 (1997)
92. C. Jarzynski, Phys. Rev. E **56**, 5018 (1997)
93. G. Crooks, Phys. Rev. E **60**, 2721 (1999)
94. T. Hatano, S. Sasa, Phys. Rev. Lett. **86**, 3463 (2001)
95. U. Seifert, Phys. Rev. Lett. **95**, 040602/1 (2005)
96. T. Sagawa, M. Ueda, Phys. Rev. Lett. **104**, 090602/1 (2010)
97. G. Gallavotti, Chaos **8**, 384 (1998)
98. R. Harris, G. Schütz, J. Stat. Mech. **7**, P07020/1 (2007)
99. U. Seifert, Eur. Phys. J. B **64**, 423 (2008)
100. C. Jarzynski, Eur. Phys. J. B **64**, 331 (2008)
101. G. Wang, E. Sevick, E. Mittag, D. Searles, D. Evans, Phys. Rev. Lett. **89**, 050601/1 (2002)
102. F. Ritort, Poincaré Seminar **2**, 195 (2003)
103. S. Ciliberto, S. Joubaud, A. Petrosyan, J. Stat. Mech. **2010**(12), P12003/1 (2010)
104. S. Toyabe, T. Sagawa, M. Ueda, E. Muneyuki, M. Sano, Nat. Phys. **6**, 988 (2010)
105. A. Alemany, M. Ribezzi, F. Ritort, AIP Conf. Proc. **1332**(1), 96 (2011)
106. C. Beck, E. Cohen, Physica A **344**, 393 (2004)
107. T. Ohkuma, T. Ohta, J. Stat. Mech. **10**, P10010/1 (2007)
108. T. Mai, A. Dhar, Phys. Rev. E **75**, 061101/1 (2007)
109. S. Chaudhury, D. Chatterjee, B. Cherayil, J. Stat. Mech.: Theor. Exp. **10**, P10006/1 (2008)
110. H. Touchette, E. Cohen, Phys. Rev. E **76**, 020101(R)/1 (2007)
111. H. Touchette, E. Cohen, Phys. Rev. E **80**, 011114/1 (2009)
112. M. Esposito, K. Lindenberg, Phys. Rev. E **77**, 051119/1 (2008)
113. M. Sellitto, Phys. Rev. E **80**, 011134/1 (2009)
114. J. Kurchan, J. Stat. Mech.: Theor. Exp. **2007**(07), P07005/1 (2007)
115. R. Kubo, M. Toda, N. Hashitsume, *Statistical Physics*, 2nd edn. Solid State Sciences, vol. 2 (Springer, Berlin, 1992)
116. R. van Zon, E. Cohen, Phys. Rev. E **67**, 046102/1 (2003)
117. R. van Zon, E. Cohen, Phys. Rev. Lett. **91**, 110601/1 (2003)
118. H. Risken, *The Fokker-Planck Equation*, 2nd edn. (Springer, Berlin, 1996)
119. R. Kubo, Rep. Prog. Phys. **29**, 255 (1966)
120. F. Zamponi, F. Bonetto, L. Cugliandolo, J. Kurchan, J. Stat. Mech.: Theor. Exp. **09**, P09013/1 (2005)
121. R. Harris, A. Rákos, G. Schütz, Europhys. Lett. **75**, 227 (2006)
122. D. Evans, D. Searles, L. Rondoni, Phys. Rev. E **71**, 056120/1 (2005)
123. J. Porra, K.G. Wang, J.Masoliver, Phys. Rev. E **53**, 5872 (1996)
124. E. Lutz, Phys. Rev. E **64**, 051106/1 (2001)
125. A. Taloni, A. Chechkin, J. Klafter, Phys. Rev. Lett. **104**, 160602/1 (2010)

126. D. Panja, J. Stat. Mech.: Theor. Exp. **06**, P06011/1 (2010)
127. M. Sellitto, Fluctuations of entropy production in driven glasses (1998). Preprint arXiv:q-bio.PE/0404018
128. F. Zamponi, G. Ruocco, L. Angelani, Phys. Rev. E **71**, 020101(R)/1 (2005)
129. D. Lauffenburger, A.F.Horwitz, Cell **84**, 359 (1996)
130. T. Lämmermann, M. Sixt, Curr. Opin. Cell Biol. **21**(5), 636 (2009)
131. P. Friedl, K. Wolf, J. Cell Biol. **188**, 11 (2010)
132. G. Dunn, A. Brown, J. Cell Sci. Suppl. **8**, 81 (1987)
133. C. Stokes, S.W. D.A. Lauffenburger, J. Cell Sci. **99**, 419 (1991)
134. R. Hartmann, K. Lau, W. Chou, T. Coates, Biophys. J. **67**, 2535 (1994)
135. A. Upadhyaya, J. Rieu, J. Glazier, Y. Sawada, Physica A **293**, 549 (2001)
136. L. Li, S. Norrelykke, E. Cox, PLoS ONE **3**, e2093/1 (2008)
137. H. Takagi, M. Sato, T. Yanagida, M. Ueda, PLoS ONE **3**, e2648/1 (2008)
138. H. Bödeker, C. Beta, T. Frank, E. Bodenschatz, Europhys. Lett. **90**, 28005/1 (2010)
139. E. Barkai, R. Silbey, J. Phys. Chem. B **104**, 3866 (2000)
140. R. Gorenflo, F. Mainardi, in *Fractals and Fractional Calculus in Continuum Mechanics*, ed. by A. Carpinteri, F. Mainardi. CISM Courses and Lecture Notes, vol. 378 (Springer, Berlin, 1997), pp. 223–276
141. D. Martin, M. Forstner, J. Käs, Biophys. J. **83**, 2109 (2002)
142. W. Schneider, W. Wyss, J. Math. Phys. **30**, 134 (1989)
143. S. Eule, R. Friedrich, F. Jenko, D. Kleinhans, J. Phys. Chem. B **111**, 11474 (2007)
144. C. Semmrich, T. Storz, J. Glaser, R. Merkel, A. Bausch, K. Kroy, Proc. Natl. Acad. Sci. **104**, 20199 (2007)
145. O. Bénichou, C. Loverdo, M. Moreau, R. Voituriez, Rev. Mod. Phys. **83**(1), 81 (2011). doi:10.1103/RevModPhys.83.81
146. G. Viswanathan, M. da Luz, E. Raposo, H. Stanley, *The Physics of Foraging* (Cambridge University Press, Cambridge, 2011)

Chapter 2
Directed Transport in a Stochastic Layer

Alexei Vasiliev

Abstract We consider a problem of transport in a spatially periodic potential under the influence of a slowly time-dependent unbiased periodic external force. Using methods of the adiabatic perturbation theory we show that for a periodic external force of general kind the system demonstrates directed (ratchet) transport in the chaotic domain on very long time intervals and obtain a formula for the average velocity of this transport. Two cases are studied: the case of the external force of small amplitude and the case of the external force with amplitude of order one.

2.1 Introduction

In recent years, studies of transport phenomena in nonlinear systems have been attracting a growing interest. In particular, a large and constantly growing number of papers are devoted to dynamics in systems which allow for directed (on average) motion under unbiased external forces and are referred to as ratchet systems. (The name comes from the famous Feynmann's lecture [1] on impossibility to obtain a directed motion and usable work with a system in the state of thermodynamic equilibrium.) Intensive study of ratchet systems was motivated by problems of motion of Brownian particles in spatially periodic potentials, unidirectional transport of molecular motors in biological systems, and recognition of "ratchet effects" in quantum physics (see review [2] and references therein). Generally speaking, ratchet phenomena occur due to lack of symmetry in the spatially periodic potential and/or the external forcing. It is interesting, however, to understand microscopic mechanisms leading to these phenomena. A possible approach is to neglect dissipation and noise terms arriving at a Hamiltonian system with deterministic forcing.

A. Vasiliev (✉)
Space Research Institute, Profsoyuznaya 84/32, Moscow 117997, Russia
e-mail: valex@iki.rssi.ru

X. Leoncini and M. Leonetti (eds.), *From Hamiltonian Chaos to Complex Systems: A Nonlinear Physics Approach*, Nonlinear Systems and Complexity 5,
DOI 10.1007/978-1-4614-6962-9_2,

Thus, one can make use of results obtained and methods developed in the theory of Hamiltonian chaos. Many papers studying chaotic transport in such Hamiltonian ratchets appeared in the last years (see, e.g., [3–9]).

Roughly speaking, Hamiltonian ratchets are related to an equation of the kind,

$$\ddot{q} + \frac{\partial U}{\partial q} = f(t),$$

with $2\pi-$periodic potential $U(q+2\pi) = U(q)$ and time-periodic external force $f(t+\tilde{T}) = f(t)$ with zero time average:

$$\int_0^{\tilde{t}} f \mathrm{d}t = 0.$$

Typically, a phase space of such a system contains invariant tori carrying regular motions and domains where motion is chaotic (stochastic layers). A most interesting fact found numerically (see the references above) is that in general for a phase trajectory in a stochastic layer there exists a nonzero limit:

$$V_q = \lim_{t\to\infty} q(t)/t \neq 0,$$

which means that there is directed transport (sometimes referred to as ratchet current) in stochastic layer in such systems. This phenomenon has been widely investigated, yet only few explicit analytical results were obtained. In particular, in [7] the ratchet current is estimated in the case when there are stability islands in the chaotic domain in the phase space of the system. The borders of such islands are "sticky" [10] and this stickiness together with desymmetrization of the islands is responsible for the occurrence of the ratchet transport.

We consider the problem of motion of a particle in a periodic potential under the influence of unbiased time-periodic external forcing. In numerics, we take $U(q) = \omega_0^2 \cos q$, where q is the coordinate and $\omega_0 = \text{const}$. Thus the equations are the same as in the paradigmatic model of a nonlinear pendulum under the action of external torque with zero time average. We study the case when the external forcing is time periodic with a large period of order ε^{-1}, $0 < \varepsilon \ll 1$, and use results and methods of the adiabatic perturbation theory. If ε is small enough, there are no stability islands in the domain of chaotic dynamics (see [11]). Thus, the mechanism of ratchet transport in this system differs from one suggested in [7].

The main objective of this chapter is to find a formula for the average velocity $V_q = \langle \dot{q} \rangle$ of a particle in the chaotic domain on very large time intervals. We study two cases: of the external force with the amplitude of order 1 and of the external force of small amplitude of order ε. We show that in the first case chaos develops as a result of multiple passages through a resonance. Each passage produces a small variation (a jump) of the value of the adiabatic invariant of the system. These jumps result in effective mixing and uniform distribution of the adiabatic invariant along a trajectory in the chaotic domain. On the other hand, direction and value of velocity depend on the immediate value of the action. Thus, to find the average velocity of transport on time intervals of order or larger than the mixing time, we find formulas for displacement in q at a given value of the action and then integrate them over the interval of values of the action corresponding to the chaotic domain. The situation is similar in the case of small external force. In this case, a typical phase trajectory repeatedly crosses a separatrix on the phase portrait. At each crossing the adiabatic invariant undergoes a quasi-random jump (see [12, 13]). Like in the first case, these jumps produce chaotic dynamics in the domain of separatrix crossings. In both cases, we demonstrate that for an external force of general kind (i.e. with zero time average but lowered time symmetry, cf. [6]), there is directed transport in the chaotic domain and obtain an analytic formula for the average velocity V_q of this transport. In both cases the width of the chaotic domain in the phase space is large: in the case of forcing of order one the width is $\sim \varepsilon^{-1}$, and in the case of small forcing the width is ~ 1. This is a common situation in systems with adiabatic chaos. Indeed, chaos in such systems develops in the domain filled with phase trajectories that repeatedly cross the resonance or the separatrix (see for particular examples, e.g., [13–18]). Thus the total phase flux due to the directed transport in the considered system is also large.

The chapter is based on results obtained in [19] by Leoncini, Neishtadt, and the author.

2.2 External Forcing of Order One

In this section we study the case when the external forcing is not small. The equation of motion has the form

$$\ddot{q} + \frac{\partial U}{\partial q} = f(\tau),$$

where a dot denotes t-derivative, $0 < \varepsilon \ll 1$ is a small parameter, $\tau = \varepsilon t$ is called the "slow time", function $f(\tau)$ is periodic with period T, i.e. $f(\tau + T) = f(\tau)$, and has zero time average. The system can be rewritten in the form

$$\dot{q} = p, \ \dot{p} = -\frac{\partial U}{\partial q} + f(\tau), \ \dot{\tau} = \varepsilon. \tag{2.1}$$

This is a Hamiltonian system with time-dependent Hamiltonian:

$$H = \frac{p^2}{2} + U(q) - f(\tau)q.$$

One can see from the second equation in (2.1) that magnitude of momentum p can reach values of order ε^{-1}. Make the canonical transformation of variables $(p, q) \mapsto (\bar{p}, \bar{q})$ with generating function $W = (\bar{p} - \varepsilon^{-1}F(\varepsilon t))q$, where $F(\tau)$ is defined as $F(\tau) = -\int_0^\tau f(x)\mathrm{d}x + C$. Here C is a constant which we are free to choose.

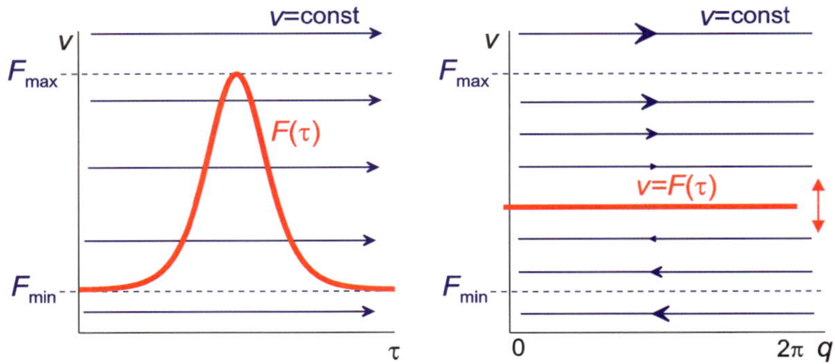

Fig. 2.1 *Left panel*: phase portrait of system (2.4). The *bold line* represents one period of function $F(\tau)$. *Right panel*: dynamics on (q,v)-plane

To simplify the following discussion, we take it large enough to make F positive at all values of τ and assume that function F has only one minimum and one maximum on its period. Note that $\bar{q} \equiv q$. After this transformation, Hamiltonian of the system acquires the form (bars over q are omitted):

$$H = \frac{(\bar{p} - \varepsilon^{-1}F(\tau))^2}{2} + U(q). \tag{2.2}$$

Introduce rescaled momentum $v = \varepsilon\bar{p}$ and rescaled time $\theta = \varepsilon^{-1}t$. We denote the derivative with respect to θ with prime and thus obtain:

$$q' = v - F(\tau),\ v' = -\varepsilon^2\frac{\partial U}{\partial q},\ \tau' = \varepsilon^2. \tag{2.3}$$

This is a system in a typical form for application of the averaging method. Variable q is fast, and variables v and τ are slow. Take into account that U is a 2π-periodic function of q and thus its q-derivative has zero q-average. We average over fast variable q and obtain the averaged system:

$$v' = 0,\ \tau' = \varepsilon^2. \tag{2.4}$$

Thus, v is constant along a phase trajectory of the averaged system and is an adiabatic invariant of the exact system. The approximation $v = \text{const}$ is called adiabatic. The averaged system describes the dynamics adequately everywhere in the phase space except for a small neighborhood of the resonance at $v - F(\tau) = 0$, where the "fast" variable q is not fast.

In Fig. 2.1, left, a phase portrait of system (2.4) is shown. The horizontal lines are phase trajectories of the averaged system. Along every trajectory $v = \text{const}$. The bold line represents one period of function $F(\tau)$. When a phase trajectory of the averaged system crosses the resonance $v = F(\tau)$, value of the adiabatic invariant undergoes a quasi-random jump of typical order $\sqrt{\varepsilon^2} = \varepsilon$ (see, e.g., [20, 21] and references therein). A jump of v at the resonance crossing can be expressed as $\Delta v = \varepsilon G(q_*)$, where $q_* \in (0, 2\pi)$ is the value of $q \bmod 2\pi$ at the resonance crossing in the adiabatic approximation, and $G(q_*) \sim 1$ is a smooth function on the interval $(0, 2\pi)$. Magnitude of the jump should be considered as a quasi-random value, because a small variation of initial conditions results generally in large, of order one, variation of q_*. Consider two successive resonance crossings, corresponding to $q_* = q_1$ and $q_* = q_2$. It follows from the second equation in (2.4) that time interval (in terms of θ) between these crossings is a value of order ε^{-2}. A small variation δq_1 in q_1 produces variation of order ε in Δv. This latter variation after a long time interval $\sim \varepsilon^{-2}$ results in large variation of q_2: $\delta q_2 \sim \delta q_1/\varepsilon$. Therefore, jumps of the adiabatic invariant at successive resonant crossings can be considered as statistically independent random values.

Another representation of dynamics is shown in Fig. 2.1, right, on the plane (q, v). The bold line correspond to the position of the resonance $v = F(\tau)$. It slowly moves upwards and downwards in the picture, oscillating between $v = F_{\max}$ and $v = F_{\min}$, i.e. the maximal and the minimal values of $F(\tau)$. Note that $\dot{q} = 0$ on the line $v = F(\tau)$, $\dot{q} > 0$ above this line, and $\dot{q} < 0$ under this line. The region swept by this line in its slow motion is the region where resonance crossings occur. Uncorrelated jumps of adiabatic invariant v result in stochastization of dynamics in this region. The dynamics can be considered as a random walk between level lines of v. On a period of function $F(\tau)$ (after two resonance crossings) v changes by a value of order ε. Hence, after $N \sim \varepsilon^{-2}$ separatrix crossings, v varies by a value of order one. As a result, in time of order $t_{\text{diff}} \sim \varepsilon^{-3}$, the value of adiabatic invariant is distributed in all the range of values corresponding to the domain of resonance crossings. Phase trajectories of the averaged system that cross the resonance correspond to values of $\tilde{p}$ belonging to the interval $(F_{\min}, F_{\max})$. Therefore the chaotic domain of the exact system is, in the main approximation, a strip $F_{\min} \le v \le F_{\max}$. It is reasonable to assume that distribution of values of v in the stochastic layer is uniform. Captures into the resonance followed by escapes from the resonance (see [20, 21]) are also possible in this system. However, probability of a capture is small, of order ε, and hence impact of these phenomena on the transport is small.

To check these conclusions, we take $U(q) = -\omega_0^2 \cos q$ and $F(\tau) = A\left(1 + 2\exp\left[-\alpha(\sin\tau)^2\right]\right)$. A plot of $F(\tau)$ is shown in Fig. 2.2, left; the plot of the corresponding function $f(\tau)$ is shown in Fig. 2.2, right (recall that $f = \mathrm{d}F/\mathrm{d}\tau$). In Fig. 2.3 we represent a sample of Poincaré section of a long phase trajectory of (2.2) and the corresponding histogram of $\bar{p}$ for this trajectory. The plots show that the distribution of $\bar{p}$ is close to the uniform one.

To find the mean velocity V_q in q-coordinate, we take into account that v is uniformly distributed in the chaotic domain. Thus V_q is velocity at a fixed value of v averaged over all v-s within the stochastic layer, i.e. over interval $(F_{\min}, F_{\max})$. In other words, one can find the value of displacement Δq on one period of

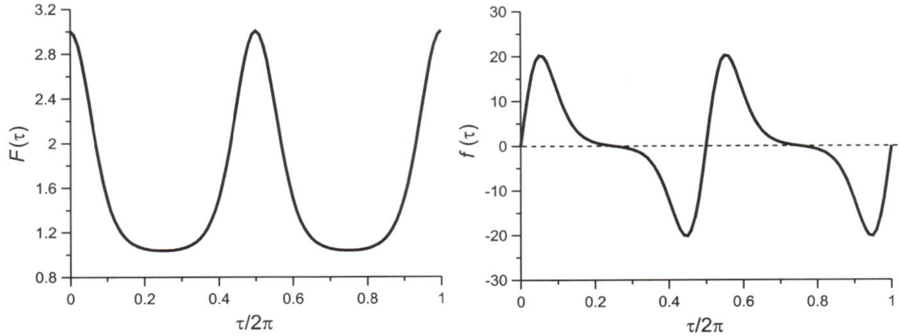

Fig. 2.2 *Left panel*: plot of the function $F(\tau)$ used in numerics. *Right panel*: plot of the external forcing $f(\tau)$

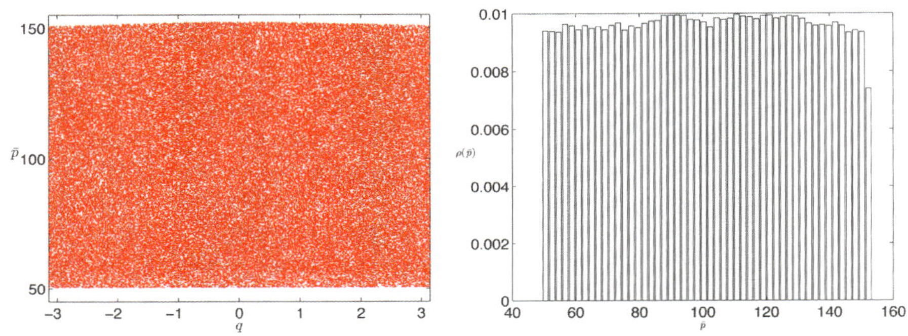

Fig. 2.3 *Left panel*: Poincaré section at $\tau = 0 \bmod 2\pi$ of a long phase trajectory ($5 \cdot 10^4$ dots) of system (2.2). All the points are mapped onto the interval $q \in (-\pi, \pi)$. $U(q) = -\omega_0^2 \cos q$, $F(\tau) = A\left(1 + 2\exp\left[-\alpha(\sin\tau)^2\right]\right)$ with $A = 0.5, \alpha = 4, \varepsilon = 0.01, \omega_0 = 1$. *Right panel*: histogram of $\bar{p}$ along the same phase trajectory

perturbation, then average it over the range of adiabatic invariant v corresponding to the chaotic domain, and find the average velocity of transport. Thus we find

$$\Delta q = \int_0^{T/\varepsilon} p\mathrm{d}t = \int_0^{T/\varepsilon} \left(\bar{p} - \varepsilon^{-1}F(\tau)\right)\mathrm{d}t = \frac{1}{\varepsilon^2}\int_0^T (v - F(\tau))\mathrm{d}\tau. \tag{2.5}$$

To find V_q, we have to integrate this expression over v from $F_{\min}$ to $F_{\max}$ and divide the result by $(F_{\max} - F_{\min})$ and by the length of the period of the external forcing T/ε. Thus we obtain

$$V_q = \frac{\varepsilon}{T(F_{\max} - F_{\min})}\int_{F_{\min}}^{F_{\max}} \Delta q\,\mathrm{d}v. \tag{2.6}$$

Table 2.1 Numerically found values of εV_q corresponding to various values of parameters ε, α in system (2.2) for $F(\tau) = A\left(1+2\exp\left[-\alpha(\sin\tau)^2\right]\right)$ (four upper rows, $A = 0.5$, $\omega_0 = 1$)

	$\alpha = 1$	$\alpha = 2$	$\alpha = 4$
$\varepsilon = 0.1$	0.046	0.128	0.253
$\varepsilon = 0.05$	0.046	0.112	0.225
$\varepsilon = 0.01$	0.0353	0.1050	0.2044
$\varepsilon = 0.005$	0.0369	0.1081	0.1916
$\varepsilon V_q^{\text{theor}}$	0.0389	0.1018	0.2006

In the bottom row theoretical values $\varepsilon V_q^{\text{theor}}$ obtained according to (2.8) are shown

Substituting Δq from (2.5) and integrating, one straightforwardly obtains

$$V_q = \frac{1}{2T\varepsilon}\int_0^{2\pi}(F_{\max}+F_{\min}-2F(\tau))\,\mathrm{d}\tau. \tag{2.7}$$

Note that formula (2.7) can be rewritten in a more elegant form as

$$V_q = \frac{1}{\varepsilon}\left(\frac{F_{\max}+F_{\min}}{2} - \langle F(\tau)\rangle\right), \tag{2.8}$$

where the angle brackets denote time average. The results of numerical checks of the formula are represented in Table 2.1. To obtain values presented in the table we integrated the system with Hamiltonian (2.2) on a long time interval $\Delta t = 2\pi\cdot 10^6/\varepsilon$.

Remarkably, formula (2.8) is the same for any smooth 2π-periodic potential (not necessarily harmonic). The potential may also depend periodically on time with the same period as that of the external force.

2.3 Small External Forcing

In the case considered in the previous section, chaotization of motion in the stochastic layer was a result of multiple resonance crossings. In the case to be studied in this section, the chaos is due to *separatrix* crossings. This produces somehow different estimates of the diffusion time. Besides, it results in a more complicated formula for the mean transport velocity.

2.3.1 Main Equations: Diffusion of the Adiabatic Invariant

Consider now the case when the external forcing is small, of order ε. The Hamiltonian equations of motion are

$$\dot{q} = p, \ \dot{p} = -\frac{\partial U}{\partial q} + \varepsilon f(\tau), \ \dot{\tau} = \varepsilon. \tag{2.9}$$

The time-dependent Hamiltonian function is

$$H = \frac{p^2}{2} + U(q) - \varepsilon f(\tau) q. \tag{2.10}$$

Consider for definiteness $U(q) = -\omega_0^2 \cos q$ (qualitative results do not depend on this choice). Similar to the previous section, we make a canonical transformation of variables $(p,q) \mapsto (\bar{p},\bar{q})$ using generating function $W_1 = (\bar{p} - F(\varepsilon t))q$, where $F(\tau)$ was defined in Sect. 2.2. Thus, $F(\tau)$ is again a periodic function defined up to an additive constant, which we are free to choose. To make the following presentation more clear, we choose this constant in such a way that the minimal value of F is $F_{\min} > 4\omega_0/\pi$. Note that $\bar{q} \equiv q$. After this transformation of variables, Hamiltonian of the system acquires the form (bars over q are omitted):

$$H = \frac{(\bar{p} - F(\tau))^2}{2} - \omega_0^2 \cos q. \tag{2.11}$$

This is a system explicitly depending on the slow time τ. A standard approach to study such a system is to consider it first at frozen τ, i.e. at $\tau = \text{const}$. Phase portrait of the system at a frozen value of τ (we call it the unperturbed system) is shown in Fig. 2.4. There is a separatrix on the portrait. It divides the phase space into the domains of direct rotations (above the upper branch of the separatrix), oscillations (between the separatrix branches), and reverse rotations (below the lower branch of the separatrix). Introduce the "action" I associated with a phase trajectory of the unperturbed system on this portrait. In the domains of rotation, I equals an area between the trajectory, the lines $q = -\pi$, $q = \pi$, and the axis $\bar{p} = 0$, divided by 2π; in the domain of oscillations, this is an area surrounded by the trajectory divided by 2π. It is known that I is an adiabatic invariant of (2.11): far from the separatrix its value is preserved along a phase trajectory with the accuracy of order ε on long time intervals (see, e.g., [21]).

Location of the separatrix on the $(q,\bar{p})$-plane depends on the value of $F(\tau)$. As τ slowly varies, the separatrix slowly moves up and down, and phase points cross the separatrix and switch its regime of motion from direct rotations to reverse rotations and vice versa. Recall known results on variation of the adiabatic invariant when a phase point crosses the separatrix. The area surrounded by the separatrix is constant, and hence, capture into the domain of oscillations is impossible in the first approximation (in the exact system, only a small measure of initial conditions correspond to phase trajectories that spend significant time in this domain; thus their influence on the transport is small). To be definite, consider the situation when the separatrix on the phase portrait slowly moves down. Thus, phase points cross the separatrix and change their mode of motion from reverse rotation to direct rotation. Let the action before the separatrix crossing at a distance of order 1 from the

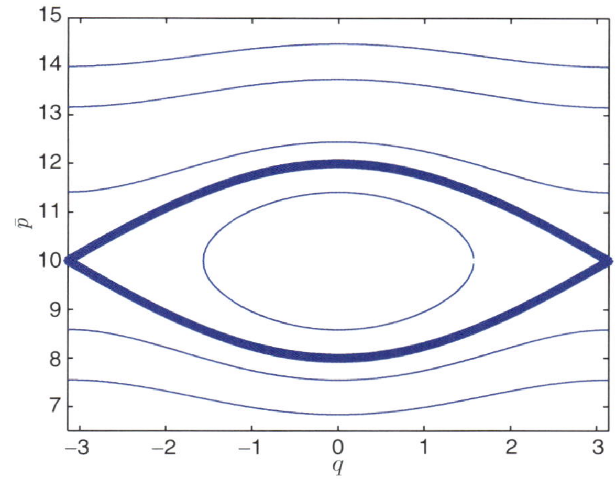

Fig. 2.4 Phase portrait of system (2.11) at frozen τ. The *bold line* is the separatrix

separatrix be $I = I_-$ and let the action after the crossing (also at a distance of order 1 from the separatrix) be $I = I_+$. In the first approximation, we have $I_+ = I_- + 8\omega_0/\pi$, i.e. the action increases by the value of the area inside the separatrix divided by 2π (see, e.g., [22, 23]). We shall call this change in the action a "geometric jump". If the separatrix contour slowly moves up and a phase point goes from the mode of direct rotation to the mode of reverse rotation, the corresponding value of the action decreases by the same value $8\omega_0/\pi$. Thus, in this approximation, the picture of motion looks as follows. While a phase point is in the domain of reverse rotation, the value of I along its trajectory stays constant: $I = I_-$. After transition to the domain of direct rotation, this value changes by the value of the geometric jump. The transition itself in this approximation occurs instantaneously. After the next separatrix crossing, the adiabatic invariant changes again by the value of the geometric jump, with the opposite sign, and returns to its initial value I_-. We call this approximation adiabatic.

In the next approximation, the value of action at the separatrix crossing undergoes a small additional jump. Consider for definiteness the case when the separatrix contour on the phase portrait moves down, and I_- and I_+ are measured when it is in its uppermost and lowermost positions, accordingly. Results of [12, 13] imply the following formula for the jump in the adiabatic invariant:

$$
\begin{aligned}
2\pi(I_+ - I_-) &= 16\omega_0 + 2a(1-\xi)\varepsilon\Theta\ln(\varepsilon\Theta) \\
&\quad + a\varepsilon\Theta\ln\frac{2\pi(1-\xi)}{\Gamma^2(\xi)} - 2b\varepsilon\Theta(1-\xi),
\end{aligned}
$$

where $a = \omega_0^{-1}$, $b = \omega_0^{-1}\ln(32\omega_0^2)$, and $\Theta = 2\pi F'(\tau_*)$. Here F' is the τ-derivative of F, τ_* is the value of τ at the separatrix crossing found in the adiabatic approximation, and $\Gamma(\cdot)$ is the gamma-function. Value ξ is a so-called pseudo-phase

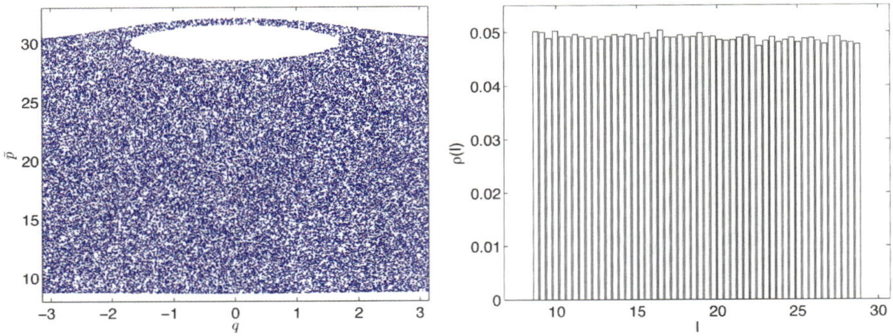

Fig. 2.5 *Left panel*: Poincaré section at $\tau = 0 \bmod 2\pi$ of a long phase trajectory ($5 \cdot 10^4$ dots). All the points are mapped onto the interval $q \in (-\pi, \pi)$. $F(\tau) = A\left(1 + 2\exp\left[-\alpha(\sin\tau)^2\right]\right)$ with $A = 10, \alpha = 16, \varepsilon = 0.005, \omega_0 = 1$. The empty region in the chaotic sea corresponds to phase points eternally locked in the domain of oscillations; they never enter the chaotic domain and do not participate in the transport. *Right panel*: histogram of I on the segment $(I_{\min}, I_{\max} - 8\omega_0/\pi)$ along the same phase trajectory

of the separatrix crossing; it strongly depends on the initial conditions and can be considered as a random variable uniformly distributed on interval $(0, 1)$ (see, e.g., [13]). Thus, value of the jump in the adiabatic invariant at the separatrix crossings has a quasi-random component of order $\varepsilon\ln\varepsilon$.

Similar to the case considered in Sect. 2.2, accumulation of small quasi-random jumps due to multiple separatrix crossings produces diffusion of adiabatic invariant (see, e.g., [13]). On a period of $F(\tau)$ (after two separatrix crossings), the action changes by a value of order $\varepsilon\ln\varepsilon$. Hence, after $N \sim \varepsilon^{-2}(\ln\varepsilon)^{-2}$ separatrix crossings, the adiabatic invariant varies by a value of order one. As a result, in time of order $t_{\text{diff}} \sim \varepsilon^{-3}(\ln\varepsilon)^{-2}$, the value of adiabatic invariant is distributed in all the range of values corresponding to the domain where phase points cross the separatrix on the phase plane; its distribution is close to the uniform one. We have checked this fact numerically for the same sample function $F(\tau)$ as in Sect. 2.2 at various parameter values. Poincaré sections and distribution histograms of I in all the cases look similar; see an example in Fig. 2.5.

2.3.2 Average Velocity of the Transport

Our aim is to find a formula for average velocity V_q along a phase trajectory on time intervals of order t_{diff} or larger. We first only take into consideration the geometric jumps, and afterwards, to obtain the final result, we take into account the mixing due to small quasi-random jumps. To simplify the consideration, assume again that function $F(\tau)$ has one local minimum $F_{\min}$ and one local maximum $F_{\max}$ on the interval $(0, T)$. The main results are valid without this assumption.

Introduce $\tilde{I}$, defined in the domains of rotation, as follows: it equals the area bordered by the trajectory, the line $\bar{p} = F(\tau)$, and the lines $q = -\pi$, $q = \pi$, divided by 2π. Thus, $\tilde{I} = |F(\tau) - I|$. Frequency of motion in the domains of rotation is $\omega(\tilde{I})$, where $\omega(\tilde{I})$ at $\tilde{I} > 4\omega_0/\pi$ is the frequency of rotation of a standard nonlinear pendulum with Hamiltonian:

$$H_0 = p^2/2 - \omega_0^2 \cos q,$$

expressed in terms of its action variable $\tilde{I}$. We do not need an explicit expression for function $\omega(\tilde{I})$. From Hamiltonian (2.11) we find $\dot{q} = \bar{p} - F(\tau)$. Consider a phase trajectory of the system frozen at $\tau = \bar{\tau}$ in a domain of rotation. Let the value of action on this trajectory be $I = I_0$ and the period of rotation be T_0 (note that $T_0 = 2\pi/\omega$ by definition). Then the value of $\dot{q}$ averaged over a period of rotation equals

$$\int_0^{T_0} \frac{|\dot{q}|}{T_0} \mathrm{d}t = 2\pi/T_0 = \omega(|F(\bar{\tau}) - I_0|).$$

Now consider a long phase trajectory in the case of slowly varying τ. Let on the interval (τ_1, τ_2) a phase point of (2.11) be below the separatrix contour. In the adiabatic approximation, the value I_0 of the adiabatic invariant along its trajectory is preserved on this interval. Hence, at $\tau \in (\tau_1, \tau_2)$ we have

$$2\pi F(\tau) - 2\pi I_0 \geq 8\omega_0, \tag{2.12}$$

and the equality here takes place at $\tau = \tau_1$ and $\tau = \tau_2$. In the process of motion on this time interval, q changes (in the main approximation) by a value:

$$\Delta q_-(I_0) = -\frac{1}{\varepsilon} \int_{\tau_1}^{\tau_2} \omega(F(\tau) - I_0) \mathrm{d}\tau. \tag{2.13}$$

On the interval $(\tau_2, \tau_1 + T)$ the phase trajectory is above the separatrix contour, and the value of the adiabatic invariant equals $\hat{I}_0 = I_0 + 8\omega_0/\pi$ due to the geometric jump. On this interval we have

$$2\pi F(\tau) - 2\pi I_0 \leq 8\omega_0. \tag{2.14}$$

In the process of motion on this time interval, q changes by a value:

$$\Delta q_+(I_0) = \frac{1}{\varepsilon} \int_{\tau_2}^{\tau_1 + T} \omega(|F(\tau) - \hat{I}_0|) \mathrm{d}\tau. \tag{2.15}$$

Total displacement in q on the interval $(\tau_1, \tau_1 + T)$ equals $\Delta q(I_0) = \Delta q_-(I_0) + \Delta q_+(I_0)$, and the average velocity on this interval is $\varepsilon \Delta q(I_0)/T$.

Consider now the motion on a long enough time interval $\Delta t \sim t_{\mathrm{diff}}$. Due to the diffusion in the adiabatic invariant described above, on this time interval, values of I_0, defined as a value of I when the phase point is *below* the separatrix contour, cover the interval $(I_{\min}, I_{\max} - 8\omega_0/\pi)$. Here $I_{\min} = F_{\min} - 4\omega_0/\pi$ and $I_{\max} = F_{\max} + 4\omega_0/\pi$. Assume that the distribution of I on this interval is uniform. To find the average velocity, we integrate $\varepsilon\Delta q(I_0)/T$ over this interval. Integrating (2.13) over I_0 and changing the order of integration we find

$$\int_{I_{\min}}^{I_{\max}-8\omega_0/\pi} \Delta q_- \mathrm{d}I_0 = -\frac{1}{\varepsilon}\int_0^T \mathrm{d}\tau \int_{I_{\min}}^{F(\tau)-4\omega_0/\pi} \omega(F(\tau)-I_0)\mathrm{d}I_0$$
$$= -\frac{1}{\varepsilon}\int_0^T \mathrm{d}\tau \int_{4\omega_0/\pi}^{F(\tau)-I_{\min}} \omega(\eta)\mathrm{d}\eta.$$

Now we take into account the equality

$$\omega(\tilde{I}) = \frac{\partial H_0(\tilde{I})}{\partial \tilde{I}}$$

(recall that $H_0(\tilde{I})$ is the Hamiltonian of a nonlinear pendulum as a function of its action variable) and obtain

$$-\int_0^T \mathrm{d}\tau \int_{4\omega_0/\pi}^{F(\tau)-I_{\min}} \omega(\eta)\mathrm{d}\eta = -\int_0^T (H_0(F(\tau)-I_{\min}) - H_0^s)\,\mathrm{d}\tau, \qquad (2.16)$$

where H_0^s is the value of H_0 on the separatrix. Similarly, integrating (2.15) we obtain

$$\int_{I_{\min}}^{I_{\max}-8\omega_0/\pi} \Delta q_+ \mathrm{d}I_0 = \frac{1}{\varepsilon}\int_0^T (H_0(I_{\max}-F(\tau)) - H_0^s)\,\mathrm{d}\tau. \qquad (2.17)$$

Adding (2.16) to (2.17) and dividing by $T(F_{\max} - F_{\min})/\varepsilon$ we find the expression for the average velocity V_q of transport on long time intervals:

$$V_q = \frac{1}{T(F_{\max} - F_{\min})}$$
$$\times \int_0^{2\pi} (H_0(I_{\max} - F(\tau)) - H_0(F(\tau) - I_{\min}))\,\mathrm{d}\tau. \qquad (2.18)$$

In (2.18), $H_0(I)$ can be found as the inverse function to $\tilde{I}(h)$, which defines action as a function of energy in domains of rotation of a nonlinear pendulum. For the latter function, the following formula holds (see, e.g., [24]):

$$\tilde{I}(h) = \frac{4}{\pi}\omega_0 \kappa \mathscr{E}(1/\kappa), \ \kappa \geq 1, \qquad (2.19)$$

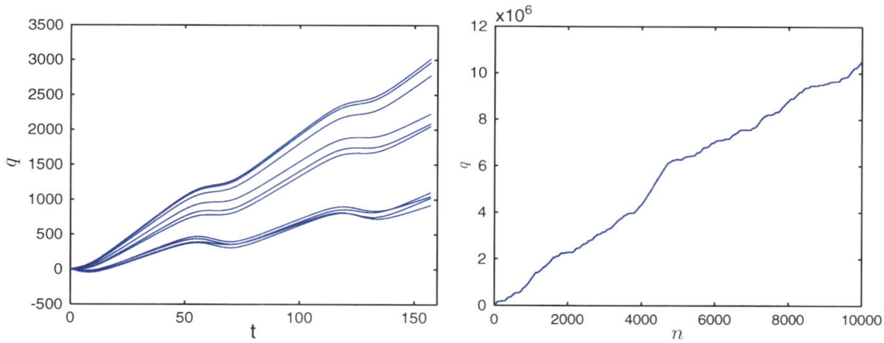

Fig. 2.6 *Left panel*: q against t for ten different initial conditions (comparatively short time interval), $\alpha = 4, \varepsilon = 0.05$. *Right panel*: q against the number of periods of the external force for a sample trajectory (10^4 periods), $\alpha = 16, \varepsilon = 0.05$. Parameter $A = 10$ in both cases

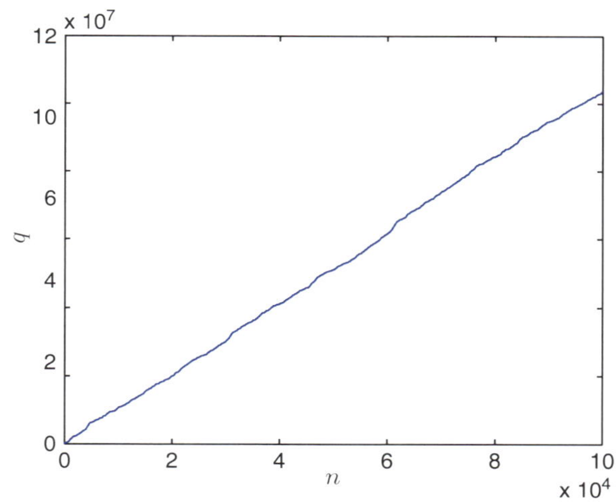

Fig. 2.7 Coordinate q against the number of periods of the external force for the same trajectory as in Fig. 2.6 (10^5 periods), $\alpha = 16, \varepsilon = 0.05, A = 10$

where $\kappa^2 = (1 + h/\omega_0^2)/2$, $\mathscr{E}(\cdot)$ is the complete elliptic integral of the second kind. If function $F(\tau)$ has several local extremes on the interval $(0, 2\pi)$, $F_{\min}$ and $F_{\max}$ in (2.18) are the smallest and largest values of F, respectively.

It can be seen from (2.18) that for function $F(\tau)$ of general type V_q is not zero, and hence there is the directed transport in the system. We checked this formula numerically for the sample function $F(\tau) = A\left(1 + 2\exp\left[-\alpha(\sin\tau)^2\right]\right)$, $\alpha > 0$ at various values of parameters ε and α. Typical plots of q against time t are shown in Figs. 2.6 and 2.7.

The results of numerical checks of formula (2.18) are represented in Table 2.2. To find numerical values of V_q presented in the table, we integrated the system with Hamiltonian (2.11) on a long time interval $\Delta t = 2\pi \cdot 10^6/\varepsilon$ with a constant time

Table 2.2 Numerically found values of V_q corresponding to various values of parameters ε, α (four upper rows, $A = 10, \omega_0 = 1$) and theoretical values V_q^{theor} obtained according to (2.18) (the bottom row)

	$\alpha = 4$	$\alpha = 8$	$\alpha = 16$
$\varepsilon = 0.1$	4.721	6.756	8.363
$\varepsilon = 0.05$	4.446	6.681	8.076
$\varepsilon = 0.01$	4.298	6.211	7.442
$\varepsilon = 0.005$	4.598	6.702	8.202
V_q^{theor}	4.393	6.679	8.110

step of $\pi/100$ (fifth order symplectic scheme [25]). Use of a symplectic scheme for long time simulations of Hamiltonian systems is necessary in order to ensure that creeping numerical error do not end up washing off the invariant tori bounding the chaotic domain. The table demonstrates satisfactory agreement between the formula and the numerics.

Finally, we note that formula (2.18) can be used also in the case of arbitrary (nonharmonic) spatially periodic time-independent potential in place of the term $-\omega_0^2 \cos q$ in (2.10) and (2.11). Of course, in this case function H_0 is different from the Hamiltonian of the nonlinear pendulum, but it always can be found, at least numerically.

2.4 Summary

To summarize, we have considered the phenomenon of the directed transport in a spatially periodic potential adiabatically influenced by a slow periodic in time-unbiased external force. We have shown that for the external force of a general kind the system exhibits directed transport on long time intervals. Direction and average velocity of the transport in the chaotic domain are independent of initial conditions and determined by properties of the external force. We studied two different cases: the case of small amplitude of the external force and the case, when this amplitude is a value of order one. We have obtained an approximate formula for average velocity of the transport and checked it numerically. The final formulas (2.8) and (2.18) are valid for any smooth periodic potential (not necessarily harmonic one).

Acknowledgements I am thankful to Neishtadt and Leoncini, who coauthored paper [19]. The work was partially supported by the RFBR grant 13-01-00251.

References

1. R.P. Feynmann, R.B. Leighton, M. Sands, *The Feynman Lectures on Physics*, Chap. 46, vol. 1 (Addison-Wesley, Reading, 1966)
2. P. Reimann, Phys. Rep. **361**, 57–265 (2002)
3. P. Jung, J.G. Kissner, P. Hänggi, Phys. Rev. Lett. **76**, 3436 (1996)
4. J.L. Mateos, Phys. Rev. Lett. **84**, 258 (2000)
5. O. Yevtushenko, S. Flach, K. Richter, Phys. Rev. E **61**, 7215 (2000)
6. S. Flach, O. Yevtushenko, Y. Zolotaryuk, Phys. Rev. Lett. **84**, 2358 (2000)
7. S. Denisov, S. Flach, Phys. Rev. E **64**, 056236 (2001)
8. S. Denisov et al., Phys. Rev. E **66**, 041104 (2002)
9. D. Hennig, L. Schimansky-Geier, P. Hänggi, Eur. Phys. J. **B 62**, 493–503 (2008)
10. G.M. Zaslavsky, Phys. Rep. **371**, 461–580 (2002)
11. A.I. Neishtadt, A.A. Vasiliev, Chaos **17**, 043104 (2007)
12. J. Tennyson, J.R. Cary, D.F. Escande, Phys. Rev. Lett. **56**, 2117–2120 (1986)
13. A.I. Neishtadt, Sov. J. Plasma Phys. **12**, 568–573 (1986).
14. Y. Elskens, D.F. Escande, Nonlinearity **4**, 615–667 (1991)
15. D.L. Vainshtein, A.A. Vasiliev, A.I. Neishtadt, Chaos **6**, 514–518 (1996).
16. A.I. Neishtadt, D.L. Vainshtein, A.A. Vasiliev, Physica D **111**, 227–242 (1998)
17. A.I. Neishtadt, A.A. Vasiliev, Nucl. Instr. Meth. Phys. Res. **A 561**, 158–165 (2006)
18. D.L. Vainchtein, J. Widloski, R.O. Grigoriev, Phys. Rev. Lett. **99**, 094501 (2007)
19. X. Leoncini, A. Neishtadt, A. Vasiliev, Phys. Rev. **E 79**, 026213 (2009).
20. A.I. Neishtadt, Celestial Mech. Dynam. Astron. **65**, 1–20 (1997).
21. V.I. Arnold, V.V. Kozlov, A.I. Neishtadt, *Mathematical Aspects of Classical and Celestial Mechanics (Encyclopedia of Mathematical Sciences 3)* (Springer, Berlin, 2006)
22. B.V. Chirikov, Sov. Phys. Doklady **4**, 390–393 (1959)
23. A.I. Neishtadt, J. Appl. Math. Mech. **39**, 594–605 (1975)
24. R.Z. Sagdeev, D.A. Usikov, G.M. Zaslavsky, *Nonlinear Physics: From Pendulum to Turbulence and Chaos* (Harwood Academic, Chur, 1992)
25. R.I. McLachlan, P. Atela, Nonlinearity **5**, 541 (1992)

Part II
From Chaos to Kinetics: Application to Hot Plasmas

Chapter 3
On the Nonlinear Electron Vibrations in a Plasma

Didier Bénisti

Abstract Many applications, including the control of parametric instabilities detrimental for inertial confinement fusion, which motivates the present work, require an accurate kinetic description of the electron vibrations in a plasma, henceforth called *electron plasma waves*. This issue actually gave rise to a countless number of papers, even beyond the plasma physics community, due to some fascinating effects like Landau damping, which is the most famous example of collisionless dissipation. However, very few theoretical results are available when the wave is so intense that it deeply traps a significant fraction of the electrons in its potential, and these results are mostly restricted to academic situations. By contrast, in this chapter we provide a description of nearly monochromatic electron plasma waves valid from the linear to the strongly nonlinear regime, using hypotheses general enough to address a real physics situation like stimulated Raman scattering in a fusion plasma. Completely new theoretical results are obtained regarding the collisionless dissipation and the dispersion relation of an electron plasma wave, whose accuracy was tested against very careful kinetic simulations of stimulated Raman scattering.

3.1 Introduction

This chapter describes the electron vibrations in a plasma, treated as collective processes involving space scales much larger than the average distance between two electrons. As is intuitively obvious, for dense enough plasmas like those considered in this chapter, collective effects dominate over individual ones, so that rapidly fluctuating fields resulting from pairwise electron interactions, which may be viewed as *collisions*, will henceforth be neglected. Moreover, when studying phenomena

D. Bénisti (✉)
CEA, DAM, DIF 91297 Arpajon, France
e-mail: didier.benisti@cea.fr

X. Leoncini and M. Leonetti (eds.), *From Hamiltonian Chaos to Complex Systems: A Nonlinear Physics Approach*, Nonlinear Systems and Complexity 5,
DOI 10.1007/978-1-4614-6962-9_3,

occurring over very short timescales, such as stimulated Raman scattering which originally motivated the present work, the ions may be considered as motionless, and this hypothesis will be used throughout this chapter. We therefore investigate here pure electron collective vibrations, which actually are nearly electrostatic waves that we call *electron plasma waves* (EPWs).

When an electron plasma wave propagates inside an initially Maxwellian plasma, its electrostatic energy decays and is converted into kinetic energy, although collisions are negligible. Hence, the propagation of an electrostatic wave in a collisionless plasma is a *dissipative* process. The best-known example of collisionless dissipation is Landau damping, i.e., the exponential decay of a freely propagating wave. This quite unexpected result was first derived more than six decades ago in [20] from the resolution of the linearized Vlasov-Poisson equations, and then proved mathematically very recently in [25] by Mouhot and Villani who tackled the nonlinear regime. As shown quite clearly in [2, 25], the homogenization of the distribution function due to the *nearly ballistic* electron motion is at the origin of Landau damping. Hence, although this effect remains after linear theory has broken down, it only exists provided that the initial wave amplitude is small enough, or only manifests itself for short times, before the electron motion has become "too nonlinear." As first shown by O'Neil in [26], for a nearly monochromatic and harmonic wave, the latter notion can be translated into the simple criterion that $\int_0^t \omega_B(t')dt'$ must be much less than 2π, where ω_B is the so-called bounce frequency, i.e., the frequency of a deeply trapped orbit. As regards the EPW propagation in the regime when $\int_0^t \omega_B(t')dt' > 2\pi$, this has been an issue for several decades, to which we provide definite results in this chapter.

Actually, when considering wave-particle interaction, the opposite limit of the Landau regime, when the interaction is very weak and the electrons are nearly freely streaming, is the adiabatic regime when the wave amplitude, $\mathscr{E}_p$, is very large and evolves slowly in space and time, so that the typical period of a "frozen orbit" (corresponding to a fixed value of $\mathscr{E}_p$) is much smaller than the typical timescale of variation of the wave amplitude, as experienced by the electrons. In such a regime, the electron motion is "enslaved" to the variations of $\mathscr{E}_p$, so that the distribution function may be directly related to the wave amplitude. Consequently, in the academic situation considered by O'Neil in [26] when the EPW amplitude is uniform, there is no way for the distribution function to evolve in time and, therefore, collisionless dissipation, and in particular Landau damping, cannot exist. In this chapter, we consider the more general situation of a wave packet whose amplitude may vary in the three space directions and show that, even in the adiabatic regime, the EPW experiences collisionless dissipation, which we relate to electron trapping and quantify. In particular, we establish that the wave is not Landau damped but that the nonlinear, and nonlocal, variations of the EPW group velocity entail the shrinking of the plasma wave packet, both in the longitudinal and transverse directions, and therefore the decrease of the electrostatic energy.

Moreover, we show here the quite unexpected result that the transition between the Landau and adiabatic regimes is very abrupt and occurs when $\int_0^t \omega_B(t')dt' \approx 6$. This allows us to provide an envelope equation for a nearly monochromatic plasma wave packet, which is valid in a three-dimensional geometry, from the linear to the strongly nonlinear regime. Moreover, our equation applies to a wave which either freely propagates or is laser driven, and addressing a driven wave is important for many reasons. First, a wave which has undergone collisionless dissipation may only have grown beyond the noise level if it has been driven. Then, to address such a phenomenon as Landau damping where one needs to specify the "initial condition," i.e., the electron distribution and the corresponding electrostatic field which is about to experience damping, one clearly needs to calculate how the plasma has been *driven* to such an initial condition. The same is clearly true to discuss collisionless dissipation in the strongly nonlinear regime, once Landau's theory has broken down. Moreover, addressing a driven wave is essential to correctly derive the nonlinear dispersion relation of an EPW that has grown in an initially Maxwellian plasma, which has also been an issue for several decades. Indeed, it is quite clear that only if it is driven may an EPW grow in initially Maxwellian plasma and remain sustained for a time long enough to see its collisionless damping rate, $\nu_{\rm NL}$, significantly reduced compared to the Landau value. However, all the previous results on the nonlinear dispersion relation of an EPW we know of rely on the hypotheses that $\nu_{\rm NL} = 0$ and that the wave has always been freely propagating, which leads to conclusions that will be widely discussed in chapter. In particular, when the wave amplitude grows, its frequency is known to downshift, and we will investigate here, both theoretically and numerically, how accounting for the drive may lead to values for the frequency shift different from previously published ones. Moreover, in this chapter, we will study in detail the ability to laser drive a large-amplitude plasma wave, and we will compare our theoretical results against those from Vlasov simulations. This will allow us to discuss the prediction made by Holloway and Dorning in [17] that an electron plasma wave cannot not exist in the strongly nonlinear regime when $\nu_{\rm NL} \approx 0$ if $k_p\lambda_{\rm D} > 0.53$, where k_p is EPW wave number and $\lambda_{\rm D}$ is the Debye length defined by Eq. (3.4).

This chapter is organized as follows. Section 3.2 describes the perturbative response to a slowly varying electrostatic wave whose amplitude only depends on time. From these results, we derive in Sect. 3.3 an envelope equation for the EPW, whose range of validity is furthermore extended to the non-perturbative regime. The results of Sect. 3.3 are discussed physically in Sect. 3.4 using a variational approach, which allows us to generalize the previous envelope equation to waves whose amplitudes vary in the three space dimensions. Section 3.5 is devoted to the dispersion relation of a laser-driven plasma wave, while Sect. 3.6 summarizes and concludes our work.

3.2 Perturbative Motion of Electrons Acted Upon by an Electrostatic Wave

In this section, we introduce the formalism that will be used throughout this chapter and apply it to the derivation of the perturbative electron response to a slowly varying electrostatic wave. Although a perturbative analysis is not enough to provide a correct description of the strongly nonlinear regime we want to address here, the results obtained in this section form the cornerstone of our theory.

3.2.1 General Formalism

Let us start by formulating the main hypotheses and main ideas our theoretical developments rest on. Since our analysis was originally motivated by the modeling of SRS, we will henceforth consider a (laser) driven electrostatic wave, so that the total longitudinal force acting upon the electrons is the sum of that due to the electrostatic wave and of that due to the drive. We moreover assume that the corresponding electrostatic and driving fields, E_{el} and E_{drive}, write in terms of a slowly varying envelope and of an eikonal, namely,

$$E_{\mathrm{el}} = -i(\mathscr{E}_p/2)\mathrm{e}^{i\varphi_p} + c.c., \tag{3.1}$$

$$E_{\mathrm{drive}} = (\mathscr{E}_d/2)\mathrm{e}^{i(\varphi_p-\delta\varphi)} + c.c., \tag{3.2}$$

where $\mathscr{E}_p$ and $\mathscr{E}_d$ are positive amplitudes such that $|\mathscr{E}_{p,d}^{-1}\partial_x\mathscr{E}_{p,d}| \ll k_p \equiv \partial_x\varphi_p$ and $|\mathscr{E}_{p,d}^{-1}\partial_t\mathscr{E}_{p,d}| \ll \omega_p \equiv -\partial_t\varphi_p$. In practice, ω_p is of the order of the so-called plasma frequency:

$$\omega_{pe} \equiv \sqrt{n\mathrm{e}^2/\varepsilon_0 m}, \tag{3.3}$$

n being the electron density, m its mass, and $-e$ its charge, while k_p is of the order of the Debye length:

$$\lambda_{\mathrm{D}} \equiv v_{\mathrm{th}}/\omega_{pe}, \tag{3.4}$$

where v_{th} is the thermal speed. For the parameters considered in this chapter, $\omega_{pe} \sim 10^{15}\,\mathrm{s}^{-1}$ while $\lambda_{\mathrm{D}} \sim 10^{-8}\,\mathrm{m}$.

We moreover assume that $\delta\varphi \ll \varphi_p$ (otherwise the electrostatic wave cannot be coherently driven). As for the total field $E \equiv E_{\mathrm{el}} + E_{\mathrm{drive}}$, it writes

$$E \equiv -i(E_0/2)\mathrm{e}^{i\psi} + c.c., \tag{3.5}$$

with $E_0 \equiv \sqrt{\mathscr{E}_p^2 + \mathscr{E}_d^2 - 2\mathscr{E}_p\mathscr{E}_d\sin(\delta\varphi)}$ and, clearly,

$$E_0\mathrm{e}^{i(\psi-\varphi_p)} = \mathscr{E}_p + i\mathscr{E}_d\mathrm{e}^{-i\delta\varphi}. \tag{3.6}$$

This total field may be viewed as an "effective" electrostatic wave, and we will henceforth study the motion of electrons acted upon by this *effective* wave. We will actually be mainly interested in the charge density, ρ, induced by this wave, which we will therefore write

$$\rho \equiv (\rho_0/2)\mathrm{e}^{i\psi} + c.c., \tag{3.7}$$

where ρ_0 is a slowly varying complex amplitude defined by the requirement (derived from Gauss' law) that

$$k_p\mathscr{E}_p - i\partial_x\mathscr{E}_p = (\rho_0/\varepsilon_0)\mathrm{e}^{i(\psi-\varphi_p)}. \tag{3.8}$$

At this stage, it is actually very convenient to introduce

$$\chi \equiv -\frac{\rho_0}{\varepsilon_0 k_p E_0}, \tag{3.9}$$

which, for purely time varying field amplitudes, reduces to the electron susceptibility. Then, by making use of Eq. (3.6), one easily finds that Eq. (3.8) translates into

$$k_p\mathscr{E}_p - i\partial_x\mathscr{E}_p = -k_p\chi(\mathscr{E}_p + i\mathscr{E}_d\mathrm{e}^{-i\delta\varphi}). \tag{3.10}$$

The real and imaginary parts of Eq. (3.10), respectively, yield

$$(1+\chi_r)\mathscr{E}_p = \mathscr{E}_d[\chi_i\cos(\delta\varphi) - \chi_r\sin(\delta\varphi)], \tag{3.11}$$

$$\chi_i\mathscr{E}_p - k_p^{-1}\partial_x\mathscr{E}_p = \mathscr{E}_d[-\chi_r\cos(\delta\varphi) - \chi_i\sin(\delta\varphi)], \tag{3.12}$$

where $\chi_r \equiv \mathrm{Re}(\chi)$ and $\chi_i \equiv \mathrm{Im}(\chi)$. From the previous equations it is clear that calculating χ_r and χ_i is enough to derive the nonlinear properties of a plasma wave (namely its dispersion relation, group velocity, and rate of conversion from electrostatic to kinetic energy), as well as how efficiently such a wave may be driven. One of the main purpose of this chapter is precisely to show how χ_r and χ_i may be derived directly from the investigation of the electron motion.

Before proceeding in the derivation of χ_r and χ_i, let us discuss the physics of Eqs. (3.11) and (3.12), and let us write them under a more convenient form. Equation (3.11) is the dispersion relation of the electron plasma wave (when $\mathscr{E}_d = 0$, one recovers the usual dispersion relation for a freely propagating wave, $1+\chi_r = 0$). In order to write it in a more convenient way, we use Eq. (3.12) to find

$$\chi_i = \frac{-\chi_r(\mathscr{E}_d/\mathscr{E}_p)\cos(\delta\varphi) + (k_p\mathscr{E}_p)^{-1}\partial_x\mathscr{E}_p}{1+(\mathscr{E}_d/\mathscr{E}_p)\sin(\delta\varphi)}. \tag{3.13}$$

Now, as will be made clear in a few lines, the dispersion relation for the driven plasma wave may be solved by making use of the adiabatic approximation, i.e., at 0-order in the space and time variations of the wave amplitudes. Then, neglecting the space derivative of $\mathscr{E}_p$ in Eq. (3.13) and plugging the corresponding value of χ_i into Eq. (3.10) yields

$$1 + \alpha_d\chi_r = 0, \tag{3.14}$$

with

$$\alpha_d \equiv \frac{1+2(\mathscr{E}_d/\mathscr{E}_p)\sin(\delta\varphi) + (\mathscr{E}_d/\mathscr{E}_p)^2}{1+(\mathscr{E}_d/\mathscr{E}_p)\sin(\delta\varphi)}. \tag{3.15}$$

Equation (3.12) is the envelope equation for the plasma wave which, when using Eq. (3.13) to estimate its right-hand side, reads

$$\chi_i \mathscr{E}_p - k_p^{-1}\partial_x \mathscr{E}_p = -\mathscr{E}_d \frac{\chi_r \cos(\delta\varphi) + (k_p\mathscr{E}_p)^{-1}\partial_x\mathscr{E}_p \sin(\delta\varphi)}{1+(\mathscr{E}_d/\mathscr{E}_p)\sin(\delta\varphi)} \approx \mathscr{E}_d \cos(\delta\varphi), \quad (3.16)$$

because $|(k_p\mathscr{E}_p)^{-1}\partial_x\mathscr{E}_p| \ll 1$ and $\mathscr{E}_d/\mathscr{E}_p \ll 1$. The relative values of $\mathscr{E}_d$ and $\mathscr{E}_p$ will be discussed in great detail in the next sections; however, one may notice from Eq. (3.12) that $\mathscr{E}_d/\mathscr{E}_p$ is of the order of χ_i, which is either of the order of the SRS growth rate or of the Landau damping rate, normalized to the plasma frequency, which are supposed to be small quantities.

Note also that we neglected the term proportional to $\partial_x\mathscr{E}_p$ in the right-hand side of Eq. (3.16) but not in its left-hand side. This is because, unless $\cos(\delta\varphi) \ll \sin(\delta\varphi)$ and the laser drive is essentially ineffective, $|\chi_r \cos(\delta\varphi)| \gg |(k_p\mathscr{E}_p)^{-1}\partial_x\mathscr{E}_p \sin(\delta\varphi)|$. By contrast, χ_i is essentially proportional to the space and time derivatives of $\mathscr{E}_p$, especially in the nonlinear regime once Landau damping has vanished, so that $(k_p\mathscr{E}_p)^{-1}\partial_x\mathscr{E}_p$ is not negligible compared to χ_i. Another way to understand our approximations is to remark that keeping the term proportional to $\partial_x\mathscr{E}_p$ in the right-hand side of Eq. (3.16) amounts to changing $(-k_p^{-1}\partial_x\mathscr{E}_p)$ into $(-k_p^{-1}\partial_x\mathscr{E}_p)[1-(\mathscr{E}_d/\mathscr{E}_p)\sin(\delta\varphi)]$ in the left-hand side of this equation, and $|(\mathscr{E}_d/\mathscr{E}_p)\sin(\delta\varphi)| \ll 1$.

Note that Eqs. (3.14) and (3.16) are, of course, valid whether the EPW is driven or not.

In order to derive χ we now need to relate it more specifically to the electron motion. To do so *we henceforth specialize, in all this section, to the case when the field amplitudes, and therefore ρ_0 and $\delta\varphi$, only depend on time*, and our results will be generalized to allow for three-dimensional (3-D) space variations of the fields in Sect. 3.4. Clearly, from Eq. (3.7), if ρ_0 only depends on time, then

$$\rho_0 = 2 \times \frac{1}{2\pi}\int_{-\pi}^{\pi} \rho e^{-i\psi} d\psi. \quad (3.17)$$

Moreover, from the very definition of the electron distribution function, f,

$$\rho = -n_0 e \left[\int_{-\infty}^{+\infty} f dv - 1\right], \quad (3.18)$$

where n_0 is the unperturbed electron density, so that

$$\rho_0 = -2n_0 e \times \frac{1}{2\pi}\int_{-\infty}^{+\infty}\int_{-\pi}^{\pi} f e^{-i\psi} d\psi dv \quad (3.19)$$

$$\equiv -2n_0 e \langle e^{-i\psi}\rangle, \quad (3.20)$$

where $\langle . \rangle$ stands for a local, in space, statistical averaging. Then, using Eq. (3.9) for χ, we find

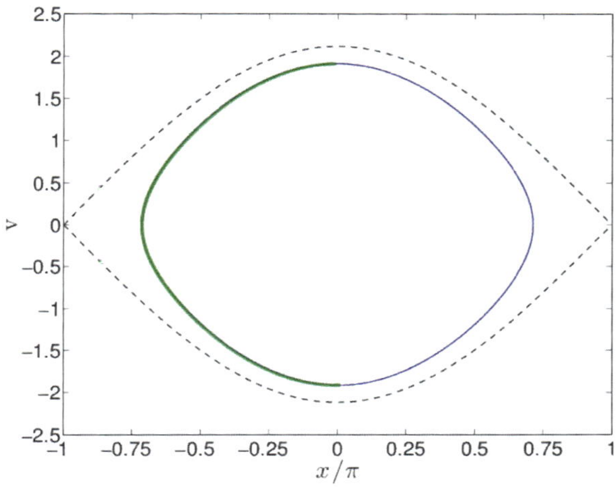

Fig. 3.1 Orbit, calculated between times $t_0 = 11,626.6$ and $t_1 = 11,635.3$, for the dynamics $\mathrm{d}x/\mathrm{d}t = v$, $\mathrm{d}v/\mathrm{d}t = -E_0 e^{\gamma_0 t}\sin(x)$, with $E_0 = 10^{-5}$ and $\gamma_0 = 10^{-3}$, and corresponding to the initial position and velocity, $x_0 = 0$ and $v_0 = 1$. The *thin solid line* (right part) is the actual orbit of the trapped electron. The *thick curve* (left part) is the symmetric image, with respect to the v-axis, of that part of the orbit lying on the half plane, $x > 0$. The *black dashed curve* is the virtual separatrix corresponding to the amplitude at $t = t_0$

$$\chi = \frac{2n_0 e\langle e^{-i\psi}\rangle}{\varepsilon_0 k_p E_0}, \tag{3.21}$$

which shows that χ is proportional to $\langle e^{-i\psi}\rangle$.

We now use heuristic arguments to take advantage of the latter result, and assume that calculating $\langle e^{-i\psi}\rangle$ amounts to averaging $e^{-i\psi}$ along all the electrons orbits in phase space. Then, because the wave amplitude varies very slowly in time, these orbits are nearly symmetric with respect to the velocity axis (see, e.g., Fig. 3.1). One may therefore calculate $\langle\cos(\psi)\rangle$ by assuming that these orbits are exactly symmetric with respect to the v-axis, which amounts to making the adiabatic approximation. By making use of this approximation, one may therefore derive a *non-perturbative* estimate (with respect to the wave amplitude) of $\langle\cos(\psi)\rangle$ that will be detailed in Sect. 3.5. χ_r will therefore be evaluated at 0-order in the variations of the fields amplitudes, which explains why the same approximation was made in order to cast the dispersion relation Eq. (3.11) in the form given by Eq. (3.13).

As regards $\langle\sin(\psi)\rangle$, if the electron orbits were exactly symmetric with respect to the velocity axis, then, because $\sin(\psi)$ is an odd function, averaging over such orbits would just yield $\langle\sin(\psi)\rangle = 0$ (which explains why the adiabatic estimate of χ_i is just $\chi_i = 0$). Now, clearly, these orbits are all the more symmetric as the typical timescale of variation of E_0 is large compared to the time it takes for ψ, or the polar angle in phase space, to change by 2π. The latter time is very close to $2\pi/\omega_B$ for a trapped orbit far enough from the virtual separatrix, where $\omega_B \equiv$

$\sqrt{eE_0k_p/m}$ (m being the electron mass) is the so-called bounce frequency. Hence, as shown in Fig. 3.1, when $\omega_B \gg E_0^{-1}\mathrm{d}E_0/\mathrm{d}t$, the orbits of "deeply" trapped electrons are nearly symmetric with respect to the v-axis, and such electrons contribute very little to $\langle\sin(\psi)\rangle$ and therefore to χ_i. Hence, their contribution will henceforth be disregarded. These "deeply" trapped electrons were found in [3] to be such that their initial velocities v_0 fulfill the condition $|v_0 - v_\phi| \leq V_l$, where

$$V_l \equiv \max\left[0; \frac{4}{\pi}\sqrt{\frac{eE_0}{k_pm}}\left(1 - \frac{3}{\int_0^t \omega_B \mathrm{d}t'}\right)\right]. \tag{3.22}$$

This condition may be understood the following way, $|v_0 - v_\phi| \leq \frac{4}{\pi}\sqrt{eE_0/k_pm}$ is the condition for an electron to be trapped, as derived by making use of the adiabatic approximation (see Sect. 3.5.1 for details). Accounting for the extra factor $(1 - 3/\int \omega_B \mathrm{d}t')$ in Eq. (3.22) amounts to defining an electron as "deeply trapped" provided that it has experienced a substantial fraction (about one half) of its trapped orbit. The latter condition is similar to the condition for an efficient phase mixing, as invoked by O'Neil in [26] to explain the nonlinear reduction of the Landau damping rate.

In conclusion, the function $\langle\sin(\psi)\rangle$ will be calculated the following way:

$$\langle\sin(\psi)\rangle \equiv \int_{|v_0 - v_\phi| \geq V_l} S(v_0, E_0) f_0(v_0) \mathrm{d}v_0, \tag{3.23}$$

where $f_0(v_0)$ is the electron distribution function in the limit of a vanishing field amplitude and $S(v_0, E_0)$ is the contribution to $\langle\sin(\psi)\rangle$ of those electrons whose unperturbed velocity is v_0.

3.2.2 Perturbative Analysis

In this section, we provide a first estimate of the function $S(v_0, E_0)$ in Eq. (3.23) using a perturbative analysis of the electron motion that we now detail.
The motion of electrons in the electrostatic field Eq. (3.5) is given by the following equations:

$$\mathrm{d}x/\mathrm{d}t = v, \tag{3.24}$$

$$\mathrm{d}v/\mathrm{d}t = (ie/2m)E_0 \mathrm{e}^{i\psi(x,t)} + c.c., \tag{3.25}$$

which derive from the Hamiltonian,

$$\mathscr{H} = v^2/2 + V(x,t), \tag{3.26}$$

where $-\partial_x V = (ie/2m)E_0 e^{i\psi(x,t)} + c.c.$ The perturbative technique consists in defining a new set of variables, x' and v', such that in these variables the particle motion is nearly unperturbed, i.e., v' remains nearly constant and, up to the accuracy of the perturbative scheme, may be identified with the initial velocity. The change in variables is defined by using a generative function $F(x,v',t)$ (see [14] for details) and is

$$x' = x + \partial F/\partial v', \tag{3.27}$$

$$v = v' + \partial F/\partial x. \tag{3.28}$$

In these new variables, the new Hamiltonian is

$$\mathscr{H}' = \mathscr{H} + \partial F/\partial t \tag{3.29}$$

$$= v'^2/2 + v'\partial_x F + (\partial_x F)^2/2 + V + \partial F/\partial t. \tag{3.30}$$

One would like to choose F so that $\mathscr{H}' = v'^2/2$, and v' is a constant of motion. This is usually done by perturbation, meaning that one uses the following expansion, $F = \sum_n E_0^n F_n$, such that v' is a constant up to terms of the order to E_0^{n+1}. Note though that, usually, the perturbation series does not converge, as will be discussed below.

3.2.2.1 First-Order Results

A first-order perturbative analysis amounts to choosing F so that

$$v'\partial_x F + \partial_t F = -V(x,t), \tag{3.31}$$

which would make v' constant up to terms of order E^2. Equation (3.31) is easily solved in

$$F(x,v',t) = -\int_0^t V[x - v'(t-t'),t']\mathrm{d}t'. \tag{3.32}$$

Assuming $v' \approx v_0$, we then find that, $\delta v \approx v - v' = \partial_x F$ is given by

$$\delta v = (ie/2m)\int_0^t E_0(t')e^{i\psi[x-v_0(t-t'),t']}\mathrm{d}t' + c.c. \tag{3.33}$$

A first-order calculation of the electron motion is equivalent to a linear analysis. However, since, as is clear from Eq. (3.23), we disregard the contribution of the deeply trapped electrons to derive χ_i, our result for χ_i will be different from the linear one.

From the very definition of the charge density Eq. (3.18), it is clear that if we write the electron velocity as $v = v_0 + \delta v$, then

$$\rho = -n_0 e \int f_0(v_0) \frac{\partial \delta v}{\partial v_0} \mathrm{d}v_0. \tag{3.34}$$

Plugging the value found at first order for δv into the latter expression for ρ yields

$$\rho_0(x,t)\mathrm{e}^{i\psi} = \frac{-in\mathrm{e}^2}{m} \int f_0(v_0) \frac{\partial}{\partial v_0} \left\{ \int_0^t E_0(t')\mathrm{e}^{i\psi[x - v_0(t-t'),t']} \mathrm{d}t' \right\} \mathrm{d}v_0. \tag{3.35}$$

Calculating the previous time integral by parts and at zero order in the space and time variations of k_p and ω_p, we find

$$\frac{\partial}{\partial v_0} \int_0^t E_0 \mathrm{e}^{i\psi} \mathrm{d}t' = k_p^{-2} \left[\frac{ik_p E_0}{(v_0 - v_\phi)^2} - \frac{2(\mathrm{d}E_0/\mathrm{d}t)}{(v_0 - v_\phi)^3} \right] \mathrm{e}^{i\psi}, \tag{3.36}$$

up to terms of the order of $\left[k_p^{-3}(v_0 - v_\phi)^{-4}(\mathrm{d}^2E_0/\mathrm{d}t^2)\right]$. Clearly, the previous estimate only makes sense for large enough values of $(v_0 - v_\phi)$ and, actually, plugging Eq. (3.36) into Eq. (3.35) would lead, when the integration is carried out over all velocities v_0, to a singular expression. Nevertheless, it is always possible to define E_0 in such a way that ω_p has a nonzero imaginary part, ω_p^i, and in the limit $\omega_p^i \to 0$ the integral,

$$I \equiv \int_{|v_0 - v_\phi| \geq V_l} 2 \frac{f_0(v_0) - f_0(v_\phi) - (v_0 - v_\phi) f_0'(v_\phi)}{k^3 (v_0 - v_\phi)^3} \mathrm{d}v_0, \tag{3.37}$$

converges whatever the value of V_l (it converges to its Cauchy principal part, which is well defined, when $V_l = 0$). In order to take advantage of this result, using the definition Eq. (3.9) for χ, the expression Eq. (3.35) for ρ_0, and the fact that the contribution to χ_i of the deeply trapped electrons may be disregarded, we now write χ_i as

$$\chi_i = \frac{\omega_{pe}^2}{k_p E_0} \mathrm{Re}[(\chi_1 + \chi_2)\mathrm{e}^{-i\psi}], \tag{3.38}$$

where $\omega_{pe} \equiv \sqrt{n_0 \mathrm{e}^2/\varepsilon_0 m}$ is the plasma frequency, and

$$\chi_1 \equiv ik_p \int_{|v_0 - v_\phi| \geq V_l} [f_0(v_0) - f_0(v_\phi) - (v_0 - v_\phi) f_0'(v_\phi)] \int_0^t (t' - t) E_0(t') \mathrm{e}^{i\psi} \mathrm{d}t' \mathrm{d}v_0, \tag{3.39}$$

$$\chi_2 \equiv ik_p f_0'(v_\phi) \int_0^t (t' - t) E_0(t') \int_{|v_0 - v_\phi| \geq V_l} (v_0 - v_\phi) \mathrm{e}^{i\psi(x - v_0(t-t'),t')} \mathrm{d}v_0 \mathrm{d}t'. \tag{3.40}$$

Now, from our previous discussion, it is valid to calculate the time integral in χ_1 by parts as in Eq. (3.36), which yields

$$\mathrm{Re}(\chi_1 \mathrm{e}^{-i\psi}) \approx -k_p I(\mathrm{d}E_0/\mathrm{d}t), \tag{3.41}$$

where I is given by Eq. (3.37). When $V_l = 0$, $-\omega_{pe}^2 I \equiv -\partial_\omega \chi_r^{\text{lin}}$, where χ_r^{lin} is the linear value of the real part of the electron susceptibility, as derived, for example, in [15], calculated by making use of the adiabatic approximation (i.e., at 0-order in the variations of the wave amplitude). The value of $-\omega_{pe}^2 I$ does not vary much with the wave amplitude unless V_l is of the order of, or larger than, the thermal velocity, v_{th}.

As for χ_2, since

$$\int_{-\infty}^{+\infty} (v_0 - v_\phi) \mathrm{e}^{ik_p(v_0-v_\phi)(t'-t)} \mathrm{d}v_0 = (-2i\pi/k_p^2)\partial_{t'}\delta(t'-t), \tag{3.42}$$

where $\delta(t)$ is the Dirac distribution, and since at 0-order in the variations of k_p and ω_p we may replace $\psi(x,t)$ by $(k_p x - \omega_p t)$, we find that when $V_l = 0$

$$\chi_2 = -\frac{2\pi}{k_p} f_0'(v_\phi) \mathrm{e}^{i\psi} \int_0^t (t'-t) E_0(t') \partial_{t'} \delta(t'-t) \mathrm{d}t' \tag{3.43}$$

$$= \frac{-\pi}{k_p} E_0(t) f_0'(v_\phi) \mathrm{e}^{i\psi}. \tag{3.44}$$

Hence a first-order perturbation analysis yields, when $V_l = 0$ (which corresponds to the linear limit),

$$\chi_i = \partial_\omega \chi_r^{\text{lin}} \frac{(\mathrm{d}E_0/\mathrm{d}t)}{E_0} - \frac{\pi\omega_{pe}^2}{k_p^2} f_0'(v_\phi). \tag{3.45}$$

Now, because $\mathscr{E}_d \ll \mathscr{E}_p$, $E_0^{-1}(\mathrm{d}E_0/\mathrm{d}t) \approx \mathscr{E}_p^{-1}(\mathrm{d}\mathscr{E}_p/\mathrm{d}t)$ so that

$$\chi_i \approx \partial_\omega \chi_r^{\text{lin}} \frac{(\mathrm{d}\mathscr{E}_p/\mathrm{d}t)}{\mathscr{E}_p} - \frac{\pi\omega_{pe}^2}{k_p^2} f_0'(v_\phi). \tag{3.46}$$

Plugging this value of χ_i into Eq. (3.16) yields the following envelope equation for the electron plasma wave (when $\mathscr{E}_p$ depends on time only):

$$\mathrm{d}\mathscr{E}_p/\mathrm{d}t + \nu_{\text{L}}\mathscr{E}_p = \mathscr{E}_d \cos(\delta\varphi)/\partial_\omega \chi_r^{\text{lin}}, \tag{3.47}$$

where $\nu_{\text{L}} \equiv -\pi\omega_{pe}^2 f_0'(v_\phi)/(k_p^2 \partial_\omega \chi_r^{\text{lin}})$ is the Landau damping rate, derived in [20]. Since the latter equation is valid whether the wave is driven or not, it unambiguously shows that, if an EPW is driven to a level significantly larger than that due to electrostatic fluctuations, yet small enough for $V_l = 0$, and if this EPW is then left freely propagate, it will necessarily damp at the rate derived by Landau. The non-Landau damping predicted by Belmont et al. in [2] cannot be found in that case.

Let us now address the nonlinear regime, $V_l > 0$. When $[k_p V_l]^{-1}$ is much less than the typical time, τ_0, over which E_0 varies, calculating the time integral in Eq. (3.35) by parts is valid, which yields

$$
\begin{aligned}
\chi_i &\approx \omega_{pe}^2 \frac{(\mathrm{d}E_0/\mathrm{d}t)}{E_0} \int_{|v_0-v_\phi|\geq V_l(t)} \frac{-2f_0(v_0)}{k_p^3(v_0-v_\phi)^3}\mathrm{d}v_0 \\
&\equiv \frac{(\mathrm{d}E_0/\mathrm{d}t)}{E_0}\partial_\omega \chi_r^{\mathrm{eff},1},
\end{aligned} \tag{3.48}
$$

where $\chi_r^{\mathrm{eff},1}$ is the real part of an "effective" susceptibility, which does not account for the contribution of the deeply trapped electrons, and which is derived from a first-order perturbative analysis of the electron motion and at 0-order in the variations of the wave amplitude (which amounts to making use of the adiabatic approximation). Plugging this value of χ_i into equation Eq. (3.16) yields

$$
\mathrm{d}\mathscr{E}_p/\mathrm{d}t = \mathscr{E}_d \cos(\delta\varphi)/\partial_\omega \chi_r^{\mathrm{eff},1}, \tag{3.49}
$$

which shows that there is no damping term in the envelope equation for the EPW when k_pV_l has become so large compared to τ_0^{-1} that an integration by parts yields an accurate estimate for the change in velocity, δv, of the electrons which significantly contribute to χ_i. Note again that $V_l > 0$ when $\int_0^t \omega_B \mathrm{d}t' > 3$, i.e., when the first trapped electrons have completed about one-half of their trapped orbit, so that our result is consistent with that published by O'Neil in [26] on the nonlinear reduction of Landau damping for a wave with constant and uniform amplitude.

We therefore successfully derived an explicit expression for χ_i, both when $V_l = 0$ and when $k_pV_l \gg \tau_0^{-1}$. Now, it would be very convenient to have a practical formula for χ_i, and especially for the term χ_2 (since the value of χ_1 does not change much) valid whatever V_l. First note that, in the limit $k_pV_l \gg \tau_0^{-1}$, calculating the time integral in Eq. (3.40) by parts easily yields

$$
\mathrm{Re}(\chi_2 \mathrm{e}^{-i\psi}) \approx \frac{-4f_0'(v_\phi)}{k_p^2 V_l}\frac{\mathrm{d}E_0}{\mathrm{d}t}. \tag{3.50}
$$

In the opposite limit, $k_pV_l \ll \tau_0^{-1}$, we may write $\chi_2 \equiv (-E_0/k_p)f_0'(v_\phi)\mathrm{e}^{i\psi}[\pi+\delta\chi_2]$, with

$$
\delta\chi_2 \equiv \frac{k_p}{E_0}\int_0^t (t'-t)E_0(t')\partial_{t'}G(t'-t)\mathrm{d}t', \tag{3.51}
$$

where

$$
\begin{aligned}
G(t'-t) &= \int_{-V_l}^{V_l} \mathrm{e}^{ik_p(v_0-v_\phi)(t-t')}\mathrm{d}v_0 \\
&= \frac{2\sin[k_p(V_l-v_\phi)(t-t')]}{k_p(t'-t)}.
\end{aligned} \tag{3.52}
$$

Clearly, the timescale of variations of G is V_l^{-1}, while $\partial_{t'}G|_{t'=t} = 0$, and $\partial_{t'}^2 G|_{t'=t} = 2k_p^2V_l^3/3$. Integrating the right-hand side of Eq. (3.51) three times by parts then yields

$$\delta\chi_2 \approx -\frac{4(k_pV_l)^3}{3E_0}\int_0^t\int_0^{t'}\int_0^{t''}E_0(u)\mathrm{d}u\mathrm{d}t''\mathrm{d}t'. \tag{3.53}$$

These results for χ_2 are now to be compared to those obtained by assuming $E_0(t) = \mathscr{E}_0 \mathrm{e}^{\Gamma t}$, where $\mathscr{E}_0$ and Γ are constants. In this case, it is straightforwardly found

$$\mathrm{Re}(\chi_2\mathrm{e}^{-i\psi}) = \frac{-f_0'(v_\phi)E_0}{k_p}\left[\pi - 2\tan^{-1}\left(\frac{k_pV_l}{\Gamma}\right) + \frac{2\Gamma k_pV_l}{\Gamma^2+(k_pV_l)^2}\right]. \tag{3.54}$$

Using this formula, one recovers results similar to those found in the general case, i.e., when $k_pV_l \gg \Gamma$,

$$\mathrm{Re}(\chi_2\mathrm{e}^{-i\psi}) \approx \frac{-4f_0'(v_\phi)}{k_p^2V_l}\Gamma E_0, \tag{3.55}$$

and when $k_pV_l \ll \Gamma$, $\mathrm{Re}(\chi_2\mathrm{e}^{-i\psi}) = (-E_0/k_p)f_0'(v_\phi)[\pi+\delta\chi_2]$ with

$$\delta\chi_2 \approx -\frac{4}{3}\left(\frac{k_pV_l}{\Gamma}\right)^3. \tag{3.56}$$

It is worth noting here that, when $E_0 = \mathscr{E}_0\mathrm{e}^{\Gamma t}$, Γ may be interpreted as $\Gamma = E_0^{-1}(\mathrm{d}E_0/\mathrm{d}t)$ or as $\Gamma = E_0/\int_0^t E_0(u)\mathrm{d}u$ (when $t \gg \Gamma^{-1}$). Using the definition $\Gamma = E_0^{-1}(\mathrm{d}E_0/\mathrm{d}t)$, Eq. (3.55) is exactly the same as Eq. (3.50), while Eqs. (3.53) and (3.56) compare better if one uses for Γ, $\Gamma = E_0/\int_0^t E_0(u)\mathrm{d}u$. Then, one may think of using Eq. (3.54) as a practical formula for $\mathrm{Re}(\chi_2\mathrm{e}^{-i\psi})$ valid whatever V_l and whatever the variations of E_0, with Γ continuously changing from $E_0/\int_0^t E_0(t')\mathrm{d}t'$ when $k_pV_l \ll \tau_0^{-1}$ to $E_0^{-1}(\mathrm{d}E_0/\mathrm{d}t)$ when $k_pV_l \gg \tau_0^{-1}$, where τ_0 is the typical time of variations of E_0. We therefore propose to use Eq. (3.54) for $\mathrm{Re}(\chi_2\mathrm{e}^{-i\psi})$ in the general case, with

$$\Gamma \equiv \frac{E_0(t) - E_0[t-\pi/(k_pV_l)]}{\int_{t-\pi/(k_pV_l)}^t E_0(u)\mathrm{d}u}, \tag{3.57}$$

which has the desired properties, $\Gamma \approx E_0/\int_0^t E_0(t')\mathrm{d}t'$ when $k_pV_l \ll \tau_0^{-1}$ and $\Gamma \approx E_0^{-1}(\mathrm{d}E_0/\mathrm{d}t)$ when $k_pV_l \gg \tau_0^{-1}$. In Sect. 3.4 we will see that using Eq. (3.54) for $\mathrm{Re}(\chi_2\mathrm{e}^{-i\psi})$ with Γ given by Eq. (3.57) will provide a practical and accurate analytic formula for the nonlinear counterpart of the Landau damping rate of an SRS-driven plasma wave.

3.2.2.2 Higher-Order Results

We worked out the perturbation analysis of the electron motion, when $E_0 = \mathscr{E}_0\mathrm{e}^{\Gamma t}$, up to the 11th order. The corresponding tedious calculations will of course not be reproduced here, nor will be the high-order formulas for $\langle\sin(\psi)\rangle$, which may nevertheless be found in [3]. Here, we will just outline the limits and advantages of using high-order results.

First of all, although rigorous estimates remain to be done, the small parameter, ε, of the perturbative expansion appears to be of the order of $\varepsilon = \omega_B^2/[\Gamma^2+(k_pV_l)^2]$ and is therefore indeed small when $\omega_B \ll \Gamma$, while in the opposite limit $\omega_B \gg \Gamma$, $\omega_B \approx k_pV_l$ and $\varepsilon \approx 1$. Hence, unlike in the situation where one tries to calculate perturbatively the motion of all electrons (i.e., when $V_l = 0$) and where ε would be a *big* parameter whenever $\omega_B \gg \Gamma$, here we are in a limit situation where $\varepsilon \to 1$ when $\omega_B/\Gamma \to \infty$. On a more physical basis, the less symmetric orbits (with respect to the v-axis) are those close to the virtual separatrix so that, as ω_B increases, the relative contribution to $\langle\sin(\psi)\rangle$ from electrons lying on those orbits becomes more important, and it is well known that the motion close to the separatrix is not perturbative, with respect to the wave amplitude. We therefore expect our perturbative estimate of $\langle\sin(\psi)\rangle$ to eventually break down but to remain valid beyond the regime $\omega_B \ll \Gamma$.

This is exactly what we observe when comparing our perturbative estimate of $\langle\sin(\psi)\rangle$ to that derived from test particle simulations, as shown in Fig. 3.2. These simulations consist in calculating the motion of electrons acted upon by an exponentially growing electrostatic wave of constant phase velocity, and $\langle\sin(\psi)\rangle$ is estimated numerically by using the formula:

$$\langle\sin(\psi)\rangle = \sum_{n=1}^{N} f_0(v_{0,n})\sin(\psi_n), \tag{3.58}$$

where N is the total number of electrons in the simulation, $v_{0,n}$ is the initial velocity of the nth electron, and $\psi_n \equiv k_px_n - \omega_pt$, where x_n is the position of electron #n at time t. In the case of Fig. 3.2, f_0 is a Maxwellian, and the electrons are initially uniformly distributed in space (over 25 different positions for each initial velocity) and in velocity (over 1,000 different velocities ranging from $-10v_{\text{th}}$ to $10v_{\text{th}}$, where v_{th} is the thermal velocity), the wave phase velocity is $v_\phi = 3v_{\text{th}}$, and its growth rate is $\Gamma = \omega_{pe}/10$. From Fig. 3.2 it may be seen that, when using an 11th order analysis, good agreement (relative discrepancy less than 15 %) between the theoretical and numerical values of $\langle\sin(\psi)\rangle$ is found up to $\omega_B/\Gamma \approx 15$, while when using a first-order analysis, such a good agreement may only be found when $\omega_B/\Gamma < 3$. Hence, using higher-order expansions yields a good estimate for $\langle\sin(\psi)\rangle$ up to larger values of ω_B/Γ. However, clearly, a non-perturbative theory is needed to address the electron response when the wave bounce frequency is much larger than its growth rate. Non-perturbative values of $\langle\sin(\psi)\rangle$, accurate whenever $\omega_B/\Gamma > 3$, will be provided in the next section. Then, as will be shown in Sect. 3.3, by "connecting" perturbative and non-perturbative values of $\langle\sin(\psi)\rangle$, one may get a very good estimate of this function whatever the wave amplitude, and, going to higher order in the perturbative expansion just increases the accuracy of the theoretical result.

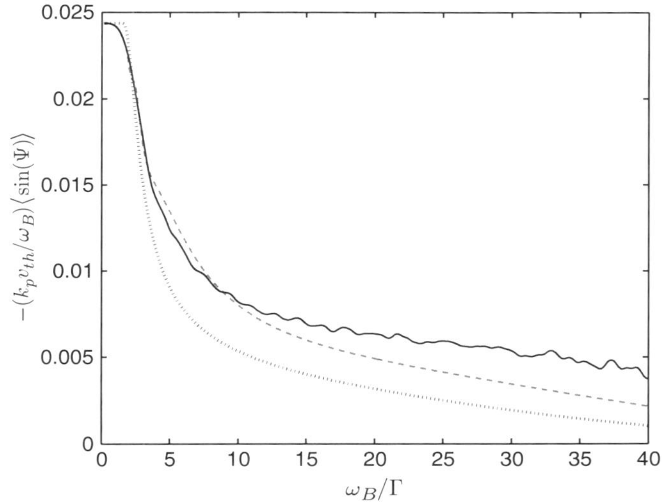

Fig. 3.2 $-(k_p v_{\mathrm{th}}/\omega_B)\langle\sin(\psi)\rangle$ versus ω_B/Γ as calculated by using a first-order perturbation analysis (*dotted line*), an 11th order analysis (*dashed line*), and from test particle simulations (*solid line*)

Before ending this section we need to make two important remarks. First, one could generalize the high-order formulas obtained when $E_0 = \mathscr{E}_0 \mathrm{e}^{\Gamma t}$ by using Eq. (3.57) for Γ. Second, in the limit $\omega_B \gg \Gamma$ one finds $\chi_i \approx E_0^{-1}(\mathrm{d}E_0/\mathrm{d}t)\partial_\omega \chi_r^{\mathrm{eff},n}$ where $\chi_r^{\mathrm{eff},n}$ is the real part of an "effective" susceptibility, which does not account for the contribution of the deeply trapped electrons and which is derived from an nth order perturbative analysis of the electron motion and at 0-order in the variations of the wave amplitude (which amounts to making use of the adiabatic approximation). Hence the result obtained at first order generalizes to higher orders. The corresponding tedious proof of this result will not be provided here but will be illustrated numerically in the next section.

3.3 Envelope Equation for a Purely Time-Dependent Wave Amplitude

In this section, we generalize the perturbative results obtained previously to derive a *non-perturbative* expression for χ_i, from which we deduce a very accurate envelope equation for the electron plasma wave, including an explicit analytic expression for the nonlinear counterpart of its Landau damping rate. We restrict here to the situation when the wave amplitude only depends on time and generalize our results in the next section by making us of a variational approach.

Let us first derive a theoretical estimate for the imaginary part of χ [defined by Eq. (3.9)], which we obtain, once again, by making use of heuristic arguments.

Heuristically, one would like to think of the wave growth (or decay) rate Γ as the imaginary part of the wave frequency and use the following Taylor expansion:

$$\chi_i(\omega_p + i\Gamma) \approx \chi_i(\omega_p + i0) + \Gamma \partial_\omega \chi_r. \tag{3.59}$$

As will be discussed in detail in the next section, such an expansion is actually not valid because of the collisionless dissipation induced by trapping. However, since the deeply trapped electrons contribute very little to χ_i, the former difficulty may be alleviated by withdrawing the contribution of these electrons in the expansion Eq. (3.59), which then writes

$$\chi_i(\omega + i\Gamma) \approx \chi_i^{\text{eff}}(\omega + i0) + \Gamma \partial_\omega \chi_r^{\text{eff}}, \tag{3.60}$$

where the superscript "eff" means, as in Sect. 3.2, that the contribution of the deeply trapped electrons has been disregarded. When plugging the latter expression for χ_i into the envelope equation (3.16), one clearly sees that the term $\chi_i^{\text{eff}}(\omega + i0)$ accounts for collisionless damping. Now, from O'Neil's work of [26], this term is expected to decrease and become negligible as $\int \omega_B \mathrm{d}t$ increases. Moreover, from an nth order perturbative analysis, we found in Sect. 3.2 that, indeed, when $\int \omega_B \mathrm{d}t$ is large, $\chi_i \approx \Gamma \partial_\omega \chi_r^{\text{eff},n}$ where $\chi_r^{\text{eff},n}$ is the nth order estimate of χ_r^{eff} calculated by making use of the adiabatic approximation. Now, clearly, there is no need to resort to a perturbation analysis to evaluate $\partial_\omega \chi_r^{\text{eff}}$ adiabatically, and how to do this is explained in detail in [3] and briefly recalled in the Appendix. Then, for large values of $\int \omega_B \mathrm{d}t$, we are naturally led to the following *non-perturbative* estimate for χ_i:

$$\chi_i \approx E_0^{-1}(\mathrm{d}E_0/\mathrm{d}t)\partial_\omega \chi_r^{\text{eff}} \approx \mathscr{E}_p^{-1}(\mathrm{d}\mathscr{E}_p/\mathrm{d}t)\partial_\omega \chi_r^{\text{eff}}. \tag{3.61}$$

3.3.1 Exponentially Growing Wave

As in Sect. 3.2, in order to test the accuracy of the latter expression, we compare the values of $\langle \sin(\psi) \rangle$ deduced from Eq. (3.61) to results from test particle simulations, for an exponentially growing wave. As shown in Fig. 3.3a, there exists a range in ω_B/Γ where the values of $\langle \sin(\psi) \rangle$ deduced from Eq. (3.61) match the perturbative ones, whether one uses a first-order or an 11th order expansion. However, the range in ω_B/Γ is larger and the matching is better when using a higher-order analysis. Moreover, as may be seen in Fig. 3.3b, using Eq. (3.61) yields very accurate values for $\langle \sin(\psi) \rangle$ whatever $\omega_B/\Gamma \geq 3$.

Another interesting feature illustrated in Fig. 3.3a is the very abrupt convergence of χ_i towards $\Gamma \partial_\omega \chi_r^{\text{eff}}$ when $\omega_B/\Gamma \geq 3$. Since, when $\omega_B/\Gamma \leq 3$, a perturbative

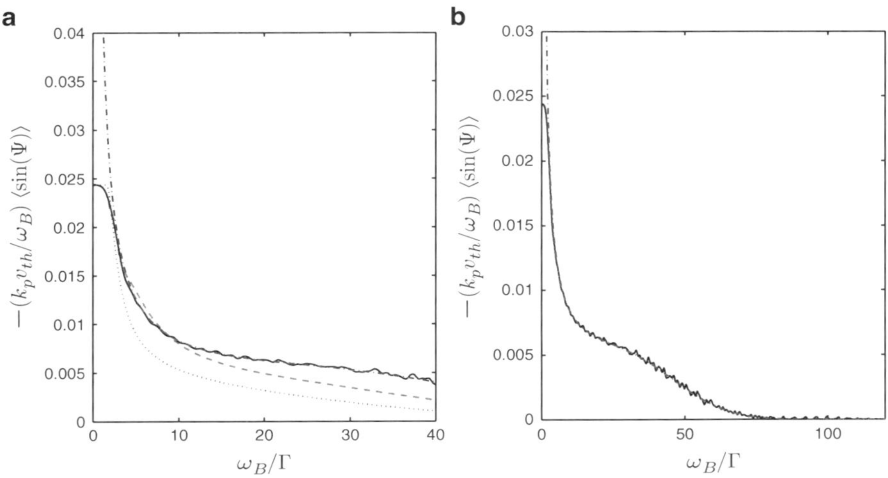

Fig. 3.3 $-(k_p v_{\rm th}/\omega_B)\langle\sin(\psi)\rangle$ versus ω_B/Γ as calculated by using a first-order perturbation analysis (*dotted line*) and an 11th order perturbative analysis (*dashed line*), from Eq. (3.61) (*dash–dotted line*) and from test particle simulations (*solid line*) for the same conditions as in Fig. 3.2

estimate of χ_i is quite accurate, one is therefore naturally led to the following estimate for χ_i:

$$\chi_i \approx \chi_i^{\rm per} \times [1 - Y(\omega_B/3\Gamma)] + \Gamma\partial_\omega\chi_r^{\rm eff} \times Y(\omega_B/3\Gamma), \tag{3.62}$$

where $\chi_i^{\rm per}$ is the perturbative value of χ_i given in Sect. 3.2 and where $Y(x)$ grows from 0 to 1 as x increases. Moreover, since the convergence of χ_i to $\Gamma\partial_\omega\chi_r^{\rm eff}$ is quite abrupt when $\omega_B \geq 3\Gamma$, we choose $Y(x)$ so that it rises very quickly from 0 to 1 as x becomes larger than unity, namely, we choose

$$Y(x) = \tanh^5[(\mathrm{e}^x - 1)^3]. \tag{3.63}$$

As may be seen in Fig. 3.4a, Eq. (3.62) yields an excellent estimate for χ_i when $\chi_i^{\rm per}$ is calculated at the 11th order, while, as shown in Fig. 3.4b, using a first-order perturbation analysis already provides a good accuracy.

3.3.2 *Generalized Expression for χ_i*

For an exponentially growing wave, and for large enough values of Γt, $2\omega_B/\Gamma \approx \int_0^t \omega_B \mathrm{d}t'$ which is a well-known parameter to measure the degree of nonlinearity of the electron motion (see [26] for example). Hence, it was expected that, for

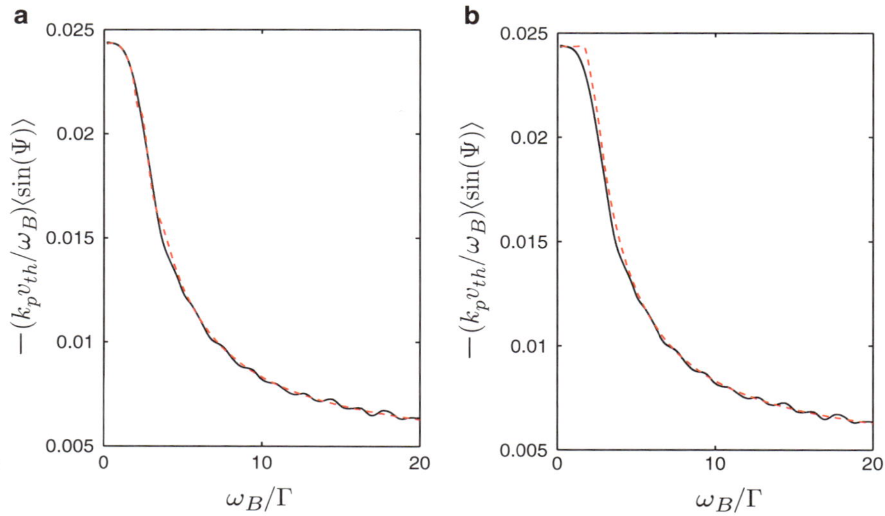

Fig. 3.4 $-(k_p v_{\text{th}}/\omega_B)\langle\sin(\psi)\rangle$ versus ω_B/Γ as calculated from test particle simulations (*solid line*) and from Eq. (3.62) (*dashed line*) with, panel (**a**), χ_i^{per} calculated at the 11th order and, panel (**b**), χ_i^{per} calculated at first order

an exponentially growing wave, perturbative results would break down for large enough values of ω_B/Γ. Moreover, in order to make the results derived in the previous section valid whatever the time evolution of the wave amplitude, we are naturally led to generalize Eq. (3.62) into

$$\chi_i \approx \chi_i^{\text{per}} \times \left[1 - Y\left(\int \omega_B \mathrm{d}t/6\right)\right] + E_0^{-1}(\mathrm{d}E_0/\mathrm{d}t)\partial_\omega \chi_r^{\text{eff}} \times Y\left(\int \omega_B \mathrm{d}t/6\right), \quad (3.64)$$

where χ_i^{per} is derived from the values obtained for a purely time growing wave by using for Γ the value given by Eq. (3.57) of Sect. 3.2. There, χ_i^{per} was expressed in terms of f_0, defined as the electron distribution function in the limit of a vanishing wave amplitude. We now make explicit what this means.

3.3.3 Symmetric Detrapping

If E_0 has kept on increasing with time, f_0 is nothing but the unperturbed distribution function, assumed to be a Maxwellian. If E_0 has reached a large enough value to induce nonlinear electron motion before decreasing back to nearly 0, a perturbative analysis of the electron motion from $t = 0$ is no longer valid when E_0 is, again, very small. However, one may calculate the electron motion perturbatively from $t = +\infty$ by invoking the time-reversal invariance of the dynamics. Then, f_0 is the distribution

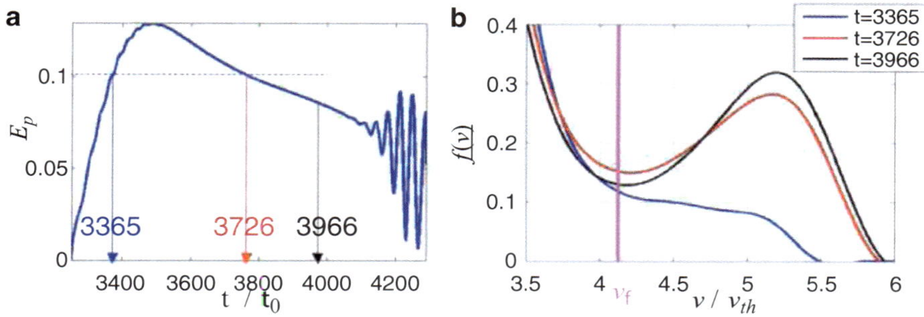

Fig. 3.5 Results from Vlasov simulations of stimulated Raman scattering showing, panel (**a**), the plasma wave amplitude (in its reference frame and in arbitrary units) as a function of time (normalized to the laser period), panel (**b**), the space-averaged electron distribution function at the three different times indicated by the *arrows* in panel (**a**). Note that, as the EPW amplitude decreases, the space-averaged distribution function becomes more symmetric with respect to v_ϕ. Note moreover that, although $\mathscr{E}_p$ is the same at times $t/\tau_0 = 3,365$ and $t/\tau_0 = 3,726$, the space-averaged distribution functions at these two times are very different from each other. Hence, the electron distribution function depends not only on the instantaneous wave amplitude but also on the maximum one

function in the limit $t \to +\infty$ which, as shown in [3] and as illustrated in Fig. 3.5, results from the electrons symmetric detrapping with respect to the wave phase velocity, v_ϕ. As a result, in the interval $|v - v_\phi| > \max(V_l)$, $f_0(v, t = +\infty)$ assumes the same values as the initial, unperturbed distribution function, while in the interval $|v - v_\phi| \leq \max(V_l)$, $f_0(v, t = +\infty)$ is nearly symmetric with respect to v_ϕ. Then, electrons whose initial velocity lies within the latter interval contribute very little to χ_i. This means that, once deeply trapped, electrons no longer contribute significantly to χ_i, even after being detrapped. This implies, in particular, that Landau damping is not restored after the wave amplitude has decreased back to very small values and explains why using $\int \omega_B \mathrm{d}t$ as an argument of the function Y is much more appropriate than using $\omega_B/|\Gamma|$.

In practice, when the electron motion has reached such a nonlinear regime that symmetric detrapping needs to be accounted for, $\int \omega_B \mathrm{d}t$ is so important that the perturbative term in Eq. (3.64) is negligible, while only the untrapped electrons (whose distribution function is unperturbed) contribute to $\partial_\omega \chi_r^{\text{eff}}$. Hence, one may always use for f_0 the unperturbed distribution function (assumed here to be a Maxwellian) in all the previously derived expressions and replace V_l defined by Eq. (3.22) of Sect. 3.2.1 by $\max_{t'<t}(V_l)$.

3.3.4 Nonlinear Landau Damping Rate

We now want to express χ_i, given by Eq. (3.64), in such a way that when plugged back into Eq. (3.16), this equation may indeed be considered as an envelope equation, i.e., that it reads

$$d\mathscr{E}_p/dt + \nu_{NL}\mathscr{E}_p = \mathscr{E}_d \cos(\delta\varphi)/\partial_\omega \chi_r^{env}, \tag{3.65}$$

where ν_{NL} would be the nonlinear counterpart on the Landau damping rate, (henceforth more simply termed nonlinear Landau damping rate), and χ_r^{env} would be the real effective susceptibility to be used in the envelope equation for the EPW.

When $\int \omega_B dt \gg 6$, from Eq. (3.64) it is clear that $\nu_{NL} \approx 0$ while $\partial_\omega \chi_r^{env} \approx \partial_\omega \chi_{eff}^r$. As for χ_i^{per}, we will only use here its first-order expression since we saw earlier that it already yields very accurate results (examples of results obtained with χ_i^{per} calculated at the 11th order will be given in Fig. 3.6 and in Sect. 3.4.2). Then,

$$\chi_i^{per} \approx \frac{-\omega_{pe}^2 f_0'(v_\phi)}{k_p^2}\left[\pi - 2\tan^{-1}\left(\frac{k_p V_p}{\Gamma}\right) + \frac{2\Gamma k_p V_l}{\Gamma^2 + (k_p V_l)^2}\right]$$
$$+\partial_\omega \chi_r^1 E_0^{-1}(dE_0/dt), \tag{3.66}$$

where Γ is defined by Eq. (3.57), $-\partial_\omega \chi_r^1/\omega_{pe}^2$ is the integral (3.37), and V_l is the maximum, for $t' < t$, of the expression given in Eq. (3.22) of Sect. 3.2. From the latter expression for χ_i^{per}, one would like to use $\partial_\omega \chi_r^{env} = \partial_\omega \chi_r^1$, and for ν_{NL} the first term of Eq. (3.66) divided by $\partial_\omega \chi_r^1$, when $\int \omega_B dt \ll 6$.

Then, in order to get expressions for ν_{NL} and $\partial_\omega \chi_r^{env}$ valid whatever $\int \omega_B dt$, we would only need to connect the previous estimates obtained when $\int \omega_B dt \ll 6$ to those valid when $\int \omega_B dt \gg 6$ the following way:

$$\partial_\omega \chi_r^{env} = \partial_\omega \chi_r^1 \times \left[1 - Y\left(\int \omega_B dt/6\right)\right] + \partial_\omega \chi_r^{eff} \times Y\left(\int \omega_B dt/6\right), \tag{3.67}$$

$$\nu_{NL} = \frac{-\omega_{pe}^2 f_0'(v_\phi)}{k_p^2 \partial_\omega \chi_r^1}\left[\pi - 2\tan^{-1}\left(\frac{k_p V_l}{\Gamma}\right) + \frac{2\Gamma k_p V_l}{\Gamma^2 + (k_p V_l)^2}\right]$$
$$\times \left[1 - Y\left(\int \omega_B dt/6\right)\right]. \tag{3.68}$$

However, ν_{NL} as defined by Eq. (3.68) is, at first sight, much more complicated an operator than a damping rate. It may nevertheless by considered as such because it assumes nearly constant values before dropping to 0, as shown in Figs. 3.6 and 3.8. Note that $\nu_{NL} \approx 0$ whenever $\int \omega_B dt \geq 6$, i.e., after the first trapped electrons have completed about one trapped orbit. Hence, the physics of the nonlinear reduction of the collisionless damping rate is the same as in situation considered by O'Neil and is due to the trapping of the nearly resonant electrons which, on the average, no longer give or take energy from the wave as they get phase mixed in the wave trough.

Concomitant with the drop in ν_{NL} is a sudden increase of $\partial_\omega \chi_r^{env}$ because the term χ_2 defined by Eq. (3.40) (or its higher-order counterpart), responsible for Lan-

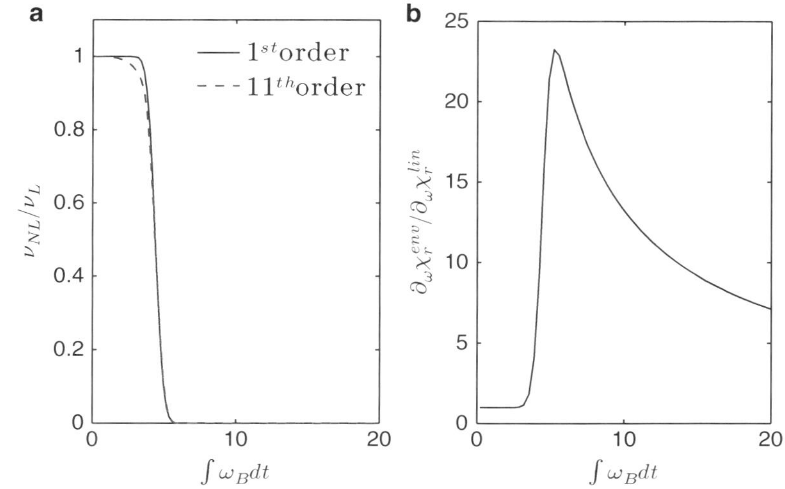

Fig. 3.6 Panel (**a**), collisionless damping rate, $\nu_{\rm NL}$, normalized to its linear value as calculated by using Eq. (3.68) (*solid line*) and as derived from an 11th order expansion for $\chi_i^{\rm per}$ (*dashed line*) versus $\int \omega_B {\rm d}t$. Panel (**b**), $\partial_\omega \chi_r^{\rm env}$ normalized to its linear value as calculated from Eq. (3.67) versus $\int \omega_B {\rm d}t$. Both $\nu_{\rm NL}$ and $\partial_\omega \chi_r^{\rm env}$ are calculated for a plasma wave whose phase velocity is $v_\phi = 3v_{\rm th}$ and growth rate is $\Gamma = 2 \times 10^{-2} \omega_{pe}$

dau damping when $\int \omega_B {\rm d}t \ll 6$, becomes nearly proportional to $\mathscr{E}_p^{-1}({\rm d}\mathscr{E}_p/{\rm d}t)$ when $\int \omega_B {\rm d}t \gg 6$ and therefore renormalizes $\partial_\omega \chi_r^{\rm env}$ once Landau damping has become negligible. Moreover, the rather crude way we model χ_2, as either proportional to $\mathscr{E}_p$ when $\int \omega_B {\rm d}t < 6$ or proportional to $\mathscr{E}_p^{-1}({\rm d}\mathscr{E}_p/{\rm d}t)$ whenever $\int \omega_B {\rm d}t > 6$, is vindicated by the abrupt convergence of $\mathscr{E}_p^{-1}({\rm d}\mathscr{E}_p/{\rm d}t)\partial_\omega \chi_r^{\rm env}$ towards χ_i when $\int \omega_B {\rm d}t > 6$, as illustrated in Fig. 3.3.

3.4 Variational Approach and Generalization to a Space-Dependent Wave Amplitude

In this section, by making use of a variational approach, we reinterpret physically the envelope equation (3.65) derived for a purely time varying wave amplitude. This allows us to generalize this equation to account for a space-dependent wave amplitude, first in one dimension where we show very careful comparisons between our theoretical predictions and results from Vlasov simulations of stimulated Raman scattering, and then in three dimensions.

3.4.1 Physical Discussion of the Previous Results Using a Variational Approach

In this section, we want to make clear how collisionless dissipation enters in the envelope equation (3.65), especially when $\nu_{\mathrm{NL}} = 0$, to make $\partial_\omega \chi_r^{\mathrm{env}} \neq \partial_\omega \chi_r$, which explains why the Taylor expansion Eq. (3.59) of Sect. 3.3 one would naively expect is not valid.

Let us therefore consider the situation when $\omega_B \gg \Gamma$, so that the electron orbits are very close to the frozen ones (corresponding to a fixed wave amplitude), and are completed within very short times. In this situation, the electron motion is nearly adiabatic and "enslaved" to the variations of the wave amplitude, so that the rate of variation of the kinetic energy is just proportional to the wave growth rate, Γ. Moreover, still in this situation, it is quite clear that the kinetic energy of the electrons which have never been trapped is only a function of the instantaneous wave amplitude. It increases when the wave grows and is converted back into electrostatic energy when the wave decays. Now, whatever Δv, adiabatic electrons with initial velocities $v_\phi \pm \Delta v$, where $v_\phi \equiv \omega/k$ is the EPW phase velocity, are all trapped at the same time, and their trapping leads to a jump in the kinetic energy by a quantity proportional to

$$\Delta K = m[f_0(v_\phi - \Delta v) - f_0(v_\phi + \Delta v)]v_\phi \Delta v, \tag{3.69}$$

where f_0 is the electron distribution function in the limit of a vanishing wave amplitude, as defined in Sects. 3.3.2 and 3.3.3. Note that ΔK is nonzero only because $f_0(v_\phi - \Delta v) \neq f_0(v_\phi + \Delta v)$ and, actually, ΔK is positive when f_0 is a decreasing function of velocity, as is the case for a Maxwellian. Now, as shown in Sect. 3.3.3, if $\mathscr{E}_p$ is decreasing, the electrons are detrapped nearly symmetrically with respect to the phase velocity. Consequently, detrapping would lead to a change in the electron kinetic energy by $\Delta K' = -\Delta K$ but, now, with $f_0(v_\phi - \Delta v) = f_0(v_\phi + \Delta v)$. Hence, $\Delta K' = 0$, the kinetic energy gained through trapping is not converted back into electrostatic energy when the electrons are detrapped. We therefore conclude that, *when the electron motion is nearly adiabatic, only trapping may lead to an irreversible increase of the kinetic energy and therefore to the collisionless dissipation of the electrostatic energy.*

This may be viewed in a more formal way by referring to the work by Yampolsky and Fisch who, using considerations based on energy conservation, found the following envelope equations for $\mathscr{E}_p$:

$$\partial_\omega \chi_r (d_t + \nu)\mathscr{E}_p = \mathscr{E}_d \cos(\delta\varphi), \tag{3.70}$$

where ν is directly related to the rate of kinetic energy gained by the electrons. As discussed before, ν should be proportional to the EPW growth rate in the limit when $\omega_B \gg \gamma$, which is exactly what Yampolsky and Fisch found in [30]. Hence, when $\omega_B \gg \gamma$, one may write $\nu \partial_\omega \chi_r E_p \equiv \partial_\omega \chi_r' \partial_t E_p$, where χ_r' is a dimensionless function of the EPW amplitude, and Eq. (3.1) becomes

$$(\partial_\omega \chi_r + \partial_\omega \chi_r')\mathrm{d}\mathscr{E}_p/\mathrm{d}t = \mathscr{E}_d \cos(\delta\varphi). \tag{3.71}$$

Since, as shown in [7], the results obtained by Bénisti et al. and Yampolsky and Fisch match over a range of wave amplitudes where the condition $\omega_B \gg \gamma$ holds, Eq. (3.1) may be identified with Eq. (3.65). This shows that, when $\omega_B \gg \gamma$, $\nu/\gamma \approx -\partial_\omega \chi_r^{\mathrm{tr}}/\partial_\omega \chi_r$, where $\chi_r^{\mathrm{tr}}(t)$ is that part of χ_r only due to the electrons which have been trapped at a given time $t' \leq t$. This formally relates collisionless dissipation to trapping, as expected from our previous discussion.

Using this result, we can now "construct" the EPW envelope equation using arguments based on energy conservation. A very well-known method to derive a nonlinear envelope equation, that automatically guarantees the conservation of the electric field energy for a freely propagating wave, is the variational approach developed by Whitham in [29]. When the wave amplitude only depends on time, Whitham's theory trivially yields

$$\mathrm{d}\left(\partial_\omega \chi_r \mathscr{E}_p\right)/\mathrm{d}t = 0. \tag{3.72}$$

We now need to account for the change in electrostatic energy due to the drive, which amounts to replacing the right-hand side of Eq. (3.72) with $\mathscr{E}_d \cos(\delta\varphi)$. We also need to allow for collisionless dissipation which, in the perturbative regime, just amounts to a Landau-like damping, as shown in the previous section. Hence, if we neglect the time variation of the wave frequency, we find, in the perturbative regime when $\omega_B \ll \Gamma$ (or, equivalently, when $\int_0^t \omega_B \mathrm{d}t' \ll 1$),

$$\partial_\omega \chi_r (\mathrm{d}\mathscr{E}_p/\mathrm{d}t + \nu_{\mathrm{NL}} \mathscr{E}_p) = \mathscr{E}_d \cos(\delta\varphi), \tag{3.73}$$

where ν_{NL} is given by Eq. (3.68) with $Y = 0$. Moreover, as discussed before, in the strongly nonlinear regime when $\omega_B \gg \Gamma$ (or, equivalently when $\int_0^t \omega_B \mathrm{d}t' \gg 1$), collisionless dissipation is only due to trapping and is accounted for in the EPW equation by the term $-\partial_\omega \chi_r^{\mathrm{tr}} \partial_t \mathscr{E}_p$, which leads to the following envelope equation:

$$(\partial_\omega \chi_r - \partial_\omega \chi_r^{\mathrm{tr}})\mathrm{d}\mathscr{E}_p/\mathrm{d}t = \mathscr{E}_d \cos(\delta\varphi), \tag{3.74}$$

where $(\partial_\omega \chi_r - \partial_\omega \chi_r^{\mathrm{tr}}) = \partial_\omega \chi_r^{\mathrm{eff}}$ as defined in Sect. 3.3. Now, using the result of the previous section that the transition between a perturbative and an adiabatic-like regime is very abrupt and occurs when $\int_0^t \omega_B \mathrm{d}t' \approx 6$, one easily recovers Eq. (3.65) by connecting Eqs. (3.73) and (3.74) using the function $Y(\int \omega_B \mathrm{d}t)$.

3.4.2 *One-Dimensional Variation of the Wave Amplitude*

We now take advantage of the previous method in order to very easily generalize Eq. (3.65) to a space-dependent wave amplitude.

3.4.2.1 Theoretical Results

In this section, we just reproduce the procedure of Sect. 3.4.1 for a wave whose amplitude depends on time, t, and on the space variable, x.

We thus start with the results derived from Whitham's theory, which would yield the following envelope equation (see [29] for details):

$$\partial_t\left(\partial_\omega \chi_r \mathscr{E}_p\right) - \partial_x\left(\partial_k \chi_r \mathscr{E}_p\right) = 0. \tag{3.75}$$

Just like in Sect. 3.4.1, account is taken on the effect of the drive by simply replacing the right-hand side of Eq. (3.75) by $\mathscr{E}_d \cos(\delta\varphi)$. As for collisionless dissipation, in order to correctly allow for it, we take advantage of the sharp transition between the perturbative and adiabatic regimes, which occurs after the trapped electrons have completed about one orbit, i.e., when $\int \omega_B \mathrm{d}t' \approx 6$, where the integral is now calculated in the frame moving at the EPW phase velocity, v_ϕ, with respect to the laboratory frame:

$$\int \omega_B \mathrm{d}t' \equiv \int_0^t \omega_B(x - v_\phi t', t')\mathrm{d}t'. \tag{3.76}$$

In the perturbative regime, when $\int_0^t \omega_B \mathrm{d}t' < 6$, collisionless dissipation amounts to a Landau-like damping so that, neglecting the time and space variations of the EPW wave number and frequency, we find

$$\frac{\partial \chi_r}{\partial \omega}\left[\frac{\partial \mathscr{E}_p}{\partial t} + \nu_{\mathrm{NL}} \mathscr{E}_p\right] - \frac{\partial \chi_r}{\partial k}\frac{\partial \mathscr{E}_p}{\partial x} = \mathscr{E}_d \cos(\delta\varphi), \tag{3.77}$$

where ν_{NL} is still given by Eq. (3.68) with, now,

$$\Gamma(x,t) = \frac{E_0(x,t) - E_0\left[x - v_\phi \pi/(k_p V_l), t - \pi/(k_p V_l)\right]}{\int_{t-\pi/(k_p V_l)}^{t} E_0\left[x - v_\phi(t-u), u\right]\mathrm{d}u}. \tag{3.78}$$

In the strongly nonlinear, adiabatic-like, regime, when $\int \omega_B \mathrm{d}t > 6$, the rate of dissipation is still proportional to the trapping rate which is now proportional to the wave growth rate calculated in the wave frame, i.e., proportional to $(\partial_t + v_\phi \partial_x)\mathscr{E}_p$. As for the coefficient of proportionality, it is the same as in the previous section, since the origin of dissipation is the same, and is therefore $-\partial_\omega \chi_r^{\mathrm{tr}} \equiv \partial_\omega \chi_r - \partial_\omega \chi_r^{\mathrm{eff}}$. Hence, wherever $\int \omega_B \mathrm{d}t > 6$, we find the following envelope equation:

$$\frac{\partial \chi_r}{\partial \omega}\frac{\partial \mathscr{E}_p}{\partial t} - \frac{\partial \chi_r}{\partial k}\frac{\partial \mathscr{E}_p}{\partial x} + \frac{\partial[\chi_r^{\mathrm{eff}} - \chi_r]}{\partial \omega}\left[\frac{\partial \mathscr{E}_p}{\partial t} + v_\phi \frac{\partial \mathscr{E}_p}{\partial x}\right] = \mathscr{E}_d \cos(\delta\varphi). \tag{3.79}$$

We now make the approximation, $1 + \chi_r = 0$, and use the result that $k^2\chi_r$ is only a function of the EPW phase velocity to find $-\partial_k \chi_r = v_\phi \partial_\omega \chi_r + 2\chi_r/k_p \approx v_\phi \partial_\omega \chi_r - 2/k_p$. Then, Eq. (3.79) is

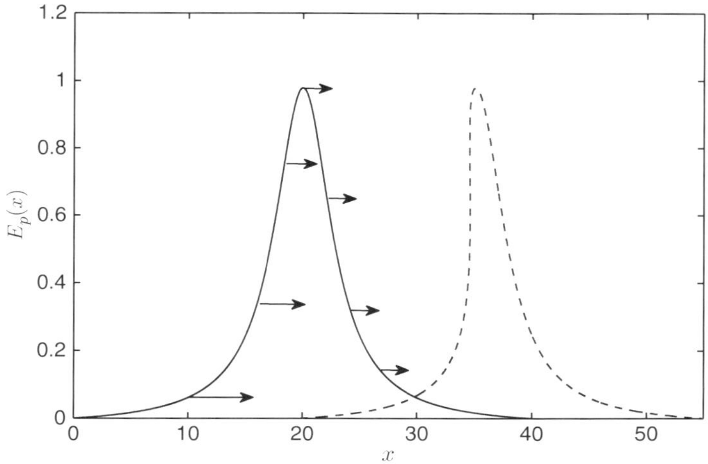

Fig. 3.7 Space profile of the plasma wave (in arbitrary units), at time t_1 (*solid line*) and at time $t_2 > t_1$ (*dashed line*), when the group velocity of the wave packet (whose amplitude is indicated by the *arrows*) decreases with the EPW amplitude at the rear side and remains fixed at its minimum nonlinear value at the front side. One clearly sees that the wave packet at time t_2 is narrower than at time t_1

$$\partial_t \mathscr{E}_p + v_g \partial_x \mathscr{E}_p = \mathscr{E}_d \cos(\delta\varphi)/\partial_\omega \chi_r^{\text{eff}}, \tag{3.80}$$

with

$$v_g \equiv v_\phi - \frac{2}{k_p \partial_\omega \chi_r^{\text{eff}}}. \tag{3.81}$$

In the strongly nonlinear regime, the EPW group velocity, v_g, is therefore *not* $-\partial_k \chi_r / \partial_\omega \chi_r \approx \partial \omega_p / \partial k_p$ because of the term we needed to add to the equation derived from Whitham's theory in order to account for collisionless dissipation. At this stage, one may wonder how collisionless dissipation actually manifests itself since, clearly, Eq. (3.80) would predict that the wave remains undamped. First, as shown in Fig. 3.6, $\partial_\omega \chi_r^{\text{eff}} \gg \partial_\omega \chi_r$, which just reflects the fact that, accounting for dissipation, one would find it much harder to laser drive a plasma wave. Second, $\partial_\omega \chi_r^{\text{eff}}$ is the contribution to $\partial_\omega \chi_r$ from those electrons which have *never* been trapped and is therefore a nonlocal function of the EPW amplitude, which mainly depends on $\mathscr{E}_{\max} \equiv \max_{t' \le t} \left[\mathscr{E}_p(x - v_\phi t', t')\right]$. Hence, from Eq. (3.81), so does the EPW group velocity, v_g, which actually decays with $\mathscr{E}_{\max}$ in the strongly nonlinear regime, as may be seen in Fig. 3.11. Then, because $v_\phi > v_g$, the EPW group velocity would mainly decrease in the ascending part of a large-amplitude plasma wave packet (before its maximum along the direction of propagation of the wave) and would remain nearly constant in the descending part of the pulse. This would automatically entail the shrinking of a large-amplitude, freely propagating, wave packet and, therefore, the decrease of its total electrostatic energy (see Fig. 3.7).

Now, in order to derive an envelope equation valid whatever the regime, and in the whole space domain, we just need to connect the envelope equations (3.77) and (3.80) exactly the same way as in Sect. 3.3, which yields

$$\partial_t \mathscr{E}_p + v_g \partial_x \mathscr{E}_p + \nu_{\mathrm{NL}} \mathscr{E}_p = \mathscr{E}_d \cos(\delta\varphi)/\partial_\omega \chi_r^{\mathrm{env}}, \tag{3.82}$$

where $\partial_\omega \chi_r^{\mathrm{env}}$ and ν_{NL} are respectively given by Eqs. (3.67) and (3.68), with $\int \omega_B \mathrm{d}t$ defined by Eq. (3.76) and Γ by Eq. (3.78), and where

$$v_g \equiv v_\phi - \frac{2}{k_p \partial_\omega \chi_r^{\mathrm{env}}}. \tag{3.83}$$

3.4.2.2 Comparisons with Vlasov Simulations of Stimulated Raman Scattering

In this section we now test our previous theoretical results against Vlasov simulations of stimulated Raman scattering (SRS) performed using the Vlasov code ELVIS [28]. It is out of the scope of this chapter to detail these simulations, which will further be described in Sect. 3.5.3, but we just want to stress here that, numerically, the EPW results from the interaction of a pump laser entering from vacuum on the left ($x = 0$) and of a small-amplitude counterpropagating "seed" light wave injected on the right. We therefore simulate the optical mixing of two lasers and, numerically, only backward stimulated Raman scattering is addressed.

Imaginary Part of the Electron Susceptibility

The first set of comparisons with numerical simulations we present here aims at checking the accuracy of our theoretical prediction for χ_i, i.e., of the very terms used in the envelope equation. Such comparisons are made possible due to the great precision of the noiseless results offered by Vlasov codes. Using a Hilbert transform of the fields (see, e.g., [16]), one can numerically calculate the ratio $[\mathscr{E}_d \cos(\delta\varphi) + k_p^{-1} \partial_x \mathscr{E}_p]/\mathscr{E}_p$, which from Eq. (3.16) yields a first, numerical, estimate of χ_i. From Vlasov simulations one can also extract the values of all the quantities, such as $\mathscr{E}_{\max}$, $\int \omega_B \mathrm{d}t$, Γ, ..., which enter our theoretical formula for χ_i. Using these values we calculate a second, theoretical, estimate for χ_i. Both these estimates are compared in Fig. 3.8a. The simulation results of Fig. 3.8 correspond to a plasma with electron temperature, $T_e = 5\,\mathrm{keV}$, and electron density $n = 8.9 \times 10^{20}\,\mathrm{cm}^{-3}$. The total length of the simulation box is $L = 270\lambda_l$, where $\lambda_l = 0.351\,\mu\mathrm{m}$ is the laser wavelength, and the data of Fig. 3.8 were measured at $x = 154\lambda_l$. The laser intensity is $I_l = 4 \times 10^{15}\,\mathrm{W/cm}^2$ while the seed intensity is $I_s = 10^{-5} I_l$ and the seed wavelength is $\lambda_s = 0.609\,\mu\mathrm{m}$. As can be seen in Fig. 3.8a, there is a very good agreement between the theoretical and numerical values of χ_i, especially as regards the decrease of χ_i from its linear value.

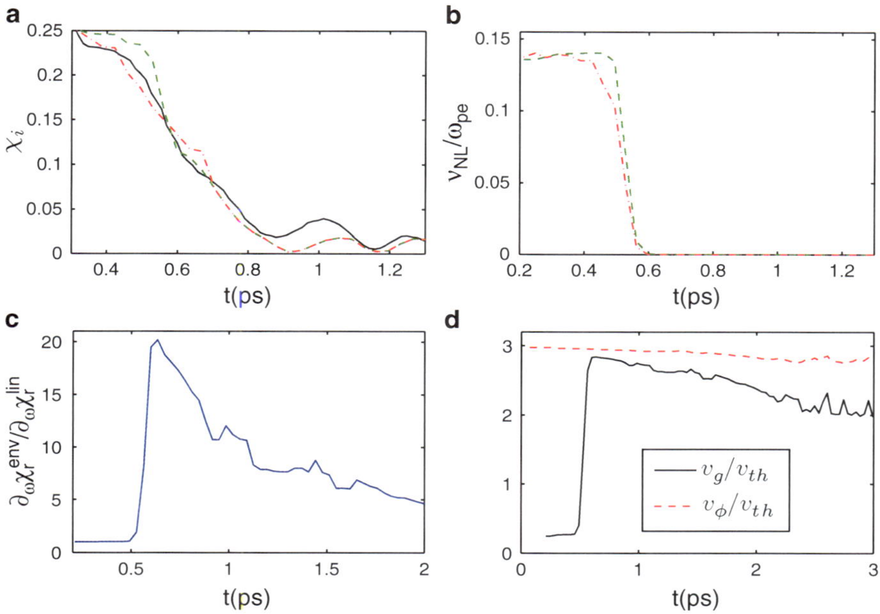

Fig. 3.8 Panel (**a**), χ_i calculated numerically (*solid line*) and theoretically using for χ_i^{per} a first order (*dashed line*) or an 11th order (*dash–dotted line*) perturbation analysis, panel (**b**), the nonlinear Landau damping rate normalized to the plasma frequency from a first order (*dashed line*) or an 11th order (*dash–dotted line*) perturbation analysis, panel (**c**), $\partial_\omega \chi_r^{\text{env}}$ normalized to its linear value, and, panel (**d**), the EPW group velocity (*solid line*) and phase velocity (*dashed line*) normalized to the thermal one

From Eqs. (3.67) and (3.68) we derive ν_{NL} and $\partial_\omega \chi_r^{\text{env}}$, whose values are plotted in Fig. 3.8b, c, and we recover mostly the same results as with test particle simulations, namely, that ν_{NL} remains nearly constant before abruptly dropping to 0 and that this drop in ν_{NL} is concomitant with a sudden rise in $\partial_\omega \chi_r^{\text{env}}$. It should be noted here that the variations of ν_{NL} plotted in Fig. 3.8b are very different from the oscillating result found by O'Neil in [26] , because we consider here slowly varying waves inducing a nearly adiabatic electron motion. As a consequence, electrons with the same initial velocity are all trapped nearly simultaneously. This is in contrast with the situation considered by O'Neil where the wave was assumed to reach instantaneously a constant and uniform amplitude, E_0. Then, by the time the EPW has grown to E_0, the electrons barely had the time to move, and electrons with the same initial velocity are not all trapped by the wave, depending on their initial position. Hence, the nonlinear mechanism leading to the decrease of ν_{NL} is much less effective in the O'Neil situation than in the one considered here. In the O'Neil case, $\nu_{\text{NL}} \approx 0$ when $\omega_B t \geq 30$, while we find $\nu_{\text{NL}} \approx 0$ when $\int \omega_B \text{d}t \geq 6$.

Nonlinear Group Velocity

Let us now use, once again, Vlasov simulations of stimulated Raman scattering in order to check our quite unexpected theoretical prediction, $v_g = v_\phi - 2/(k_p \partial_\omega \chi_r^{\text{env}}) \neq \partial \omega_p / \partial k_p$.

The simulations are the same as those presented earlier except that, in the Vlasov equation we numerically solve, we artificially multiply the ponderomotive force, $v \times B$, by a Lorentzian function. This may be viewed as a way to somehow account, in 1-D simulations, for the change in the laser intensity along its direction of propagation due to its focussing inside the plasma. For our purposes, this is also a way to impress a well-defined shape on the plasma pulse.

Once the plasma wave has reached the desired maximum amplitude, we turn the electromagnetic waves off and let the EPW pulse freely propagate, in order to measure its group velocity. This measurement is made easier by the well-defined shape impressed on the pulse while it is driven, but is hampered by electrostatic instabilities which alter this shape, and by a residual Landau-like damping. These two effects force us to find the EPW group velocity from the pulse propagation over a rather small time interval, which alters the precision of our numerical estimates. Nevertheless, as will be shown here, our numerical measurements are precise enough to discriminate between different theoretical predictions. Moreover, from our previous results, it is clear that the nonlinear Landau-like damping rate decreases along the direction of propagation of the plasma wave train (see also [5, 12]). As a result, and as explained in Fig. 3.9, the speed of propagation of the pulse maximum overestimates the EPW group velocity. However, as explained again in Fig. 3.9, an underestimate of v_g may be obtained by measuring the speed of propagation of a point corresponding to a given field amplitude and located on the right of the pulse maximum. Namely, in the notations of Fig. 3.9, $(x_B - x_A)/\delta t > v_g > (x_B - x_{B'})/\delta t$.

These two estimates are usually close to each other and close to our theoretical prediction, as shown in the example of Fig. 3.10. In the Vlasov simulation used to generate this figure, the laser intensity is $I_l = 8 \times 10^{15}$ W/cm^2, the laser wavelength is $\lambda_l = 0.351\,\mu$m, the seed intensity is $I_s = 8 \times 10^{10}$ W/cm^2, and the seed wavelength is $\lambda_s = 0.609\,\mu$m. The electron density is $n = 8.9 \times 10^{20}\,\text{cm}^{-3}$, while the electron temperature is $T_e = 5\,\text{keV}$. The linear value of the plasma wave number resulting from the optical mixing is $k_p \lambda_D \approx 0.448$, where λ_D is the Debye length defined by Eq. (3.4). The electromagnetic waves are turned off at $t \approx 1.75$ ps, and Fig. 3.10 plots the pulse amplitude at $t \approx 1.75$ ps and $t + \delta t \approx 1.86$ ps. From the decrease of the amplitude of the pulse maximum, we estimate $\nu_{\text{NL}} \approx 1.4 \times 10^{12}\,\text{s}^{-1}$, which is about 150 times less than the linear Landau damping rate, $\nu_L \approx 2.4 \times 10^{14}\,\text{s}^{-1}$. Moreover, we numerically estimate that the pulse maximum moves at velocity $v_{\max} \approx 2.4 v_{\text{th}}$, where v_{th} is the thermal velocity, while we numerically measure that the point corresponding to $\Phi = 0.05$, where $\Phi \equiv e\mathscr{E}_p/kT_e$, and located on the right of the maximum, moves at velocity $v_{\Phi_{005}} \approx 1.9 v_{\text{th}}$. Namely, using the notations of Fig. 3.10, we find $(x_{M'} - x_M)/\delta t \approx 2.4 v_{\text{th}}$ and $(x_{A'} - x_A)/\delta t \approx 1.9 v_{\text{th}}$, from which we deduce that when $\Phi_{\max} \approx 0.09$, $1.9 \leq v_g/v_{\text{th}} \leq 2.4$. This is consistent with the theoretical values reported in Fig. 3.11d predicting $v_g/v_{\text{th}} \approx 2.23$ when $\Phi_{\max} \approx 0.09$. By contrast, using for $\omega_p(\Phi, k_p)$ the very accurate values derived in

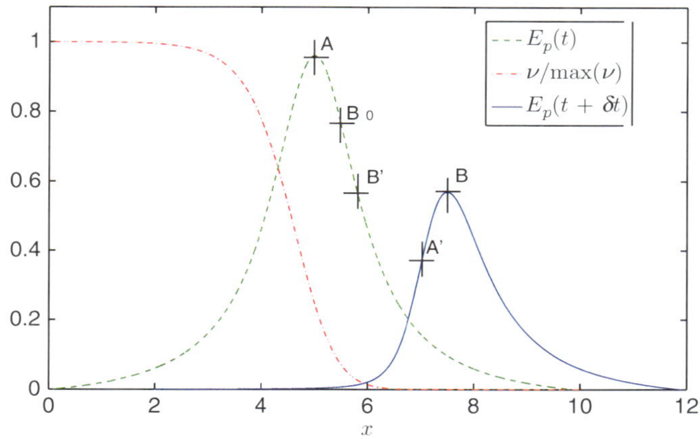

Fig. 3.9 Sketch of the propagation of a plasma pulse whose damping rate rapidly decreases with x and whose group velocity is uniform. The shape of the pulse at time t is given by the *dashed line*, and at time $t + \delta t > t$ by the *solid line*, while the shape of the damping rate is given by the *dash–dotted line*. Due to damping, the pulse maximum, A, at time t, is not located at time $t + \delta t$ at the new maximum, B, but at point A' on the left of B. Hence, for this particular example, the group velocity $v_g = (x_{A'} - x_A)/\delta t$ is less than the speed of propagation of the pulse maximum. Moreover, at time t, point B was located at B_0, i.e., on the left of point B' corresponding to the same pulse amplitude as point B. Therefore, $v_g = (x_B - x_{B_0})/\delta t > (x_B - x_{B'})/\delta t$

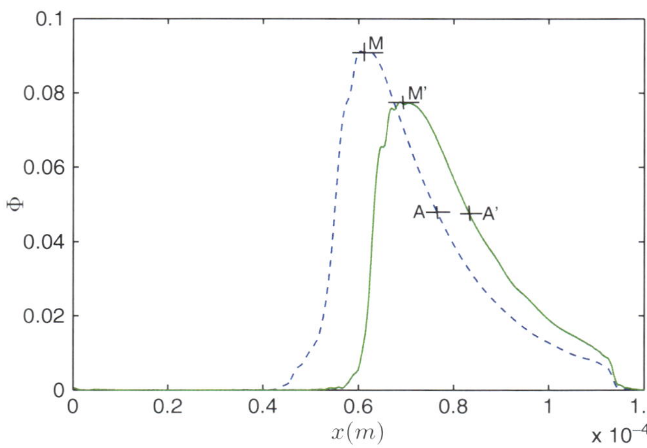

Fig. 3.10 Dimensionless plasma wave amplitude, $\Phi \equiv e\mathscr{E}_p/k_pT_e$, as a function of x at times $t \approx 1.75$ ps (*dashed line*) and $t + \delta t \approx 1.86$ ps (*solid line*), obtained from the Vlasov run of the 5 keV case of Table 3.1

the next section in order to calculate $\partial\omega_p/\partial k_p$, we find that when $0.05 \leq \Phi \leq 0.09$, $0.36v_{\text{th}} \leq \partial\omega_p/\partial k_p \leq 0.53v_{\text{th}}$. Hence, our numerical results unambiguously rule out $\partial\omega_p/\partial k_p$ as an accurate estimate of v_g.

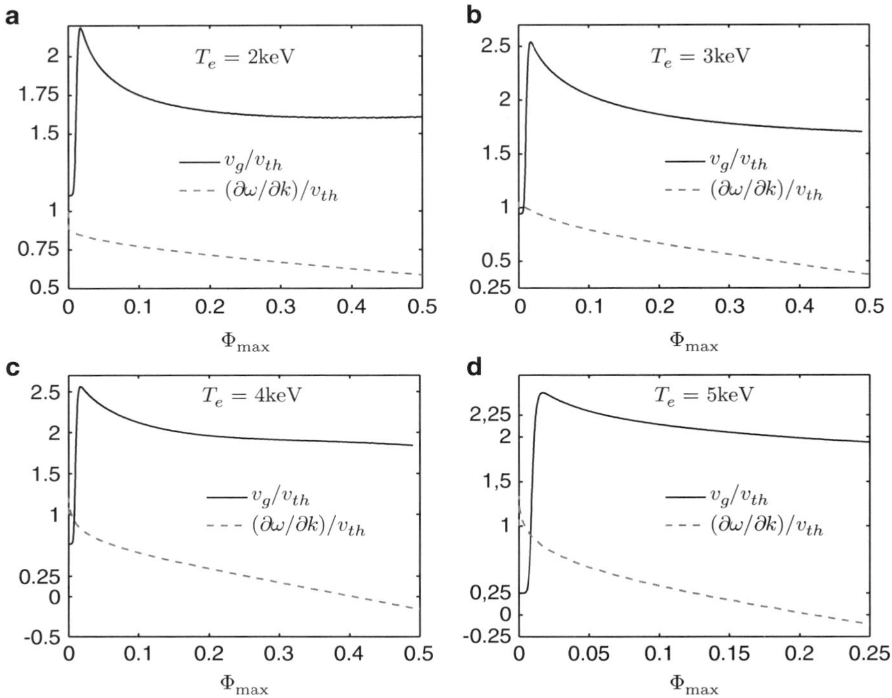

Fig. 3.11 Our theoretical predictions for the group velocity (*solid line*), and $\partial\omega_p/\partial k_p$, normalized to the thermal velocity, when the electron density is $n = 8.9 \times 10^{20}\,\text{cm}^{-3}$ and when, panel (**a**), $T_e = 2\,\text{keV}$, panel (**b**), $T_e = 3\,\text{keV}$, panel (**c**), $T_e = 4\,\text{keV}$, and panel (**d**) $T_e = 5\,\text{keV}$

Table 3.1 Values of the nonlinear group velocity, normalized to the thermal one, calculated either theoretically or numerically and compared to $\partial\omega_p/\partial k_p$, also normalized to the thermal velocity

T_e(keV)	$k\lambda_D$	Φ	v_g (theory)	v_g (numerical)	$\partial\omega_p/\partial k_p$
2	0.3	$0.25 \le \Phi \le 0.35$	$v_g \approx 1.62$	$1.5 \le v_g \le 1.65$	$0.62 \le \partial\omega_p/\partial k_p \le 0.65$
3	0.357	$0.2 \le \Phi \le 0.25$	$v_g \approx 1.83$	$1.8 \le v_g \le 2.0$	$0.61 \le \partial\omega_p/\partial k_p \le 0.66$
4	0.406	$0.2 \le \Phi \le 0.25$	$v_g \approx 1.95$	$1.9 \le v_g \le 2.6$	$0.26 \le \partial\omega_p/\partial k_p \le 0.34$
5	0.448	$0.05 \le \Phi \le 0.09$	$v_g \approx 2.23$	$1.9 \le v_g \le 2.4$	$0.36 \le \partial\omega_p/\partial k_p \le 0.53$

All results correspond to a plasma whose electron density is $8.9 \times 10^{20}\,\text{cm}^{-3}$

Figure 3.11a–d plots our theoretical predictions for v_g, together with $\partial\omega_p/\partial k_p$, as a function of Φ_{max} for T_e varying from $T_e = 2\,\text{keV}$ to $T_e = 5\,\text{keV}$, while Table 3.1 compares the numerically measured values of v_g to the theoretical ones for the four cases we investigated. For each case, v_g is numerically found to be significantly larger than $\partial\omega_p/\partial k_p$ but very close to our theoretical estimate. From these results, we therefore conclude that, indeed, $\partial\omega_p/\partial k_p$ is not the nonlinear group velocity of a plasma wave train while it seems that our theoretical prediction, $v_g = v_\phi - 2/(k_p\partial_\omega\chi_r^{\text{env}})$, is quite accurate.

Note that the comparisons we made on v_g provide an indirect numerical check of our theoretical predictions for $\partial_\omega \chi_r^{\text{env}}$. Since comparing Eq. (3.16) with Eq. (3.82) shows that $\chi_i = \partial_\omega \chi_r^{\text{env}} \left[v_{\text{NL}} + \mathscr{E}_p^{-1}(\partial_t + v_g \partial_x)\mathscr{E}_p \right] - (k_p \mathscr{E}_p)^{-1} \partial_x \mathscr{E}_p$, whose theoretical value agrees very well with the numerical one, we may consider that our numerical simulations also indirectly showed that we provided a very accurate theoretical description for v_{NL}.

Note also that, since the dispersion relation of a driven plasma wave is not exactly $1 + \chi_r = 0$ but $1 + \alpha_a \chi_r = 0$, $-\partial_k \chi_r / \partial_\omega \chi_r$ which is the linear value of v_g may significantly differ from $\partial \omega_p / \partial k_p$, all the more as α_d is large (i.e., as $k_p \lambda_{\text{D}}$ is large as will be explained in the next section), which is illustrated in Fig. 3.11 since for a given electron density $k_p \lambda_{\text{D}}$ increases with T_e.

3.4.3 Three-Dimensional Space Variation of the Wave Amplitude

We now discuss how 3-D effects may change the results derived previously, in the limit of a nearly unperturbed transverse electron motion. In case of a laser driven plasma wave, and when the laser-electric field is polarized along the y direction, one easily finds from Newton equations:

$$v_y = v_{0y} + O(eA/m), \tag{3.84}$$

$$v_z = v_{0z} + O[(eA/m)^2/c], \tag{3.85}$$

where A is the amplitude of the laser vector potential, while v_{0y} and v_{0z} are the unperturbed transverse velocities. Hence, the transverse motion may be considered as unperturbed provided that $eA/m \ll v_{\text{th}}$. This condition is fulfilled, for example, for typical laser and plasma conditions met in inertial confinement fusion [22].

Just like for the one-dimensional case, in order to derive the EPW envelope equation, we start by using Whitham's variational approach, which yields

$$\frac{\partial^2 L}{\partial t \partial \omega_p} - \nabla . \left[\nabla_{\mathbf{k}_p} L \right] = 0, \tag{3.86}$$

where the Lagrangian density is $L = \int_0^{\mathscr{E}_p} \partial_{\mathscr{E}'} L \mathrm{d}\mathscr{E}'$ and where $\partial_{\mathscr{E}'} L = 0$ is the dispersion relation of a freely propagating EPW, at 0-order in the variations of $\mathscr{E}_p$. At this order, the plasma wave may clearly be considered electrostatic, so that $\partial_{\mathscr{E}'} L = (1 + \chi_r)\mathscr{E}'$, where χ_r is the adiabatic approximation of the real part of the electron susceptibility. It assumes the same values as in 1-D and only depends on the plasma wave number through its modulus, k_p, since the EPW may be considered electrostatic and the electron distribution function is isotropic. Then, using $L \equiv \int_0^{\mathscr{E}_p} [1+\chi_r \mathscr{E}'] \mathrm{d}\mathscr{E}'$ and the consistency relation $\partial_t \mathbf{k}_p = -\nabla \omega_p$ (see [29]), Eq. (3.86) yields

$$\frac{\partial \chi_r}{\partial \omega}\frac{\partial \mathscr{E}_p}{\partial t} - \frac{\partial \chi_r}{\partial k}\frac{\partial \mathscr{E}_p}{\partial x_p} + \frac{\mathscr{E}_p}{2}\left[\frac{\partial \omega_p}{\partial t}\frac{\partial^2 \chi_r}{\partial \omega^2} - \nabla.\hat{x}_p \frac{\partial \chi_r}{\partial k} - \hat{x}_p.\nabla k_p \frac{\partial^2 \chi_r}{\partial k^2}\right] = 0, \quad (3.87)$$

where $\hat{x}_p \equiv \mathbf{k}_p/k_p$ and, therefore, $\partial_{x_p}\mathscr{E}_p \equiv (\mathbf{k}_p.\nabla\mathscr{E}_p)/k_p$. Due to the consistency relation $\partial_t \mathbf{k}_p = -\nabla\omega_p$, a transverse profile of the EPW frequency entails a transverse component in $\mathbf{k}_p$, usually directed towards the wave axis of propagation. Hence, in Eq. (3.87), we account for the self-focussing induced by wave front bowing, as discussed in [31]. Moreover, in practice, the term proportional to $\partial_t \omega_p$ is negligible and, within the paraxial approximation, we are using here, so are the longitudinal variations of the wave number and of $\hat{x}_p$. Hence, we simplify Eq. (3.87) in

$$\frac{\partial \chi_r}{\partial \omega}\frac{\partial \mathscr{E}_p}{\partial t} - \frac{\partial \chi_r}{\partial k}\left[\frac{\partial \mathscr{E}_p}{\partial x_p} + \frac{\mathscr{E}_p}{2k_p}\nabla_\perp.\mathbf{k}_p\right] = 0. \quad (3.88)$$

We now need to add to Eq. (3.88) the terms accounting for collisionless dissipation, i.e., as discussed in before, a Landau-like damping rate and a term allowing for the irreversible energy conversion from electrostatic to kinetic, induced by trapping. The expression we use for the collisionless damping rate is simplified compared to the 1-D case, and would be, if all electrons had the same transverse velocity $\mathbf{v}_\perp$:

$$\nu_{\mathrm{NL}}(\mathbf{v}_\perp) = \nu_{\mathrm{L}}\left[1 - Y\left(\int_{\mathbf{v}_\perp} \omega_B \mathrm{d}t\right)\right], \quad (3.89)$$

where

$$\int_{\mathbf{v}_\perp} \omega_B \mathrm{d}t \equiv \int_0^t \omega_B(x_{/\!/} - v_\phi t', \mathbf{x}_\perp - \mathbf{v}_\perp t', t')\mathrm{d}t'. \quad (3.90)$$

This simplification is vindicated by the fact that, in practice, $\nu_{\mathrm{NL}}(\mathbf{v}_\perp)$ has to be convoluted with the distribution function of the transverse electron velocities, making the dependence of the Landau damping rate on $\mathscr{E}_{\max}$ much smoother than in 1-D. Consequently, accounting for the variations of $\nu_{\mathrm{NL}}(\mathbf{v}_\perp)$ as a function of the EPW amplitude, given by an expression like Eq. (3.68), is not essential.

As for the rate of dissipation induced by trapping, it is now proportional to the wave growth rate, calculated in the wave frame, and as seen by the electrons. For electrons with transverse velocity $\mathbf{v}_\perp$, this is proportional to $\partial_t \mathscr{E}_p + v_\phi \partial_{x_p}\mathscr{E}_p + \mathbf{v}_\perp.\nabla_\perp \mathscr{E}_p$. Moreover, in the envelope equation for the EPW the prefactor of the latter expression is the same as in 1-D, i.e., $-\partial_{\omega_p}\chi_r^{\mathrm{tr}}$.

We therefore conclude that, when all electrons have the same transverse velocity, $\mathbf{v}_\perp$, allowing for collisionless dissipation leads to the following envelope equation for the plasma wave amplitude:

$$\frac{\partial_\omega \chi_r^{\text{env}}}{\partial \omega}\left[\frac{\partial \mathscr{E}_p}{\partial t}+\nu_{\text{NL}}(\mathbf{v}_\perp)\mathscr{E}_p\right]-\left[\frac{\partial \chi_r}{\partial k}+v_\phi\frac{\partial \chi_r^{\text{tr}}}{\partial \omega}\right]\frac{\partial \mathscr{E}_p}{\partial x_p}$$
$$-\frac{\partial \chi_r}{\partial k}\frac{\mathscr{E}_p}{2k_p}\nabla_\perp.\mathbf{k}_p-\frac{\partial \chi_r^{\text{tr}}}{\partial \omega}\mathbf{v}_\perp.\nabla_\perp\mathscr{E}_p=0, \tag{3.91}$$

where $\partial_\omega \chi_r^{\text{env}}$ is defined by Eq. (3.68) with $\int \omega_B dt$ replaced with $\int_{\mathbf{v}_\perp} \omega_B dt$.

When the transverse electron motion is not infinitely cold, one needs to average Eq. (3.91) over the distribution, $f(\mathbf{v}_\perp)$, of transverse velocities, which, as explained before, we approximate by the unperturbed one (assumed to be a Maxwellian). Using the facts that $\partial_\omega \chi_r$ does not depend on $\mathbf{v}_\perp$, that $f(\mathbf{v}_\perp)$ is isotropic, and that $\partial_k \chi_r \approx (2/k_p) - v_\phi \partial_{\omega_p} \chi_r$, averaging Eq. (3.91) over $\mathbf{v}_\perp$ yields

$$\frac{\partial \chi_{3\text{D}}^{\text{env}}}{\partial \omega}\left[\frac{\partial \mathscr{E}_p}{\partial t}+\mathbf{v}_g\nabla\mathscr{E}_p+v_0\frac{\mathscr{E}_p}{2k_p}\nabla_\perp.\mathbf{k}_p+\nu_{3D}\mathscr{E}_p\right]=0, \tag{3.92}$$

with

$$v_{g_x}=v_\phi-\frac{2}{k_p\partial_\omega\chi_{3\text{D}}^{\text{env}}}, \tag{3.93}$$

$$v_{g_{y,z}}=v_{gx}\frac{k_{y,z}}{k}+\frac{\int \partial_\omega \chi_r^{\text{env}} f(\mathbf{v}_\perp)v_{y,z}d\mathbf{v}_\perp}{\partial_\omega \chi_{3\text{D}}^{\text{env}}}, \tag{3.94}$$

$$v_0=v_\phi\frac{\partial_\omega \chi_r}{\partial_\omega \chi_{3\text{D}}^{\text{env}}}-\frac{2}{k_p\partial_\omega\chi_{3\text{D}}^{\text{env}}}, \tag{3.95}$$

$$\nu_{3D}=\frac{\int f(\mathbf{v}_\perp)\nu_{\text{NL}}(\mathbf{v}_\perp)\partial_\omega \chi^{\text{env}}(\mathbf{v}_\perp)d\mathbf{v}_\perp}{\partial_\omega \chi_{3\text{D}}^{\text{env}}}, \tag{3.96}$$

where $\partial_\omega \chi_{3\text{D}}^{\text{env}} \equiv \int f(\mathbf{v}_\perp)\partial_\omega \chi_r^{\text{env}} d\mathbf{v}_\perp$. Note that, in order to derive $\mathbf{v}_g$, we made use of the approximation, $\mathbf{v}_\perp \approx v_y\hat{y}+v_z\hat{z}$.

Just like in 1-D, the nonlocal dependence of the group velocity entails a longitudinal shrinking of the plasma wave packet. Moreover, the second term in the right-hand side of Eq. (3.94), which accounts for collisionless dissipation, induces the reduction of the transverse size of the plasma pulse with no change in its maximum amplitude, an effect that was very clearly observed numerically in [13]. As for the first term in Eq. (3.94), it allows for the self-focussing induced by wave front bowing.

Moreover, it is quite clear from Eqs. (3.89) and (3.96) that the Landau-like damping rate would decrease less rapidly as a function of the EPW amplitude for a narrower wave packet or larger electron thermal velocity, i.e., for a shorter interaction time between the electrons and the electrostatic wave.

3.5 Nonlinear Frequency Shift of an SRS-Driven Plasma Wave

In this section, we show how to solve the dispersion relation of a driven plasma wave, Eq. (3.14) of Sect. 3.2, and we stress the differences between the frequency shift, $\delta\omega_p$, of a driven plasma wave and that, $\delta\omega_{\text{free}}$, which would be found by assuming that the EPW freely propagates. We moreover discuss the physics relevance of the dispersion relation derived by Holloway and Dorning in [17], and then generalized by Rose and Russel in [27] to be applied to stimulated Raman scattering (or of the similar dispersion relation previously derived by Krapchev and Ram in [18]). It is shown, in particular, that the curves $\omega_p(k_p)$ drawn in these papers, as well as the notion of "loss of resonance" introduced in [27], have no physics reality just because, in order to draw correct conclusions about the dispersion properties of an EPW, one cannot avoid accounting for the fact that this wave needs to be driven in order to be able to grow in an initially Maxwellian plasma.

Let us now recall that the dispersion relation of a driven plasma wave, derived in Sect. 3.2.1, is

$$1 + \alpha_d \chi_r = 0, \tag{3.97}$$

where

$$\alpha_d \equiv \frac{1 + 2(\mathscr{E}_d/\mathscr{E}_p)\sin(\delta\varphi) + (\mathscr{E}_d/\mathscr{E}_p)^2}{1 + (\mathscr{E}_d/\mathscr{E}_p)\sin(\delta\varphi)}, \tag{3.98}$$

$\mathscr{E}_p$ being the amplitude of the plasma wave and $\mathscr{E}_d$ that of the drive. We restrict here to the case when the electrostatic wave grows due to the stimulated Raman scattering of a laser by a uniform plasma. Then, the driving amplitude $\mathscr{E}_d$ is well known to be (see [3, 19] for details)

$$\mathscr{E}_d = \frac{ek_p\mathscr{E}_l\mathscr{E}_s}{2m\omega_l\omega_s}, \tag{3.99}$$

$\mathscr{E}_l$ and $\mathscr{E}_s$ being, respectively, the amplitudes of the laser and scattered waves, and ω_l and ω_s the frequencies of these waves.

In order to solve Eq. (3.97), one therefore needs to derive the nonlinear values of both, α_d and χ_r, which is done in the following two sections.

3.5.1 Derivation of χ_r

Just like in the previous sections, we restrict to a situation when the electron motion is nonrelativistic, and where the SRS growth rate is small enough for the adiabatic approximation to be valid. Hence, we only derive here an adiabatic estimate of χ_r.

To do so, we shift to dimensionless variables and use ψ as a dynamical variable for the electrons, define the dimensionless time $\tau \equiv k_p v_{\text{th}} \omega_{pe}$, and normalize velocities to the thermal one, $\bar{v} \equiv v/v_{\text{th}}$. In these variables, the electron motion is given by the following equations:

$$\mathrm{d}\psi/\mathrm{d}\tau = \bar{v} - \bar{v}_\phi, \tag{3.100}$$

$$\mathrm{d}\bar{v}/\mathrm{d}\tau = -\Phi_0 \sin(\psi), \tag{3.101}$$

where $\bar{v}_\phi$ is the normalized wave phase velocity and $\Phi_0 \equiv eE_0/(k_p T_e)$. Clearly, Eqs. (3.100) and (3.101) derive from the following Hamiltonian:

$$\bar{H} = \frac{(\bar{v} - \bar{v}_\phi)^2}{2} - \Phi_0 \cos(\psi). \tag{3.102}$$

We now want to use the adiabatic theorem, proved in [10], which guarantees that the action remains nearly constant if the wave amplitude varies slowly enough. However, when using this theorem, one must be cautious to define the action, $\mathscr{A}$, so that it remains continuous when an electron initially lying on an untrapped orbit gets trapped (or vice versa). This is particularly true here because we want to derive the EPW frequency shift and therefore need to account for the nonlinear change of the wave phase velocity. Hence, we must make sure that we use action conservation for the dynamics of Hamiltonian $\bar{H}$, and not for the dynamics derived as though the wave frame were inertial (like has been done in all the papers we know of, which made use of the adiabatic approximation to derive the EPW nonlinear frequency shift). We then define the action of an untrapped electron as:

$$\mathscr{A} = \frac{1}{2\pi} \oint \bar{v} \mathrm{d}\psi, \tag{3.103}$$

where the integral is calculated along the orbit of a "frozen" wave with normalized amplitude Φ_0. This action is well known to be (see, e.g., [10])

$$\mathscr{A} = \frac{4\sqrt{\Phi_0}}{\pi\sqrt{m_0}} E(m_0) + \eta \bar{v}_\phi, \tag{3.104}$$

where $m_0 = 2\Phi_0/(\bar{H} + \Phi_0)$, $E(m_0)$ is the complete elliptic integral of second kind (see, e.g., [1]), and η is the sign of $(\bar{v}_0 - \bar{v}_\phi)$, $\bar{v}_0$ being the initial electron velocity. We assume that, initially, the wave amplitude is infinitely small. Then, the initial action is easily calculated: $\mathscr{A}(0) = \eta \bar{v}_0$. Action conservation for untrapped particles therefore writes

$$\frac{4\sqrt{\Phi_0}}{\pi\sqrt{m_0}} E(m_0) = |\bar{v}_0 - \bar{v}_\phi(\Phi_0)|. \tag{3.105}$$

When Φ_0 is a constant, it is well known that the energy of any untrapped electron is larger than Φ_0. Therefore, in phase space $(\psi, \bar{v})$, trapped and untrapped electrons are separated by a curve (the so-called separatrix), whose equation is $\bar{v} = \pm 2\sqrt{\Phi_0} \cos(\psi/2) + \bar{v}_\phi$. Then, the action of an untrapped electron infinitely close to the separatrix is $\mathscr{A}_s = 4\sqrt{\Phi_0}/\pi + \eta\bar{v}_\phi$, and the action of any untrapped electron is larger than $\mathscr{A}_s$. Going back to the case when Φ_0 has slowly grown from zero to a given value Φ^*, we obtain the following adiabatic condition for trapping: an electron with initial velocity $\bar{v}_0$ is trapped if Φ^* is larger than the value $\Phi_0(\bar{v}_0)$ defined by

$$|\bar{v}_0 - \bar{v}_\phi[\Phi_0(\bar{v}_0)]| = 4\sqrt{\Phi_0(\bar{v}_0)}/\pi, \tag{3.106}$$

where $\bar{v}_\phi[\Phi_0(\bar{v}_0)]$ is the wave phase velocity when its amplitude is $\Phi_0(\bar{v}_0)$.

As for trapped electrons, we define their action by

$$\begin{aligned}\mathscr{A} &= \frac{1}{4\pi}\oint \bar{v}\mathrm{d}\psi + \eta\bar{v}_\phi\left[\Phi_0(\bar{v}_0)\right] \\ &= \frac{4\sqrt{\Phi_0}}{\pi}[(m_1-1)K(m_1)+E(m_1)] + \eta\bar{v}_\phi\left[\Phi_0(\bar{v}_0)\right],\end{aligned} \tag{3.107}$$

where $K(m_1)$ is the complete elliptic integral of the first kind (see [1]), $m_1 = (\bar{H}+\Phi_0)/2\Phi_0$, and η is the sign of $\bar{v}_0 - \bar{v}_\phi[\Phi_0(\bar{v}_0)]$, $\bar{v}_\phi[\Phi_0(\bar{v}_0)]$ being defined by Eq. (3.106). With definition Eq. (3.107), electrons infinitely close to the separatrix have the same action, whether they are trapped or untrapped, so that $\mathscr{A}$ may indeed be considered as a preserved quantity. Note, moreover, that we do account for the nonlinear change in the EPW phase velocity in our definition of $\mathscr{A}$ *via* $\bar{v}_\phi\left[\Phi_0(\bar{v}_0)\right]$. More precisely, in case when Φ_0 has grown from zero to a given value Φ^*, one needs to account for the variations in $\bar{v}_\phi$ during the growth of Φ_0 when solving for the trapped electrons' action conservation:

$$\frac{4\sqrt{\Phi_0}}{\pi}[(m_1-1)K(m_1)+E(m_1)] = |\bar{v}_0 - \bar{v}_\phi[\Phi_0(\bar{v}_0)]|, \tag{3.108}$$

and not only assume that the EPW phase velocity has kept the constant value $\bar{v}_\phi(\Phi^*)$. Note, moreover, that allowing for the nonlinear change in $\bar{v}_\phi$ is only needed to derive the adiabatic response of the trapped electrons.

From Eq. (3.21) of Sect. 3.2, we now need to compute $\langle\cos(\psi)\rangle$ in order to derive χ_r, which we do for a given amplitude $\Phi_0 = \Phi^*$ by using the action-angle variables $(\mathscr{A},\theta)$ in order to perform the statistical averaging. Assuming that the plasma is initially Maxwellian, and noting that the action is $\mathscr{A} = \pm\bar{v}_0$ and that the Jacobian of the change of variables $(\psi,\bar{v}) \to (\mathscr{A},\theta)$ is unity, one easily finds

$$\langle\cos(\psi)\rangle = \int_{-\infty}^{+\infty}\frac{\mathrm{e}^{-\bar{v}_0^2/2}}{\sqrt{2\pi}}\left\{\int\cos\left[\psi(\bar{v}_0,\theta)\right]\frac{\mathrm{d}\theta}{2\pi}\right\}\mathrm{d}\bar{v}_0. \tag{3.109}$$

For untrapped electrons, the angle θ is given by

$$\theta = \frac{\pi}{K(m_0)}\int_0^{\psi/2}\frac{\mathrm{d}u}{\sqrt{1-m_0\sin^2 u}}. \tag{3.110}$$

Then, for untrapped electrons,

$$\int \cos\left[\psi(\bar{v}_0,\theta)\right] \frac{\mathrm{d}\theta}{2\pi} = \int_0^{2\pi} \frac{\cos(\psi)\,\mathrm{d}\psi}{4K(m_0(\bar{v}_0))\sqrt{1-m_0(\bar{v}_0)\sin^2(\psi/2)}}$$

$$= 1 + \frac{2}{m_0(\bar{v}_0)}\left[\frac{E(m_0(\bar{v}_0))}{K(m_0(\bar{v}_0))} - 1\right], \tag{3.111}$$

where $m_0(\bar{v}_0)$ solves Eq. (3.105) with $\Phi_0 = \Phi^*$.

In the case of trapped electrons,

$$\theta = \frac{\pi F(\vartheta|m_1)}{2K(m_1)}, \tag{3.112}$$

where $F(\vartheta|m_1)$ is the elliptic integral of first kind (see [1]), and ϑ is defined by $\sin(\psi/2) = \sqrt{m_1}\sin(\vartheta)$. Then, for trapped electrons,

$$\int \cos\left[\psi(\bar{v}_0,\theta)\right] \frac{\mathrm{d}\theta}{2\pi} = \int_0^{2\pi} \frac{(1-2m_1(\bar{v}_0)\sin^2\vartheta)}{4K(m_1(\bar{v}_0))} \frac{\mathrm{d}\vartheta}{\sqrt{1-m_1(\bar{v}_0)\sin^2\vartheta}}$$

$$= -1 + 2\frac{E(m_1(\bar{v}_0))}{K(m_1(\bar{v}_0))}, \tag{3.113}$$

where $m_1(\bar{v}_0)$ solves Eq. (3.108).

Putting all the pieces of the calculation together we find that, when $\Phi_0 = \Phi^*$,

$$\langle\cos(\psi)\rangle = \int_{|\bar{v}_0-\bar{v}_\phi(\Phi^*)|>4\sqrt{\Phi^*}/\pi} \frac{\mathrm{e}^{-\bar{v}_0^2/2}}{\sqrt{2\pi}}\mathrm{d}\bar{v}_0 \left\{1 + \frac{2}{m_0(\bar{v}_0)}\left[\frac{E(m_0(\bar{v}_0))}{K(m_0(\bar{v}_0))} - 1\right]\right\}$$

$$+ \int_{|\bar{v}_0-\bar{v}_\phi(\Phi^*)|<4\sqrt{\Phi^*}/\pi} \frac{\mathrm{e}^{-\bar{v}_0^2/2}}{\sqrt{2\pi}}\mathrm{d}\bar{v}_0 \left\{-1 + 2\frac{E(m_1(\bar{v}_0))}{K(m_1(\bar{v}_0))}\right\}. \tag{3.114}$$

This value of $\langle\cos(\psi)\rangle$, when plugged into Eq. (3.21), yields the expression of χ_r that we used in the dispersion relation Eq. (3.97) in order to derive the frequency shifts $\delta\omega_p$ plotted in Figs. 3.12 and 3.13.

Now, before ending this section, we want to stress the following features:

1. The range in Φ_0 for which we could find solutions to the dispersion relation Eq. (3.97) was much larger when accounting for the nonlinear variations of the EPW phase velocity than when assuming that the wave frame was inertial. This is mainly why v_ϕ must not be treated as a constant when using action conservation.
2. We numerically checked the relevance of the adiabatic approximation in [3] and found that it was valid provided that the EPW growth rate was less than about $\omega_{pe}/20$, a condition easily reached for an SRS-driven plasma wave. Actually,

the good agreement between the theoretical and numerical values of the EPW frequency shift shown in Figs. 3.12 and 3.13 illustrates the relevance of this approximation.

3.5.2 Derivation of α_d

The linear value of α_d is chosen to be that of the linearly most unstable SRS-driven mode. It is systematically larger than unity, which implies that the linear frequency of a driven plasma wave is always slightly larger than that of the freely propagating EPW with the same wave number. Physically, this is easily understood the following way. An electrostatic wave is more easily driven if it is close to a natural mode of the plasma but, also, if it is not too Landau damped. Since, as is well known, Landau damping decreases with the wave phase velocity, the frequency of the linearly most unstable mode against SRS is larger than that of a natural mode, all the more as the Landau damping rate is large. Actually, it is clear from Eq. (3.16) that $\mathscr{E}_d/\mathscr{E}_p$ is of the order of χ_i which, in the linear regime, is of the order of the Landau damping rate, ν_L, normalized to the plasma frequency. Hence, the departure of the linear value of α_d from unity is, indeed, larger for larger values of ν_L.

Now, as shown in Fig. 3.8 of Sect. 3.4, χ_i quickly decreases with the EPW amplitude and therefore so does $\mathscr{E}_d/\mathscr{E}_p$, which entails a rapid convergence of α_d towards unity, and a quick drop in ω_p. Then, although the frequency of a driven wave is close to that of a natural mode, its frequency shift (much smaller than the wave frequency itself) is significantly altered by the nonlinear variations of α_d, as is obvious from Figs. 3.12 and 3.13 showing that $|\delta\omega_{\text{free}}|$ is significantly less than $|\delta\omega_p|$.

We will actually not use Eq. (3.16) to derive the nonlinear values of α_d because this would require solving for $\delta\omega_p$ together with the SRS growth, while we intend here to derive $\delta\omega_p$ only as a function of the plasma wave amplitude $\mathscr{E}_p$. Then, to derive α_d, we resort to the envelope equation of the scattered wave, which, directly from Maxwell equations, is easily shown to be (see [6])

$$[\partial_t + v_{gs}\partial_x - i\Delta_s]\,\mathscr{E}_s = (\Gamma_s/2)\mathscr{E}_l\mathscr{E}_p e^{-i\delta\varphi}, \tag{3.115}$$

where $\Gamma_s \equiv ek_p/(2m\omega_l)$, $v_{gs} \equiv k_s c^2/\omega_s$ (c being the speed of light in vacuum and k_s the scattered wave number), and $\Delta_s \equiv [\omega_s^2 - (k_s c)^2 - \omega_{pe}^2]/2\omega_s$ represents the detuning of the scattered wave from resonance. Using the definition (3.99) for $\mathscr{E}_d$, the envelope equation (3.115) may also be written

$$\frac{\mathscr{E}_d}{\mathscr{E}_p} = \frac{k_p^2 v_{osc}^2 e^{-i\delta\varphi}}{8\omega_s(G_s - i\Delta_s)}, \tag{3.116}$$

where $G_s \equiv \mathscr{E}_s^{-1}(\partial_t + v_{gs}\partial_x)\mathscr{E}_s$ is the scattered wave growth rate calculated in its own reference frame and where $v_{osc} \equiv e\mathscr{E}_l/m\omega_l$. Assuming that v_{osc} and ω_s remain constant, one may deduce $\mathscr{E}_d/\mathscr{E}_p$ and $\delta\varphi$ (and therefore α_d) from Eq. (3.116) provided that G_s and Δ_s are known. Now, in order to derive Δ_s, we assume that k_s remains constant while the frequency of the scattered wave nonlinearly shifts by $\delta\omega_s = -\delta\omega_p$. Then, as the EPW amplitude grows and ω_p nonlinearly shifts, Δ_s increases compared to G_s, which makes $\mathscr{E}_d/\mathscr{E}_p$, and therefore the EPW frequency, drop. As for G_s, we either kept it constant or tried to account for its increase due to the nonlinear reduction of $\nu_{\rm NL}$ in a simple way (as explained in [4]), without noticing a great change in $\delta\omega_p$.

3.5.3 Comparisons with Results from Vlasov Simulations of Stimulated Raman Scattering and with Previous Theories

Let us now compare the values $\delta\omega_p$ deduced from our resolution of $1+\alpha_d\chi_r = 0$ with the values, $\delta\omega_{\rm num}$, of the EPW frequency shift inferred from simulations of stimulated Raman scattering performed with the Vlasov code ELVIS [28].

In our simulations, the space and time steps are $\Delta x/\lambda_l = c\Delta t/\lambda_l = 0.03$, where λ_l is the laser wavelength. The velocity step varies from run to run, with $0.0016 \le \Delta v/v_{\rm th} \le 0.015$. The density profile is finite, with a central, flat region from $x/\lambda_l =$ 28 to 242 (see Fig. 1 of [28]). The laser enters from vacuum on the left ($x = 0$), and a small-amplitude seed light wave is injected on the right with λ_s chosen to match the frequency of the most unstable mode. The seed intensity varied from $I_s/I_l = 10^{-5}$ to 10^{-8}, without affecting the dispersion relation. $\delta\omega_{\rm num}$ and $\Phi \equiv e\mathscr{E}_p/k_pT_e$ are obtained via the Hilbert transform (see, e.g., [16]) of the electrostatic field versus time at one x. All the simulations whose results are presented here, as well as the diagnostic used to derive the EPW amplitude and frequency from the Hilbert transform of the electrostatic field, were performed by D. Strozzi.

As illustrated in Figs. 3.12 and 3.13, we always find an excellent agreement between $\delta\omega_p$ and $\delta\omega_{\rm num}$. For all runs, the unperturbed plasma density n_0 is $8.9\times10^{20}\,{\rm cm}^{-3}$ and the laser vacuum wavelength is $\lambda_l = 0.351\,\mu$m. The values of the laser intensity, I_l, and of the electron temperature, T_e, are specified in the figure captions. The indicated value of $k_p\lambda_{\rm D}$ in these figures refers to the wave number of the linearly most unstable SRS-driven EPW for the given plasma and laser parameters. $\delta\omega_{\rm num}$ is only plotted before Φ reaches its first local time maximum. After this maximum, and near the laser entrance, one may see pulses in the time evolution of Φ. The good agreement between $\delta\omega_p$ and $\delta\omega_{\rm num}$ usually remains for the early pulses (not only for the first one) but eventually breaks down together with the validity of the adiabatic approximation. Away from the laser entrance, we numerically find that Φ increases with time until a sideband eventually grows, which is reminiscent of the result of Brunner and Valeo [9], and which then makes the notions of a central frequency, and its shift, irrelevant. For the range of intensities

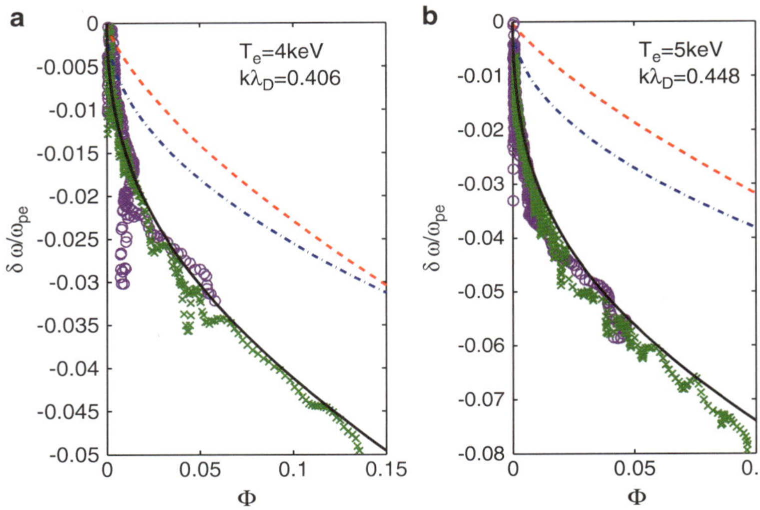

Fig. 3.12 $\delta\omega_p$ (*solid line*), $\delta\omega_{\text{free}}$ (*dashed line*), $\delta\omega_{\text{D}}$ (*dash–dotted line*) and $\delta\omega_{\text{num}}$ at $x = 77\lambda_l$ (*circles*) and at $x = 193\lambda_l$ (*crosses*) for $I_l = 2\,\text{PW/cm}^2$ and (**a**) $T_e = 4\,\text{keV}$, and (**b**) $T_e = 5\,\text{keV}$

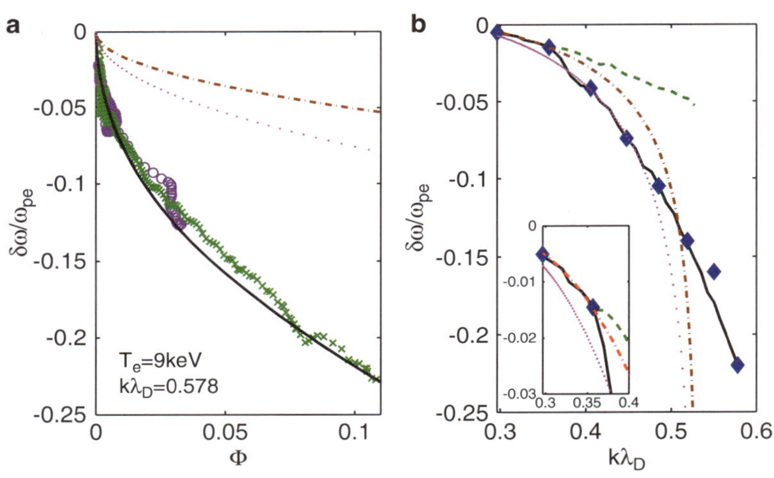

Fig. 3.13 Panel (**a**) $\delta\omega_p$ (*solid line*), $\delta\omega_{\text{D}}$ (*dash–dotted line*), $\delta\omega_{\text{MO}}$ (*dots*), and $\delta\omega_{\text{num}}$ at $x = 77\lambda_l$ (*circles*) and at $x = 232\lambda_l$ (*crosses*) for $T_e = 9\,\text{keV}$ and $I_l = 8\,\text{PW/cm}^2$. Here $\delta\omega_{\text{D}}$ and $\delta\omega_{\text{MO}}$ are calculated by using for ω_{lin} the linear frequency of the SRS-driven plasma wave. Panel (**b**) $\delta\omega_{\text{num}}$ (*diamonds*), $\delta\omega_p$ (*solid line*), $\delta\omega_{\text{free}}$ (*dashed line*), $\delta\omega_{\text{D}}$ (*dash–dotted line*), and $\delta\omega_{\text{MO}}$ (*dotted line*) *versus* $k_p\lambda_{\text{D}}$ when $\Phi = 0.1$. Each numerical result is for a distinct run with a different T_e, and $I_l = 2\,\text{PW/cm}^2$ for $T_e < 6\,\text{keV}$ ($k\lambda_{\text{D}} < 0.485$), $I_l = 4\,\text{PW/cm}^2$ for $T_e = 6\,\text{keV}$, $I_l = 6\,\text{PW/cm}^2$ for $T_e = 7\,\text{keV}$, and $I_l = 8\,\text{PW/cm}^2$ for $T_e > 7\,\text{keV}$

we investigated, $I_l \leq 10\,\text{PW/cm}^2$, and when $0.3 \leq k_p\lambda_D \leq 0.58$, we thus find that our theory breaks down mainly when, eventually, the EPW can no longer be considered nearly monochromatic. For lower values of $k_p\lambda_D$, and maybe larger intensities, a nearly monochromatic EPW may reach so large an amplitude that higher harmonics and a "DC" field need to be accounted for in order to correctly calculate the frequency shift, as recently reported in [21]. However, we never had to account for these to find a good agreement between our numerical and theoretical estimates of the EPW frequency shift.

When comparing $\delta\omega_p$ and $\delta\omega_{\text{num}}$ to $\delta\omega_{\text{free}}$, we find that $\delta\omega_{\text{free}}$ misses the initial rapid drop in $\delta\omega_p$ due to the rapid convergence of α_d towards unity, while, for larger wave amplitudes, the variations of $\delta\omega_p$ and $\delta\omega_{\text{free}}$ with Φ are similar and are mainly due to the nonlinear change in χ_r. As a result, for the examples of Fig. 3.12, $|\delta\omega_{\text{free}}|$ underestimates $|\delta\omega_p|$ by a factor close to two.

Let us now compare $\delta\omega_p$ and $\delta\omega_{\text{num}}$ to well-known previously published formulas for the frequency shift. We start with that, $\delta\omega_D$, derived by Dewar in [11] for a small-amplitude freely propagating EPW, by assuming (as we do it here) adiabatic electron motion but by neglecting the nonlinear change in the wave phase velocity when enforcing action conservation. Dewar then found

$$\frac{\delta\omega_D}{\omega_{pe}} \equiv \frac{1.09 f_0''(\bar{v}_\phi)(\omega_{\text{lin}}/\omega_{pe})\sqrt{\Phi}}{1+(k_p\lambda_D)^2-(\omega_{\text{lin}}/\omega_{pe})^2}, \tag{3.117}$$

where $f_0(\bar{v}) \equiv \exp(-\bar{v}^2/2)/\sqrt{2\pi}$, $f_0'' = d^2 f_0/d\bar{v}^2$, $\bar{v}_\phi \equiv \omega_{\text{lin}}/(kv_{Te})$, and ω_{lin} is the linear solution of $1+\chi_r = 0$, χ_r being calculated by making use of the adiabatic approximation. ω_{lin} only exists, and therefore $\delta\omega_D$ is only defined, when $k_p\lambda_D < 0.53$. As can be seen in Fig. 3.13(b), $\delta\omega_D$ yields a good estimate of $\delta\omega_p$ and $\delta\omega_{\text{num}}$ only when $k_p\lambda_D \leq 0.35$.

Another very well-known approximate formula for the frequency shift of a freely propagating wave is that derived by Morales and O'Neil in [24] by assuming that the wave amplitude grows infinitely quickly before remaining constant and uniform and by neglecting the nonlinear variations of the wave phase velocity when calculating the electron motion. The value of the frequency shift found by Morales and O'Neil is $\delta\omega_{\text{MO}} \approx (1.63/1.09)\delta\omega_D$. $\delta\omega_{\text{MO}}$ is also only defined when $k_p\lambda_D < 0.53$ and Fig. 3.13b seems to show that it is close to $\delta\omega_p$ and $\delta\omega_{\text{num}}$ only when $0.37 \leq k_p\lambda_D \leq 0.46$. This agreement is however fortuitous: the ratio $\delta\omega_p/\delta\omega_{\text{MO}}$ actually depends on Φ because $\delta\omega_p$ is not simply proportional to $\sqrt{\Phi}$.

If one were to extrapolate the values of $\delta\omega_D$ and $\delta\omega_{\text{MO}}$ beyond $k_p\lambda_D = 0.53$ by choosing for ω_{lin} the linear frequency of the SRS-driven wave, $\delta\omega_D$ and $\delta\omega_{\text{MO}}$ would be found to underestimate $\delta\omega_p$ whenever $k_p\lambda_D > 0.35$ and $k_p\lambda_D > 0.4$, respectively. An example of this is given in Fig. 3.13a.

3.5.4 Discussion of Previously Proposed Nonlinear Dispersion Relations

Several authors proposed in the past nonlinear dispersion relations for essentially undamped plasma waves which have grown in an initially Maxwellian plasma. Holloway and Dorning derived in [17] such a dispersion relation for a wave of infinitely small amplitude assumed to remain undamped in a nearly Maxwellian plasma. This result was generalized by Rose and Russell in [27] for a wave of finite amplitude which has grown infinitely quickly and by Krapchev and Ram in [18] for a wave growing slowly enough to induce adiabatic electron motion. All the corresponding curves $\omega_p(k_p)$ assume the same shape close to that derived by Holloway and Dorning, which is reproduced in Fig. 3.14, with the same peculiar property that they do not extend beyond $k_p\lambda_D \approx 0.53$, where λ_D is the Debye length defined by Eq. (3.4). This lack of solution to the dispersion relation beyond a given value of $k_p\lambda_D$ was termed a "loss of resonance" by Rose and Rusell, who deduced from this that a large-amplitude, nearly monochromatic, plasma wave with $k_p\lambda_D > 0.53$ could not exist. This assertion is nevertheless in total contradiction with the results plotted in Fig. 3.13a showing that a nearly monochromatic plasma wave with $k_p\lambda_D \approx 0.58$ can indeed grow to a large enough amplitude for its collisionless damping rate to be extremely small compared to its linear value, as we checked it by applying our theoretical estimate for ν_{NL} (3.68) to the simulations results of Fig. 3.13a.

Clearly, one can laser drive an electron plasma wave whatever its wave number and even when $k_p\lambda_D > 0.53$. Moreover, whatever the value of $k_p\lambda_D$ we investigated,

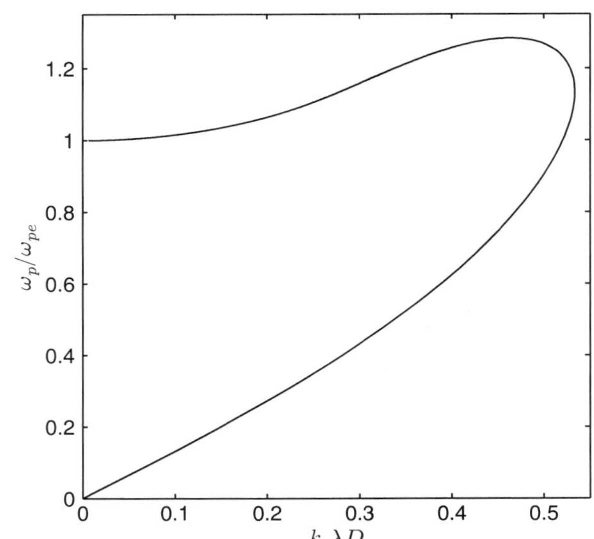

Fig. 3.14 Unrealistic dispersion relation of a plasma wave assumed to have spontaneously grown in an initially Maxwellian plasma before remaining undamped

we could always find solutions to the dispersion relation up to large values of Φ (at least $\Phi = 1$). Now, one may wonder what happens if an EPW with $k_p\lambda_D > 0.53$ is first laser driven and if the drive is then turned off to let the plasma wave freely propagate. Clearly if, when the drive is shut down, the wave amplitude is so small that ν_{NL} is close to its linear value, then the EPW frequency will assume the value derived from Landau's dispersion relation, solved, for example, in [15], but its amplitude will quickly decrease due to a strong Landau damping. If the EPW is driven to so large amplitudes that $\nu_{NL} \approx 0$, then, as shown previously, $\alpha_d \approx 1$, and the dispersion relations for the driven and freely propagating waves have become essentially the same. As a result, the EPW frequency is not affected by the shutting down of the laser. Hence, nonlinear values for the frequency of a freely and essentially undamped plasma wave, which grew in an initially Maxwellian plasma, are given by the results of the previous section, which hold even for values of $k_p\lambda_D$ significantly larger than 0.53. We therefore conclude that a dispersion relation as that plotted in Fig. 3.14 represents no physics reality.

3.6 Conclusion

We presented in this chapter a theoretical description, supported by numerical results, of the nonlinear propagation of an electron plasma wave (EPW) in a collisionless plasma and in a three-dimensional geometry. This required completely new developments in nonlinear kinetic plasma theory, leading to quite surprising results that could not be inferred from previous theoretical studies in nonlinear plasma physics, usually restricted to rather academic situations, and which have no counterpart in nonlinear optics, thus showing the uniqueness of the EPW propagation.

One of the most striking feature in the propagation of an EPW is collisionless dissipation, which we addressed in this chapter from the linear regime, where it amounts to Landau damping, to the strongly nonlinear one when the wave is undamped and dissipation would manifest itself in the shrinking of the plasma wave packet, both in the longitudinal and transverse directions.

Our results mainly follow from the direct investigation of the nonlinear electron orbits that yields a theoretical expression for the imaginary part of the electron susceptibility, χ_i, which is the key parameter to derive the nonlinear propagation of an EPW. By making use of a high-order perturbation analysis, we provide a first estimate of χ_i, which happens to be quite precise up to $\int_0^t \omega_B(t')\mathrm{d}t' \approx 15$, where ω_B is the frequency of a deeply trapped electron orbit. That perturbative results could lead to accurate estimates up to so large values of $\int_0^t \omega_B(t')\mathrm{d}t'$ was quite unexpected and is mainly due to the fact that deeply trapped electrons do not contribute much to χ_i, a property resulting from the fact that the orbits of such electrons are nearly symmetric with respect to the velocity axis. We also managed to derive quite an accurate estimate of χ_i whenever $\int_0^t \omega_B(t')\mathrm{d}t' > 6$ by noticing that χ_i should be proportional to the wave growth rate and that the coefficient of proportionality could be derived by making use of the adiabatic approximation. Since the two previous estimates of χ_i have a common domain of validity, it is just enough to "connect them" in order to derive a very precise expression for χ_i whatever $\int_0^t \omega_B(t')\mathrm{d}t'$. χ_i is further expressed in terms of the space and time derivatives of the wave amplitude, thus transforming Gauss' law into an envelope equation, which, when compared to that derived from the famous Whitham's variational approach, allows to relate collsionless dissipation in the strongly nonlinear regime to electron trapping. Then, the envelope equation derived in 1-D is very easily generalized to a three-dimensional geometry. This equation accounts for the nonlinear variations of the collisionless damping rate, ν_{NL}, and of the EPW group velocity, v_g, which are nonlocal functions of the wave amplitude. In particular, ν_{NL} is found to remain nearly constant, and close to the Landau value, before abruptly dropping to zero. As for v_g, it is found to be different from the derivative of the wave frequency with respect to its wave number and, when ν_{NL} drops, it abruptly rises from its linear value to get close to the wave phase velocity. Actually, v_g is mainly a function of the *maximum* amplitude experienced by the electrons, and, in 1-D, its nonlinear and nonlocal variations entail the shrinking of the plasma wave packet and therefore the dissipation of the electrostatic energy. In 3-D, the group velocity has nonzero components perpendicular to the wave number, which entail the transverse shrinking of the wave packet.

All the previous surprising results were actually derived for a laser-driven wave (but may be applied to a freely propagating wave by setting the drive to zero) and were carefully compared against 1-D Vlasov simulations of stimulated Raman scattering. A very good agreement was systematically found between the numerical and theoretical results, especially as regards our predictions for χ_i and v_g and, therefore, for ν_{NL}. At this point, we would like to stress the importance of addressing a driven wave, since only such a wave may grow in a Maxwellian plasma. This is particularly crucial as regards the EPW dispersion relation once ν_{NL} is strongly reduced compared to the Landau value, since erroneous conclusions were drawn from previous theories neglecting the effect of the drive. For example, Holloway and Dorning argued in [17] that an electron plasma wave with $k_p\lambda_{\mathrm{D}} > 0.53$, where k_p is EPW wave number and λ_{D} is the Debye length defined by Eq. (3.4), could not exist in the strongly nonlinear regime when $\nu_{\mathrm{NL}} \approx 0$. However, such a limitation no longer holds when account is taken of the drive, and, numerically, we could actually generate large-amplitude plasma waves with $\nu_{\mathrm{NL}} \ll \nu_{\mathrm{L}}$ and $k_p\lambda_{\mathrm{D}} > 0.53$. Moreover, when an EPW grows in an initially Maxwellian plasma, its frequency nonlinearly shifts, and we calculated this frequency shift theoretically by making use of the adiabatic approximation, by allowing for the effect of the drive, and by accounting for the fact that the reference frame moving at the phase velocity compared to the laboratory frame is not Galilean (since the wave amplitude, and therefore its frequency and phase velocity, vary with time). When doing so, we found values for the EPW frequency shift in very good agreement with those deduced from Vlasov simulations, and much larger in magnitude than previously published ones, mainly because previous theories neglected the effect of the drive. Moreover, accounting for the fact that the phase velocity is not a constant allows to find solutions to the nonlinear dispersion relation of the EPW up to wave amplitudes much larger than when relying to previous theoretical developments.

Addressing the nonlinear properties of a *driven* plasma wave is also clearly essential to tackle such an important issue as stimulated Raman scattering (SRS) in a plasma, which is currently under investigation by several groups over the world, mainly because it may be detrimental for inertial confinement fusion [22] and because it may be a very efficient and practical means to generate very intense electromagnetic radiations via the so-called backward Raman amplification [23]. A detailed description of SRS is clearly out the scope of this chapter and would require one on its own, but we nevertheless want to stress that the theoretical results reported here form the basis of an envelope code we have been developing for about 3 years in order to accurately and efficiently predict Raman reflectivity. This quantity, which pertains to the situation when SRS mainly gives rise to a *back*scattered light, is defined as the ratio between the output power of this backscattered radiation and the input power of the laser that experiences Raman scattering. SRS reflectivity is of prime importance for inertial confinement fusion because it yields the amount of energy which is indeed available to compress the fusion target. Measuring Raman reflectivity is also a direct way to quantify the efficiency of backward Raman amplification. As reported in [8], our envelope code indeed provides reflectivity levels consistent with those deduced from kinetic codes (which solve for the electron distribution function in a collisionless plasma) while being about 10^5 faster, which is needed to address such a large scale system as a fusion plasma. Hence, besides the basic issues we have been discussing in this chapter, our theoretical results are currently being used for practical applications.

Acknowledgment It is a pleasure to acknowledge Laurent Gremillet and Olivier Morice for extensive discussions, as well as David Strozzi for his numerical simulations.

3.7 Appendix: Derivation of $\partial_\omega \chi_r^{\text{eff}}$

By definition, χ_r^{eff} is calculated by summing the adiabatic contributions to χ_r of the untrapped electrons, for a given wave amplitude E_0, and by disregarding the contributions of the trapped electrons. Then, if we denote $V_{\text{tr}} \equiv \sqrt{eE_0/mk_p}$ and v_ϕ the wave phase velocity,

$$\chi_r^{\text{eff}} = \frac{2n_0 e}{\varepsilon_0 k_p E_0} \int_{|v-v_\phi|>4V_{\text{tr}}/\pi} f_0(v)\zeta\left(\left|\frac{v-v_\phi}{V_{\text{tr}}}\right|\right) \mathrm{d}v, \tag{3.118}$$

where n_0 is the unperturbed density, f_0 is the normalized electron distribution function in the limit of a vanishing wave amplitude, and where it has been shown in Sect. 3.5 [see Eq. (3.111)] that

$$\zeta\left(\left|\frac{v-v_\phi}{V_{\rm tr}}\right|\right)=1+\frac{2}{m}\left[\frac{E(m)}{K(m)}-1\right],\tag{3.119}$$

where

$$\frac{4E(m)}{\pi m}=\frac{|v-v_\phi|}{V_{\rm tr}},\tag{3.120}$$

and where $E(m)$ and $K(m)$ are, respectively, the complete elliptic integrals of second and first kind [1].

Then, from Eq. (3.118), $\partial_\omega \chi_r^{\rm eff}$ should be

$$\begin{aligned}\frac{\varepsilon_0 k_p E_0}{2n_0 e}\partial_\omega \chi_r^{\rm eff}&=k^{-1}\int_{4/\pi}^{+\infty}[f_0(v_\phi-V_{\rm tr}v')-f_0(v_\phi+V_{\rm tr}v')]\frac{\partial\zeta}{\partial v'}{\rm d}v'\\&+k^{-1}[f_0(v_\phi-4V_{\rm tr}/\pi)-f_0(v_\phi+4V_{\rm tr}/\pi)]\zeta(4/\pi).\end{aligned}\tag{3.121}$$

In Eq. (3.121), the term which is not under the integral yields the contribution to $\partial_\omega \chi_r^{\rm eff}$ of electrons such that $|v-v_\phi|=(4/\pi)V_{\rm tr}$, which are those electrons lying very close to the separatrix. The motion of these electrons is *not* adiabatic, so that their contribution to χ_r is not accurately estimated by making use of the function ζ defined by Eqs. (3.119) and (3.120). One therefore needs to replace $\zeta(4/\pi)$ in Eq. (3.121) by another constant, and this constant is found numerically in [3] to be very close to 0.27. We therefore use the following expression for $\partial_\omega \chi_r^{\rm eff}$:

$$\begin{aligned}\frac{\varepsilon_0 k_p E_0}{2n_0 e}\partial_\omega \chi_r^{\rm eff}&=k^{-1}\int_{4/\pi}^{+\infty}[f_0(v_\phi-V_{\rm tr}v')-f_0(v_\phi+V_{\rm tr}v')]\frac{\partial\zeta}{\partial v'}{\rm d}v'\\&+0.27k^{-1}[f_0(v_\phi-4V_{\rm tr}/\pi)-f_0(v_\phi+4V_{\rm tr}/\pi)].\end{aligned}\tag{3.122}$$

References

1. M. Abramowitz, I.A. Stegun (eds.), *Handbook of Mathematical Functions*, 10th edn. (Dover, New York, 1972), pp. 587–626
2. G. Belmont, F. Mottez, T. Chust, S. Hess, Phys. Plasmas **15**, 052310 (2008)
3. D. Bénisti, L. Gremillet, Phys. Plasmas **14**, 042304 (2007)
4. D. Bénisti, D.J. Strozzi, L. Gremillet, Phys. Plasmas **15**, 030701 (2008)
5. D. Bénisti, O. Morice, L. Gremillet, D.J. Strozzi, Phys. Rev. Lett. **103**, 155002 (2009)
6. D. Bénisti, O. Morice, L. Gremillet, E. Siminos, D.J. Strozzi, Phys. Plasmas **17**, 102311 (2010)
7. D. Bénisti, N.A.Y. Yampolsky, N.J. Fisch, Comparisons between nonlinear kinetic modelings of simulated Raman scattering using envelope equations. Phys. Plasmas **19**, 013110 (2012)
8. D. Bénisti, O. Morice, L. Gremillet, A. Friou, E. Lefebvre, Nonlinear kinetic modeling of stimulated Raman scattering in a multidimensional geometry. Accepted for publication to Phys. Plasmas as an invited paper of the APS/DPP meeting. Phys. Plasmas **19**, 056301 (2012)
9. S. Brunner, E.J. Valeo, Phys. Rev. Lett. **93**, 145003 (2004)
10. J.R. Cary, D.F. Escande, J.L. Tennyson, Phys. Rev. A **34**, 4256–4275 (1986)

11. R.L. Dewar, Phys. Fluids **16**, 431 (1973)
12. J.E. Fahlen, B.J. Winjum, T. Grismayer, W.B. Mori, Phys. Rev. Lett. **102**, 245002 (2009)
13. J.E. Fahlen, B.J. Winjum, T. Grismayer, W.B. Mori, Phys. Rev. E **83**, 045401(R) (2011)
14. H. Goldstein, *Classical Mechanics* (Addison-Wesley, Cambridge, 1951)
15. B.D. Fried, R.W. Gould, Phys. Fluids **4**, 139–147 (1961)
16. R.J.D. Raju, A. Sen, Phys. Plasmas **13**, 082507 (2006)
17. J.P. Holloway, J.J. Dorning, Phys. Rev. A **44**, 3856–3868 (1991)
18. V.B. Krapchev, A.K. Ram, Phys. Rev. A **22**, 1229–1242 (1980)
19. W.L. Kruer, *The Physics of Laser Plasma Interactions* (Addison-Wesley Publishing Company, Inc., Redwood City, 1988)
20. L.D. Landau, Zh. Eksp. Teor. Fiz. **16**, 574–596 (1946)
21. R.R. Lindberg, A.E. Charman, J.S. Wurtele, Phys. Plasmas **14**, 122013 (2007)
22. J.D. Lindl, P. Amendt, R.L. Berger, S. Gail Glendinning, S.H. Glenzer, S.W. Haan, R.L. Kauffman, O.L. Landen, L.J. Suter, Phys. Plasmas **11**, 339 (2004)
23. V.M. Malkin, G. Shvets, N.J. Fisch, Phys. Rev. Lett. **82**, 4448 (1999)
24. G.J. Morales, T.M. O'Neil, Phys. Rev. Lett. **28**, 417–420 (1972)
25. C. Mouhot, C. Villani, Acta. Math. **207**, 29 (2011)
26. T. OŃeil, Phys. Fluids **8**, 2255–2262 (1965)
27. H.A. Rose, D.A. Russell, Phys. Plasmas **8**, 4784–4799 (2001)
28. D.J. Strozzi, E.A. Williams, A.B. Langdon, A. Bers, Phys. Plasmas **14**, 013104 (2007)
29. G.B. Whitham, *Linear and Nonlinear Waves* (Wiley, New York, 1974)
30. N.A. Yampolsky, N.J. Fisch, Phys. Plasmas **16** 072104 (2009)
31. L. Yin, B.J. Albright, K.J. Bowers, W. Daughton, H.A. Rose, Phys. Rev. Lett. **99**, 265004 (2007)

Chapter 4
How to Face the Complexity of Plasmas?

Dominique F. Escande

Abstract This paper has two main parts. The *first part* is subjective and aims at favoring a brainstorming in the plasma community. It discusses the present theoretical description of plasmas, with a focus on hot weakly collisional plasmas. It comprises two subparts. The first one deals with the present status of this description. In particular, most models used in plasma physics are shown to have feet of clay, there is no strict hierarchy between them, and a principle of simplicity dominates the modeling activity. At any moment the description of plasma complexity is provisional and results from a collective and somewhat unconscious process. The second subpart considers possible methodological improvements, some of them specific to plasma physics and some others of possible interest for other fields of science. The proposals for improving the present situation go along the following lines: improving the way papers are structured and the way scientific quality is assessed in the referral process, developing new databases, stimulating the scientific discussion of published results, diversifying the way results are made available, assessing more quality than quantity, and making available an incompressible time for creative thinking and non-purpose-oriented research. Some possible improvements for teaching are also indicated. The suggested improvement of the structure of papers would be for each paper to have a "claim section" summarizing the main results and their most relevant connection to previous literature. One of the ideas put forward is that modern nonlinear dynamics and chaos might help revisiting and unifying the overall presentation of plasma physics.

The *second part* of this chapter is devoted to one instance where this idea has been developed for three decades: the description of Langmuir wave–electron interaction in one-dimensional plasmas by a finite-dimensional Hamiltonian. This part is more specialized and is written like a classical scientific paper. This

D.F. Escande (✉)
UMR 7345 CNRS-Aix-Marseille-Université, case 321, Av. Normandie Niemen,
FR-13397 Marseille CEDEX 20, France
e-mail: Dominique.Escande@univ-amu.fr

X. Leoncini and M. Leonetti (eds.), *From Hamiltonian Chaos to Complex Systems: A Nonlinear Physics Approach*, Nonlinear Systems and Complexity 5,
DOI 10.1007/978-1-4614-6962-9_4,

Hamiltonian approach enables recovering Vlasovian linear theory with a mechanical understanding. The quasilinear description of the weak warm beam is discussed, and it is shown that self-consistency vanishes when the plateau forms in the tail distribution function. This leads to consider the various diffusive regimes of the dynamics of particles in a frozen spectrum of waves with random phases. A recent numerical simulation showed that diffusion is quasilinear when the plateau sets in and that the variation of the phase of a given wave with time is almost non-fluctuating for random realizations of the initial wave phases. This led to new analytical calculations of the average behavior of the self-consistent dynamics when the initial wave phases are random. Using Picard iteration technique, they confirm numerical results and exhibit a spontaneous emission of spatial inhomogeneities.

Non quia difficilia sunt, non audemus, sed quia non audemus, difficilia sunt.

It is not because things are difficult that we do not dare, it is because we do not dare that things are difficult [Seneca, Epistulae morales 104, 26].

4.1 Introduction

One of the points of this chapter is that plasma physics and probably also other fields of science would benefit from a modification in the structure of scientific papers: each new paper would come with an "executive summary," longer than an abstract, providing its main results and its most relevant references. This idea is developed in Sect. 4.2.2.1 for research papers. Since this chapter must take on the usual structure of scientific papers and must start with a classical introduction, an extended summary is provided in Appendix 1 of Sect. 4.5.

This introduction is split into two parts: the introduction to the contents of the chapter and a short introduction to plasma physics for nonexperts.

4.1.1 What This Chapter Is About

For Aristotle, the study of human knowledge called theoria ("contemplation") was the highest human knowledge and happiness. This chapter is inspired by this kind of philosophy. Indeed, its first part aims at discussing the present way plasmas are theoretically described, with a focus on hot weakly collisional plasmas. It states views that are biased by the author's personal research background,[1] but which are in the spirit of favoring a collective brainstorming about how to deal with plasma complexity. Indeed, in agreement with what is told in Sect. 4.2.1.3, an authoritative

[1] Which is made explicit by a series of examples in the following.

review about plasma complexity should be written by a series of committees of experts, as was already done twice for ITER physics basis [79, 101]. The first part of this chapter (Sect. 4.2) comprises two main subparts. The first one (Sect. 4.2.1) deals with the present status of this description. The second one (Sect. 4.2.2) considers possible methodological improvements, some of them specific to plasma physics and some others of possible interest for other fields of science. To the best of the author's knowledge such a work has not yet been done, while it might be useful for the development of plasma physics. Indeed, till now, this physics has developed in a somewhat continuous way. This shows up in several ways. As to textbooks about plasma physics with a broad scope, modern ones provide synthetic views, but they do not fundamentally challenge the contents of previous ones. As to research programs, they are strongly purpose oriented, which does not help theoreticians to stop and to look backward. This is very much the case for the research on thermonuclear fusion by magnetic confinement. Indeed for more than five decades the fusion reactor has been thought as coming soon.[2]

Therefore one of the purposes of this paper is to stop and to look backward for proceeding better ahead. How do we work? How could our community improve its methodology in order to improve its efficiency and to get more satisfaction and even more joy in its practice? Although we will discuss general issues and possible ways to improve the present understanding of complex systems (in particular of plasmas), for the sake of definiteness working examples will be taken from the author's direct research experience in hot plasmas. One of the ideas put forward in Sect. 4.2.2 is that modern nonlinear dynamics and chaos might help revisiting and unifying the overall presentation of plasma physics.

The second part of this chapter (Sect. 4.3) is more specialized and is devoted to one instance where this idea has been developed for three decades: wave–particle interaction in plasmas and more specifically Langmuir wave–electron interaction in one-dimensional plasmas. In reality this second part corresponds to the author's invited talk at *Chaos, Complexity and Transport 2011*. Since this was an interdisciplinary conference, a short introduction to plasma physics is provided right after first giving a definition of "complexity" relevant to plasma physics.

"Complexity" is a word with a lot of meanings. The meaning used in this chapter is ubiquitous in modern science. However, this meaning comes with an increasing number of attributes when going from inanimate matter to living matter, to humans, and to societies. These attributes may be ordered in a sequence of levels of description. For inanimate matter the number of levels is smaller than for living systems. The coarsest level has two aspects: on the one hand the whole is more than its parts and on the second hand it displays a spontaneous self-organization.

---

[2]The TFTR tokamak was shut down in 1997. For years before, people no longer insisted into calling it Tokamak Fusion Test Reactor. When the ITER project officially started in 1992, "ITER" meant "International Tokamak Experimental Reactor." Now "ITER" is the Latin noun meaning "the way" [125]. This is so true that the KTX machine, a large reversed-field pinch, is being funded in Hefei in the frame of the Chinese ITER domestic program.

The former aspect may be very strong: the whole may be a lot beyond the sum of its parts, as occurs for open systems, be it a plasma column in a laboratory, or the human body whose matter is almost completely renewed about every two months through metabolism and repair. The latter aspect is important to tell complex systems from artifacts like computers or engines. It may come with two opposite, but possibly interrelated, features: order and chaos.

Emergence is the central feature of the level of description following the largest one. It results from self-organization and is the appearance in the system of interest of a feature (form or pattern) arising out of a multiplicity of relatively simple interactions of smaller parts. This feature cannot be anticipated from the knowledge of the parts of the system alone, even if these parts are also complex systems made up of finer scales. A typical example of emergence structure is a fluid vortex, as occurring for instance due to the motion of water particles in a pipe flow. In turn, individual vortices may interact to produce another emergent feature: turbulence. This emergence makes the water less fluid than in the laminar state: pressure drop increases, but molecules are unaffected; again the whole is a lot beyond the sum of its parts.

In living systems, next levels of description include features like cooperation and competition. There are other definitions of complexity which introduce the same attributes in a different order or other attributes which are implicitly present in the above definition: complex systems contain many interdependent constituents interacting nonlinearly, and their self-organization spans several spatial and temporal scales.

If a fluid or a plasma is described as an N-body system, its Hamiltonian is made up of the sum of all the free particle Hamiltonians, plus an interaction. Therefore the interaction is a part of the system. The same occurs when dealing with wave–particle interaction in plasmas. As will be shown in Sect. 4.3, the corresponding physics can be described by a Hamiltonian made up of a free particle part and a free wave (harmonic oscillator) part, plus a wave–particle interaction part. This description puts waves and particles on an equal footing, as occurs in modern field theory.

4.1.2 Plasma Physics

A plasma is a quasineutral system of charged particles. The plasma state is often presented as the fourth state of matter, because it can be reached by further heating matter after it experiences successively the solid, liquid, and gaseous states. With respect to these states, it has the distinctive feature of the long-range interaction of its particles because of the electromagnetic field they produce. This endows it with a ubiquitous collective behavior which shows up as waves, solitons, turbulent eddies, vortex structures, streamers, blobs, etc.

"Plasma" is a Greek word which means moldable, or without any definite shape. This lack of definite shape applies to plasma physics as well, because the plasma state is in reality contiguous to all three solid, liquid, and gaseous states, which weakens its classification as the fourth state of matter in a one-dimensional classification. This is linked to the fact that this state may be realized with densities varying over more than 30 decades and temperatures varying over more than seven decades. As a result there is a huge variety of plasma states, which sets complexity, and even complexities, at the very beginning of plasma theory. This chapter does not attempt to deal with this variety, but focuses on hot weakly collisional plasmas with relevance to astrophysics and especially thermonuclear fusion. Such plasmas are characterized by the fact that there are many particles in the Debye sphere, i.e., the sphere with a radius equal to the Debye length $\lambda_D = [(\varepsilon_0 k_B T)/(ne^2)]^{1/2}$ where ε_0 is the vacuum permittivity, k_B is the Boltzmann constant, T is the temperature, n is the density, and e is the electron charge. Therefore $n\lambda_D^3 \gg 1$, which means $T^3/n \gg 1$ with appropriate normalizations.

The plasma state is the state of matter the most spread in the universe (except for dark matter!) and has many important applications. However, despite the obvious importance of the field and the extensive knowledge on plasmas, plasma physics has got a limited scientific recognition. No Nobel prize has been awarded to its physicists.[3] It was not quoted in the one hour opening talk "A century of physics" of the APS Centennial Meeting in 1999. This lack of scientific recognition may be partly related to the slower progress toward the thermonuclear reactor than expected. However this chapter provides another clue by showing the difficulties in describing plasmas. This difficulty makes the theory of plasmas more like an impressionist painting than like a well-structured theoretical knowledge. This chapter also shows the stimulating collective research process in plasma physics to be a hindrance for an individual to bring alone a decisive progress in plasma description. Plasma complexity is at the root of this situation. It is temporal as well as spatial. Indeed it stems from the many degrees of freedom nature of plasma dynamics, which exhibits a huge variety of dynamical modes. As Kadomtsev said "Here, similar to many paintings by the prominent artist Hieronymus Bosch, there exist many levels of perception and understanding. At a cursory glance of the picture you promptly grasp the idea. But under a more scrutinized study of its second and third levels you discover new horizons of a deeper life and it turns out that your first impressions become rather shallow" [80]. The "Garden of Earthly Delights" at the Madrid Prado museum is the author's favorite experience from this point of view.[4]

[3]Alfven's Nobel citation reads: "(...) for fundamental work and discoveries in magnetohydrodynamics with fruitful applications in different parts of plasma physics". Therefore his award was not meant as presented to a plasma physicist. In reality, Alfven was in an awkward position with respect to well identified fields of physics [124].

[4]Increasing blowups are advised when using the Internet to watch a photograph of the painting.

4.2 Facing Plasma Complexity

This section aims at favoring a brainstorming in the plasma community. It is made up of two parts. The first one deals with the present status of the theoretical description of plasmas, with a focus on hot weakly collisional plasmas. The second part considers possible methodological improvements, some of them specific to plasma physics and some others of possible interest for other fields of science. Since an authoritative review about plasma complexity should be the result of a collective effort, this section is inevitably subjective. Therefore some of its statements might possibly be unwillingly polemical.

4.2.1 Present Status of the Description of Plasma Complexity

This section recalls the path used for student training, before dealing with the way theoreticians face complexity. It is recalled that models used in plasma physics, even the Vlasov equation, have feet of clay. Each plasma physicist is shown to elaborate his own global view about plasma physics from many models which do not have any strict hierarchy. *The validation of assumptions turns out to be more difficult for a complex system than for a simple one.* In agreement with Popper's falsifiability paradigm [103], at any moment the description of plasma complexity is provisional. It results from a collective and somewhat unconscious process. This makes changing views more difficult. Numerical simulations are discussed as a complex tool to face complexity.

What happens in this process is often similar to the Indian tale "The blind men and the elephant." Six blind men want to know what an elephant is and go in the courtyard of the rajah's palace to touch one. Each of them touches a different part of the animal and feels it is similar to an object he already knows: a wall, a spear, a snake, a tree, a fan, and a rope. They start arguing with one another so loudly that the rajah comes to his window, asks about the issue, and finally tells them they should put together all the pieces of information they got in order to know what an elephant is! Knowing what is a plasma comes with similar difficulties: the experimental knowledge is often scarce, but the theoretical views may be many. One must try and tie together the partial views of the experts. Though textbooks and review papers are of some help, the global knowledge about a given class of plasmas, e.g., magnetic fusion plasmas, is largely implicit, and each expert of the field has his own one.[5]

[5]An important exception is the ITER physics basis [79, 101], already mentioned in Sect. 4.1.1, that was written by a series of committees of experts. This trend toward a collective view about the tokamak has been present in the authorship of Wesson's book [119] since its first edition in 1987. It is also worth noting that most modern plasma textbooks have at least two authors.

4.2.1.1 Path for Students

Going from the simplest to the more complex in plasma physics, is a path any student must follow. At the roots of plasma physics there are basic textbook problems dealt with by elementary models. Two oppositely extreme ones are proposed to beginners: single particle and fluid models. Single-particle dynamics comes under two main aspects: motion in a magnetic field with its corresponding drifts and what is traditionally called "collisions" in plasma physics. Fluid models are introduced to describe plasmas as a whole, with two important aspects: ideal and resistive magnetohydrodynamics (MHD) on the one hand and the description of waves and instabilities on the other hand. The knowledge of waves is of paramount importance because they are the simplest instance of the ubiquitous collective motions of plasmas and because they are leading actors in plasma turbulence. Later on are introduced kinetic descriptions (Vlasov, gyrokinetic, etc., equations) which aim at describing plasmas as a whole while keeping single-particle dynamics as much as possible into account. Textbooks describe basic phenomena with the simplest possible assumptions, like homogenous plasma either isotropic or in a uniform magnetic field. Then the intricacy of actual plasmas is introduced by taking into account various gradients of magnetic field, density, temperature, etc., and new physics emerges corroborated by experimental results, which gives confidence into the models. However there are bad news.

4.2.1.2 Models Have Feet of Clay

Indeed unfortunately most models have feet of clay! Indeed very few models used by plasma physicists are derived from first principles with assumptions that are justified for the class of experimental or theoretical problem of interest (an exception is presented in Sect. 4.3). Many models are derived under given assumptions, but, because of their handiness, they are used out of their domain of proved validity or without knowing their limit of validity. Vlasov equation is considered as the reference equation for collisionless plasmas, but the derivation of this equation by a mean-field approach shows the validity of its solutions is proved up to the time of exponential divergence of nearby orbits (see for instance [113] for an introduction). For instance, this time a priori bounds the actual duration of the Bernstein–Green–Kruskal (BGK) solutions [14] of Vlasov equation. It is too stringent a time bound to describe turbulence, which is an incentive to keep plasma granularity for this purpose, as is done in Sect. 4.3. The latter approach reveals that an unstable plasma may reach a state away from the Vlasovian saturation state [64, 67] and that there are no BGK modes [41]: in reality a plasma state starting in a Vlasovian BGK state is metastable. Furthermore it was shown theoretically that a superthermal tail can be generated by beam–plasma interaction, which is missed by a Vlasovian description [120]. The discrepancy of the Vlasovian solutions can be understood as a result of the non-commutation of the infinite time and infinite number of particles limits. Indeed, consider the particles initially inside a given Debye sphere, and assume they have a chaotic motion. Later chaos stretches this sphere into a thin tube where these particles are far apart, which reveals that the plasma is no longer a fluid in phase space. Vlasov equation is also used for open systems, but its validity has not been proved formally (to the author's knowledge) in this case.

The use of basic fluid models is ubiquitous, but they cannot generally be justified from first principles for collisionless plasmas. For instance, the applicability of MHD equations to fusion plasmas may be justified if perpendicular motion dominates (see Sect. 2.5 of [17], Sect. 6 of [75], Sect. 3.5 of [76]), at least for the mass and momentum equations (see Sect. 2.4 of [69]).

As Bertrand Russell stated, "Although this may seem a paradox, all exact science is dominated by the idea of approximation" [110]. This is all the more true in plasma physics, since the nature of approximations is not clear in many theoretical descriptions. Fortunately, very often the predictions they provide are found to have experimental relevance or to be validated by more sophisticated models. This suggests that models have a larger validity than proved as yet or that their equations describe the genuine evolution of quantities that are close to those explicitly present herein.

4.2.1.3 A Global View from Many Models Without Any Strict Hierarchy

The typical complexity of plasmas rules out their description by the most general non-quantum model which would account for all its particles, with possibly their finite lifetime inside the plasma, and its actual geometry. Therefore a whole fauna of "approximate" models must be used. Theoretical description and intuition develop by successive additions of complexity on the basis of the solution of a few simple problems and of the confrontation to experiments and numerical simulations. The permanent irruption of new measurements and scientific interactions may bring a progressive complexification of a given model, or it may challenge it so much that a new one must be proposed. This evolution is often due to a community and not to an individual.

Unfortunately, starting from the academic knowledge, there is no obvious path toward the more complex for the researcher, and in particular an axiomatic approach is generally formidable (an exception to this is provided in Sect. 4.3). Indeed complexity may be added to a simple model in many different ways. It may be by increasing dimensionality, or the complexity of geometry (from slab to cylindrical and to toroidal), or the number of involved physical quantities, or the number of described species, or the number and type of transport coefficients, etc. One may go from fluid to kinetic descriptions, or from a linear to a nonlinear one, one may take inhomogeneities and fluctuations into account, etc. As plasma complexity has many dimensions, there is no strict hierarchy among these models.

A natural way to tackle a modeling issue is for the researcher to start by applying Occam's razor principle "Entities must not be multiplied beyond necessity," i.e., to choose the simplest available model in agreement with his present knowledge of the system of interest.[6] In the same spirit, he may state a priori what is the dominant physics and build a model incorporating it from first principles, e.g., conservation laws or symmetries. However, subjectivity is present in this approach: How to check whether enough complexity is included? How to compare nearby models? How to ascertain the structural stability of the description, i.e., its keeping its validity if more physics is included? Furthermore, very often implicit assumptions are made in the process. At each step of the complexification, new temporal and spatial scales may show up, as well as new dimensionless numbers. The analytical or numerical treatment of the model becomes also more tedious and possibly its experimental check as well. Therefore, going to the more complex tends to be done with caution, and there is a trend toward extracting as much physics as possible from a potentially too rough model. Furthermore, because of the lack of knowledge about the conditions of validity of a model, one cannot be sure that a more complete set of equations of this model is bound to bring an improved description of the system of interest.[7] Finally, simpler models are easier to present to colleagues or to teach to students, and their results are easier to discuss. The quote attributed to Einstein "Make everything as simple as possible, but not simpler" states an aim that is beyond one's reach for complex systems![8]

The articulation of fragmentary descriptions into a global one often is a personal and implicit process driven by interactions with the scientific community. This process is fed by the intuition developed by the knowledge of many experimental facts and simple models. Since at any moment a researcher stands in between knowing everything about nothing and nothing about everything, his global view has fuzzy outskirts. So is that of his community. Tradition plays an important role, which reminds of law reasoning and of historical, biological, and hereditary logics. Scientific schools are bound to develop in a "blind men and the elephant" process.

---

[6]A very interesting example where this process was applied repeatedly is the theory of edge localized modes (ELMs) in tokamaks. First they were considered as current driven, then as pressure driven, then back as current driven, and now as both pressure and current driven (see Chap. 3, Sect. 2.6.3 of [101] and [112]).

[7]As was told in Sect. 4.2.1.2, the applicability of MHD equations to fusion plasmas may be justified if perpendicular motion dominates, at least for the mass and momentum equations. Therefore adding more MHD equations to the latter does not mean necessarily a more accurate description. Furthermore the use of the mass conservation equation may bring an unphysical peaking of density on the axis of a pinch which is avoided by microscopic turbulence in an actual plasma (see Sect. 9.3.3 of [17]).

[8]However plasma physicists are not desperate. To the contrary going ahead by using many approximate models is like going downhill rapidly on a scree made up with small stones: this is both fast and pleasant, though none of the stone be comfortable to stand quietly! However plasma physicists may sometimes be meditative: "(...) had the range of instabilities now known to beset tokamaks been discovered by theoreticians before the experimental program was undertaken, there might have been some hesitation" (p. 562 of [119]).

4.2.1.4 Validation and Refutation of Assumptions

It is well known that "$A \Rightarrow B$" and "B right" do not prove that A be right. However, implicitly people often act as if A were right. In physics, when a theory agrees very well with experimental data and makes right predictions, it is often considered as right,[9] possibly with precise ideas about its actual real range of validity (for instance for classical mechanics when thought in the context of quantum or relativistic mechanics). However "A is right" may happen to be said even for a theory which has only partial agreement with facts or which cannot be precisely tested, because it is the only one available or because it is simple, or elegant, and taught to students. Furthermore proving that A is wrong may be difficult and may take years for a complex system, because of the lack of information about it.

The present theory of neoclassical transport is an example of reference theory which has not been tested, because it is hidden by turbulent transport which is ubiquitous in magnetized plasmas. Even for theories about anomalous transport, a clear-cut experimental check is difficult, in particular for two cases: (1) the existence of a residual stress in momentum transport [100]; (2) the existence of fractional diffusion (see [34, 35] and references therein) challenging the standard advection–diffusion picture.[10] These two checks are very difficult, since the calculation of profiles of transport coefficients belongs in the category of inverse problems which are known to come with issues of ill-conditioning or singularity. Transport codes provide a classical way to infer the profile of transport coefficients in fusion plasmas: assuming given functionals for the profiles of transport coefficients, the free parameters are iteratively adjusted to best reproduce the measurements. However this does not provide any estimate on the uncertainty of the reconstructed profiles.[11]

A case of simple and elegant theory with only a partial agreement with facts is Taylor theory of magnetic relaxation [114] applied to the reversed-field pinch[12] (RFP). When it was published in 1974, this theory brought the first theoretical attempt to explain the mysterious RFP relaxation and had several features in agreement with experiments. However, the theory was unable to explain the

[9]According to Popper's falsifiability paradigm, this theory just survives the process of refutation, but it is not protected from refutation in the future [103].

[10]This picture is quite flexible though and is justified for generic particle transport, provided there is enough randomness in the Hamiltonian describing the dynamics [57].

[11]Some progress may be expected from a new technique tailored for periodically modulated experiments [58]. This technique avoids any a priori constraint on the profiles and computes them by simply inverting a 2D matrix. It also provides the uncertainty on the reconstruction. This is done by a controllable smoothing of the experimental data, instead of the ad hoc regularization of the profile of transport coefficients operated by transport codes (see Appendix 3 of Sect. 4.7).

[12]The RFP is a magnetic configuration germane to the tokamak that produces most of its magnetic field by the currents flowing inside the plasma [53]. With respect to the tokamak and the stellarator, the RFP has a low imposed external field. It has a helical magnetic field like the stellarator, but it is more magnetically self-organized than a tokamak and much more than a stellarator.

dynamo at work in the plasma and the features of its outer part. Furthermore it described a driven ohmic system as if it were closed. Nevertheless it became the reference model, as shown in particular by reference [99] which provides a list of shortcomings of this theory in its Sect. 3.6 though. In 1999, new MHD numerical simulations gave momentum to the new paradigm of single-helicity equilibria [23, 63], already introduced in 1990 [24, 25, 65]. This was an incentive for experimentalists to look into the database of RFX, the largest RFP, and to find that quasi-single-helicity states had been present in many discharges for quite a long time. Taylor relaxation theory was not consistent with these states (see Sect. 7 of [27]), which forced to give up this paradigm [53].

Another interesting case in the frame of fluid theories is the ability of the simplest Ohm's law to enable the correct calculation of the saturation of various resistive instabilities. Indeed with general assumptions, the supplementary terms of a generalized Ohm's law do not contribute into the parallel Ohm's law used in such calculations (see Sect. 2.1 of [26]). This does not prove the validity of the simplest Ohm's law. Indeed the plasma velocity may strongly depend on the Hall contribution for instance.

Many simple models recover Landau damping in an intuitive way. As stated in Sect. 4.3.1, some of these models make wrong assumptions, though their prediction about the existence of the damping be right.

4.2.1.5 Difficulty in Being Critical and in Changing Views

A classical reference for the scientific method is provided by the four principal rules of Descartes'*Discourse on the Method* (1637). In rough words, one should apply a systematic doubt to any new statement, divide each of the difficulties under examination into as many parts as possible, go from the simplest to the more complex, and make enumerations so complete, and reviews so general, that one might be assured that nothing was omitted. As told before, each expert has his own global view about a given class of plasmas. In reality, he stands very far from the four Cartesian rules:

- He must provisionally accept for true assertions that are not clearly known to be such.
- The difficulties under examination have so many aspects that there is no way to find an adequate solution by dividing all these aspects.
- There is no way to ascend little by little to the knowledge of the more complex, by commencing with objects the simplest and easiest to know.
- As a result, there is no way to make enumerations so complete, and reviews so general, that one might be assured that nothing was omitted.

All this sounds quite deceptive, but the Cartesian dream is implicitly at work in the community working on a given class of plasmas, and induces a polarization of the heuristic efforts. However, at a given moment, this community, though being made up of physicists who aim at being critical (Cartesian doubt), must accept to think in a way similar to the legal reasoning where rationality is continuity: "In view of the accumulated knowledge, it is natural to think that..." Therefore tradition plays an important role.

An explicative model tying together the past knowledge about a given phenomenon is forcedly accepted reservedly due to the above necessary practical violation of the Cartesian rules. Therefore, challenging new experimental or theoretical results may first be felt as a mere confirmation that the as yet accepted model is not perfect, but not as a proof it is breaking down. At some moment, a new explicative model may be considered, but still keeping in mind the former one: the glass is both half-full and half-empty. Giving up on past habits is difficult to justify, since there are so many facts and factors to account for in order to provide a good theoretical description. This makes all the more difficult the painstaking change of paradigm described by Kuhn [85] and first advocated by de Broglie.[13] This mental dynamics is present, for instance, in the evolution of the theoretical view about sawteeth, ELM's (see footnote 6), and disruptions in tokamaks [77]; also about the RFP, as is now described in two steps.

Till the 2000s, Taylor relaxation theory (TRT) was so strongly accepted that another explanation of the partial magnetic relaxation observed in the RFP, proposed in 1991, was overlooked, though very simple and closer to experimental facts, as far as the current distribution is concerned: the Rusbridge theory inspired by ideas of Kadomtsev and Moffatt[14] [109].

Though the existence of long-lasting quasi-single-helicity states in RFX invalidating TRT was published in 2000 [62], it took almost a decade for the change of paradigm to become obvious with the cover story of Nature Physics in August 2009 stating: "Reversed-field pinch gets self-organized" [90]. Here, another interesting phenomenon occurred: when motivated by numerical simulations, experimentalists looked into the database of RFX and found that quasi-single-helicity states had been present in many discharges for years, but not analyzed because they were considered as atypical (out of the paradigm): in agreement with Kuhn's view [85], the previous paradigm led to a screening of the facts challenging it. It was a pity, because single helicity was predicted theoretically in 1990 (cf., Sect. 4.2.1.4), and comes with good magnetic surfaces, while multiple helicity comes with broad magnetic chaos.

[13] "The history of science shows that the progress of science has constantly been hampered by the tyrannical influence of certain conceptions that finally come to be considered as dogma. For this reason, it is proper to submit periodically to a very searching examination principles that we have come to assume without discussion [33]."

[14] The explanation goes along the following steps: (1) In the radial domain where the magnetic field is chaotic, transport is fast, and the equilibrium is almost force-free; therefore $\mathbf{J} = \mu \mathbf{B}$ where μ may be space-dependent. (2) Setting this in $\nabla \cdot \mathbf{J} = 0$, implies $\mathbf{B} \cdot \nabla \mu = 0$, which shows that μ must be constant along field lines; thus μ is constant in the chaotic radial domain. This straightforward derivation yields a result in full agreement with the fact that, in MHD simulations, μ is almost constant in most of the domain with a positive toroidal magnetic field, but not where it is reversed [22]. However this domain of almost constant value of μ was rather considered as a hint to the validity of TRT which predicts μ constant over the whole plasma radius instead [114].

4.2.1.6 Numerical Simulation: A Complex Tool to Face Complexity

Numerical simulation is a tool that enables a dramatic progress in the description of complex systems. In the past, only analytical calculations were possible, which limited strongly the set of tractable models. This reinforced the "look under the street light" syndrome. For instance, the simplest MHD model to describe RFP relaxation is made up of Navier–Stokes equation including Laplace–Lorentz force and Faraday–Ohm's law. In 1974, there was no way to simulate adequately this set of equations. As a result, the already mentioned Taylor relaxation theory [114] was built with the Ansatz that fluid motion might be neglected and that relaxation might be described in a purely magnetic way. Fifteen years later, good simulations were possible and revealed features of RFP relaxation that ruled out TRT [27]. Among them, the single-helicity states were already mentioned, but the paramount importance of fluid motion was exhibited too. This motion explains the dynamo component of the electromotive force driving the currents in the plasma, which was a mystery for the previous TRT paradigm. At present, single-helicity states are understood as a mere extension of the saturated tearing mode, in particular as far as the electric drift nature of the dynamo is concerned [18, 26]. As yet, MHD simulations remain the main way to address theoretical issues, since the analytical description of these states is just in its infancy[15] [19, 26].

In many other fields of plasma physics too, numerical simulations are the irreplaceable tool for investigating complexity.[16] This has made the issue of verification and validation of these codes a crucial one [116]. However, the role of intentionality is higher in simulations than in analytical calculations because of the choice of initial conditions and of parameters. The essentially dynamic nature of simulations mimics the experimental behavior of plasmas, but with a much higher representational ability than provided by measurements on the one hand and with a much smaller complexity than actual experiments on the other hand. Though numerical experiments have an experimental character, they often come without error bars on their predictions, in particular when they involve an underlying chaotic dynamics. The numerical coding of an analytical model often involves many uncontrolled approximations. For instance the name "Vlasov code" encompasses very different types of codes, all with non-Vlasovian features. Eulerian Vlasov solvers produce a fake diffusion in phase space that violates Liouville theorem.

---

[15] However this description already reveals that the edge current does not matter to reach shallow reversal. This is important to guide the endeavor toward improving confinement of quasi-single-helicity states: one should enable the central part of the plasma to reach a genuine ohmic equilibrium. Indeed this should induce a low resistivity central part of the discharge diminishing the loop voltage and thus the ohmic power for the same plasma current.

[16] It is hard for young physicists to imagine the age where numerical simulations were a tour de force with card punching, batch submission, and paper outputs. In 1976 the author published a one-dimensional Vlasovian simulation with 8,000 cells in phase space, while in present codes this number is larger by more than three orders of magnitude! However this lean code enabled to uncover the thermalization of a volume-created plasma due to the lack of static equilibrium [49].

Particle-in-cell codes in reality simulate a dynamics which is more Klimontovich-like than Vlasovian. They avoid fake diffusion at the expense of a higher noise because of a much smaller plasma parameter $n\lambda_D^3$ than in real plasmas. However both approaches have the advantage of avoiding the development of unphysical very thin Vlasovian filaments in phase space.

As told above, complexity may be added to a simple model in many different ways and this leads to the parallel use of codes with various abilities. However it would be very useful to check a posteriori what are the dominant terms in the calculation. As to nonlinear terms, which ones act only through their linear contribution (especially for a final "stationary" state) this might be a lead for the development of possibly analytic, reduced models.

A present trend in magnetic fusion research is to develop an integrated modeling of tokamaks by tying together a bundle of codes. This will certainly bring a new knowledge of complexity, but new techniques will be needed to check the validity of the integrated models.

Despite the growing importance of numerical simulations, analytical calculations of simple models have remained important and will stay as a reference. Indeed, an analytical calculation reveals the internal structure of a model, its dominant parameters, it has an intrinsic flexibility with respect to the parameters values, and it can be checked more easily than a numerical simulation. Analytical calculations can be used to verify numerical simulations and avoid some of their pitfalls: cancelations of large terms, problems of stability and convergence due to insufficient numerical analysis, fake boundary effects, fake dissipation, etc. Fortunately, computers also help for such calculations with computer-assisted algebra. Sometimes numerical simulations can suggest assumptions for a new analytical approach (an example is given in Sect. 4.3.5).

A final caveat might be useful. Indeed the growing power of computers and the progress in numerical analysis and data processing make numerical simulations increasingly powerful... and fascinating to students. However numerical experiments are not true experiments. Only the latter are the ultimate beacon for understanding physics. Therefore students should be encouraged to become experimentalists. This point is dealt with in Sect. 4.2.2.9.

4.2.2 Possible Methodological Improvements

We have just described the empirical way plasma physicists deal with complexity. This section considers how this description might be improved in view of the difficulties presently faced by plasma physicists. One of these difficulties, information retrieval, is specific to plasma physics and possibly more generally to the physics of complex systems. Other ones, as the inflation of publications and the growing importance of oriented programs, are a general problem of contemporary physics. Working on complex systems is a hard task, but the present trend of scientific practice makes it even harder. Therefore plasma physicists would gain very much in any improvement of this practice and might be motivated into impelling a change. Challenges like ITER and DEMO might benefit a lot from such an improvement.

Here we venture some proposals for improving the present situation along the following lines: improving the way papers are structured and the way scientific quality is assessed in the referral process, developing new databases, stimulating the scientific discussion of published results, diversifying the way results are made available, assessing quality more than quantity, and making available an incompressible time for creative thinking and non-purpose-oriented research. Some possible improvements for teaching are also indicated.

4.2.2.1 Claim Section

Many improvements may be thought of to make plasma research easier. Some of them would mean an evolution of the rules of various research organizations, which is hard to trigger on large scale in the plasma physics community. However, there is a direction which might be improved rapidly and on an experimental basis: the structure of scientific papers, in order to avoid the "can't see the forest for the trees" syndrome. Indeed, it is often difficult to get the point of a given paper from its abstract, introduction, and conclusion, in particular to find out what are the most important figures or formulas. A part of the problem comes from the haste to publish new results, which does not help in their pedagogical presentation. This also leads to misunderstandings from referees, which increases the number of referral runs. In the latter process, the paper may be clarified but possibly only in a marginal way, which leaves its understanding still difficult to future readers.

A corrective action could be each research paper, even letters, to have a claim section being a kind of executive summary, but without any stylistic effort, more a list of very synthetic claims. This section would first list the main results and then their most relevant connection to previous literature. This would provide a clear information to assess the importance, the originality, the actual scientific contribution of the paper, and about the "precedents, sources, and context of the reported work," as worded in the APS guidelines for professional conduct [123]. Salient figures or formulas would be set there to make the claims explicit. Therefore the claim section would be a lot more informative than a traditional abstract. The present chapter provides very few new scientific results, and thus a claim section would make little sense. Therefore two examples of claim section are proposed in Appendix 2 for paper [55] mentioned in Sect. 4.3.4 and in Appendix 3 for paper [58] mentioned in Sect. 4.2.1.4.

It is worth noting that the very compact way of communicating new results of such a claim section is reminiscent of the way physicists communicate the essence of their results to their colleagues in the corridor of large congresses. It is also what each reader tries to do by himself when trying to get the essence of a paper without reading it entirely. This procedure is standard in the mathematical literature. It would also improve the clarity of the papers. Writing the claim section would lead an author to state the essence of his results in a more accessible way[17] and without having to care about the literary constraints of a normal text. He would be led into a better assessment of the actual novelty of his work, which would be bound to improve both the abstract and the remainder of the manuscript.

4.2.2.2 Improving the Referral Process

When available, the list of claims could seriously help in assessing a paper. Indeed one might think the referee to be then required to assess each of these claims (right or wrong?) and to check the list of precedents, sources, and context of the reported work, before writing down the remaining of his referee report or before filling in the report form. When choosing a “false” for a given claim, he should motivate his statement.

This procedure should make the referral process more scientifically rigorous and faster. Referees would be sure not to miss the importance, the originality, or the actual scientific contribution of the first version of a manuscript, as claimed by the author. They could better help authors to adjust their initial view about this. Therefore the claim section would decrease the number of referral runs and would speed up the refereeing process. This would help referees into an ethical refereeing. Editors would benefit from a better refereeing process, which would avoid many author’s complaints, while making faster the editorial process. They would also have more factual elements to assess manuscripts and referees. Journals would benefit from the increased clarity of the contents of their published papers. The procedure might start with an experimental stage where the claim section would be optional for the authors, but not for the referee report if the claim section is available.

4.2.2.3 New Data Retrieval Technique

Finding the relevant information about a given topic in plasma physics is often a hard task. Empirically people use a mix of word of mouth, browsing textbooks and review papers, web research engines, and bibliographic databases. In reality, the capability of modern computers as far as data processing, databases, and hyperlinking are concerned might provide much better tools than those already available. An important direction of improvement might be structuring and articulating the knowledge about a given topic, especially when many papers deal with it and when it has been developing for decades. Presently web research engines and bibliographic databases give so many papers that finding the most relevant ones becomes a formidable task. In particular following quotations forward or backward in time provides an

[17]It helped for the present chapter, even under the disguise of an extended summary! The reader is invited to write a claim section for his/her next paper, in order to ascertain the interest of the method, even for a private use. This section might be put as an appendix in the paper.

exponentially growing set of papers whose majority is likely not to be informative for the topic of interest. Indeed papers may be quoted for purely technical reasons, to illustrate side comments, because they are wrong, etc.

The claim sections might provide a simple way of developing a new technique for data retrieval adapted to plasma complexity. Indeed, for each paper, this section might be set by the corresponding scientific journals or publisher into a new dedicated database accessible through the Internet where cross-referenced papers would be hyperlinked. Their hyperlinking would provide a collaboration between scientific journals and publishers, but an implicit one, requiring only marginal legal agreement. It would facilitate the assessment of the state of the art on a given topic, with respect to what is available through present bibliographical databases. Indeed, connectivity between the various papers would be more topic oriented. APS guidelines for professional conduct [123] state that "It is the responsibility of authors to have surveyed prior work in the area and to include relevant references." Presently it may be hard, even with good will, to fulfill this requirement. This might become a lot easier with the new database.

The spontaneous development of these databases would probably only incorporate new papers and old ones that are still known, but some relevant old ones might be overlooked. After the start-up phase of the bases be over, further work would be necessary to screen the past unquoted literature. Public money could be involved in this second stage, since publicizing old results is a way to save present and future research work.

4.2.2.4 Stimulating the Scientific Discussion of Published Results

With the claim sections and the corresponding hyperlinked database, researchers might be made more responsible about the contents of their papers. A feedback system should be developed to this end. In reality, there is already one: the comments sent to a journal about its published papers. Unfortunately this system is very rigid and formal, and people sometimes feel its use as unfriendly. In order to cope with this issue, one might consider broadening the way papers are commented by adding a first, friendlier step: a researcher who would disagree with the contents of a paper would get in touch with its authors directly and try to sort out the issue. If they finally agree that there is something wrong, they would publish a common short corrective communication, naturally linked to the original claim section of the original paper. A classical comment to the journal would be sent only if the authors could not agree about a common view. It would also be useful for journals to enlarge this two-step comment system in order to enable comments about publications not present in journals, but in books, articles in proceedings, etc. This new kind of scientific discussion might be encouraged by the various scientific societies linked to plasma physics.

The claim sections, the corresponding hyperlinked database, and the new comment system should lead toward talks, courses, and papers that would be more updated. They should also help to have a better view of the importance of papers. Naturally these tools will be naturally complemented by blogs, wikis, discussions in social networks, and other electronic means which are developing.

4.2.2.5 Praising Quality Not Quantity

It is well known that quality is not a consequence of quantity, but the number of publications has been gaining a growing importance in the last decades in assessing physicists and research teams. This drives inflation in the number of publications. As a result, unfortunately no physicist has the time to read the so many papers corresponding to his field. An even sadder fact is that this is not a serious issue for most of these papers! Indeed a large part of the published literature brings marginal improvements to the knowledge of a given scientific domain, which are of interest only to a small subgroup of experts. However, it is a serious issue for the most important papers which might be overlooked because they stand in a crowd. This brings the first issue: "What are the important papers?", especially when going out of one's specialty.

A collateral effect of the inflation of publications is the parallel one of the number of papers to referee. Another negative trend is the growing importance of oriented programs and of the corresponding assessment process. These two trends are strongly time-consuming. Together with the incentive to publish many papers they decrease the time to think creatively and to read.

This encourages the following failings: mental inertia, works that overlook previous results (even if the corresponding paper is quoted!), parallel physics projects without contact, absence of scientific debate, fashions and related lobbies, etc. Therefore it might be very useful to make the assessment of scientific quality really quality, and not quantity, oriented. The interlinking of claim sections might provide a way to find out better the actual importance of related works and to diminish the role of quantity in quality assessments. The same trend could be induced by diversifying the ways results are made available, as explained now.

4.2.2.6 Diversifying the Ways Results Are Made Available

One might think about diversifying the ways results are made available:

- Discussion papers might be sent to experts in the corresponding domain in order to have a chance to listen to criticism, to add possible relevant quotations, and to improve the contents before submitting a paper for publication.
- Preliminary works might be made available on databases like arXiv and modified or canceled later.
- The wording of papers might be more problematic. Sometimes this might be reflected even in the title by making it interrogative, in particular for papers about debated issues.

- The use of scientific wikis might be generalized.
- Journals might systematically propose a whole hierarchy of papers including short letters about breakthroughs, short, long, and review papers. Follow-up papers might be allowed with a short format, in particular for their introduction, and explicitly indicated in journals as such... and in publication lists! Referees would be requested to detect such papers in their review process, even if the follow-up paper is submitted to another journal than the original one. This issue becomes important with the development of numerical simulations. Indeed any modification in the simulated model or in the simulation parameters brings virtually a new result: how important is it?
- Attempts that fail are numerous and time-consuming. Natural ones are bound to be repeated by several researchers. Why not allow the publication of short communications describing such unsuccessful attempts?

The issue of publication in books is a tough one. Indeed they are generally less accessible than journals. Writing a book is a strongly time-consuming task, but it may be little rewarding from the view point of scientific communication... Finally the current trend of electronic publication sets the issue of its long-term archiving. Two new directions in scientific communication are worth mentioning: the "Article of the Future" [127] and the "Quantiki" featuring 5 min presentations of new results in quantum mechanics [128].

4.2.2.7 Time for Creative Thinking and Non-purpose-oriented Research

The idea of making available an incompressible time for creative thinking and non-purpose-oriented research is now becoming popular with the "slow science manifesto" [122]. Each researcher may try and do this, but it is important this to be recognized by scientific organizations too. More time for creative thinking, to look out of one's specialty and to try transversal views may be a big saving for science.[18] A striking example comes from magnetic fusion physics. The tokamak is known to have a density limit which is proportional to the current density: the Greenwald density limit[19] [73]. Several papers have been published to provide tentative explanations of the phenomenon. Unfortunately, to the best of the author's knowledge, none of these theories work for the RFP where the same limit is present [104, 105], but this has been overlooked by tokamak experts.

It would be very useful to have a transversal view on magnetic confinement by using the information available from various configurations for magnetic confinement. Indeed important physics issues need to be solved both for ITER and for the definition of future demonstration reactors: what is the origin of the Greenwald

[18]This is the motto of the Institute for Advanced Study in Princeton [126].

[19]This limit is an edge density limit above which the discharge cannot be sustained.

density limit, how do transport barriers form and stay,[20] what is the origin of plasma rotation, what is the effect of additional heating, how to scale reactor parameters out of smaller experiments, how dangerous fast-particle-driven MHD modes may be, what is the benefit of a helical deformation of the magnetic field, what is the role of ambipolar electric fields, etc.? In particular, understanding the density limit in magnetic confinement might enable to come closer to this limit or to overcome it and would increase considerably the reactivity of thermonuclear plasmas, which would dramatically increase the prospects of magnetic fusion. It is probable that important progress in this direction might be done by taking advantage that this limit is the same Greenwald limit in the tokamak and in the RFP, as said before.

Another topic where a transversal view would be useful is the dynamo. Indeed since (half) a dynamo is acting in the RFP, there is a natural resonance with the astrophysical dynamos. The corresponding communities have been interacting for several years, in particular in the frame of the Center for Magnetic Self-Organization in the United States. The von Karman Sodium (VKS) experiment in Cadarache came with a striking result: an incompressible fluid dynamo can drive an RFP magnetic state all by itself![21] It is striking that the incompressible turbulent flow produced by impellers leads to the same magnetic equilibrium as in a current driven pinch whose plasma is compressible. Understanding the universality of the RFP configuration might lead to a large leap forward of dynamo theory.

4.2.2.8 Improving Heuristics

The claim sections, the corresponding hyperlinked database, and the new comment system would help into improving heuristics. Indeed researchers would have a simpler and more global view of their research field. Theoreticians would be incited to make more explicit the scientific contents of their calculations, to go to the essence of phenomena, and to look for universal features and applications. "Islands of knowledge" would have a tendency toward connection. Simple models would be urged into embedding in a broader physics context and checking their structural stability when going toward more complete descriptions. It would be easier for experimentalists to be aware of theoretical results and to challenge them, since these results would not be obscured by their technical surroundings. Similarly theoreticians would get an easier access to experimental results and would have more opportunities to think about and to suggest new experiments. This increased

[20]From this point of view one might again take advantage of the analogy of the RFP with the tokamak, since in the RFP such barriers are related to shear reversal too [72].

[21]This experiment studies dynamo action in the flow generated inside a cylinder filled with liquid sodium by the rotation of coaxial soft-iron impellers (von Karman geometry). It evidenced the self-generation of a stationary dynamo when the impellers do not rotate with the same angular velocity [70,94]. The magnetic field averaged over a long enough time corresponds to a RFP magnetic state with a large $m = 0$ mode (see Fig. 7 of [94]).

interaction between theory and experiments would certainly enhance creativity and improve the quality of papers. The easier way to follow connection between papers would also help people to look out of their specialty and better feed their intuition.

4.2.2.9 Improving Teaching

Plasma physics has plenty of facets, but it is important to have a global view hereof. This has to be addressed when teaching it. Knowledge in general, and all the more that on complexity, has a multi-scale structure. Therefore, teaching should introduce from the outset the concept of complexity and try to exhibit the various scales of the structure. A caveat: analytical calculations are a powerful heuristic tool, but often make very slow the introduction of concepts. It is thus advisable to limit the cases where they are described in details.

In reality there is no single way to define the multi-scale structure of a given complexity. As a result, topics which are presented in a separated way for a given choice of the structure are linked when choosing another view. The presentation of these links may be very useful, especially for graduate students.

Limited Capabilities of Models

Since our models are generally imperfect, when teaching one should use them in a different way than in more axiomatic parts in physics. Students should be made conscious about the limited capabilities of models, especially at a graduate level. Here are a few examples.

If students are taught resistive MHD, they should be made aware that other dissipations than resistivity may be present in the plasma like viscosity or heat diffusivity; therefore the Lundquist number is just one dimensionless number among many other ones, and its importance may be challenged by other such numbers.[22]

In the last decade a series of analytical calculations computed the width of the magnetic island of a saturated tearing mode [2, 3, 56, 74, 92, 93], which revealed in particular the mechanism of the saturation [56]. However, when teaching students, one should make them aware that present calculations of the saturation solve the magnetic part of the problem but that the fluid motion part is still unsolved. This is all the more important that this fluid motion is a simple example of a dynamo, i.e., the production of an electromotive force from a fluid motion [26]. The RFP single-helicity equilibrium provides another instance which is an extension of the simplest

[22]In the fluid description of screw pinches, a classical model is provided by the combination of Faraday–Ohm's law and of Navier–Stokes equation with Lorentz force. Then the Lundquist number is an obvious parameter. One may refine this description by adding a heat transport equation, which provides a self-consistent definition of the temperature profile and accordingly of the resistivity profile. Then the Lundquist number is no longer a parameter but an output of the model.

case. There, numerical MHD simulations are necessary to provide a description of the magnetic part of the problem too. In particular, they reveal the essential role of resistivity in the nature of the equilibrium, showing that it cannot be deduced from the original closed system picture originally present in Taylor relaxation theory [114].

Teaching Physics or Calculations?

When teaching, one might think about what researchers often do when looking at a theoretical paper in order to get the essence of the physical result: they skip the calculations and possibly go to them later (normally they should check everything, but they cannot!). So do the mathematicians who separate the statement of their theorems from their proofs. It is natural for textbooks to provide the calculations related to each phenomenon, but is it necessary to systematically present them when teaching? Here are a few examples.

For instance, the description of many waves in plasmas goes through similar steps: linearization about an equilibrium and Fourier transforms in space and time. Such a calculation may be done once for a simple case but then avoided for other types of waves. Then possibly some indication may be given about the clever way to go through the set of equations, in particular as far as the physics behind approximations is concerned. This would leave more time to discuss the physics of the wave.

This is all the more true that the Fourier decomposition sometimes hides the nature of the physics underlying the wave. For instance the plasma frequency ω_p is naturally introduced as that of a harmonic oscillator corresponding to the vibration of an electron slab with respect to its neutralizing ion slab. Langmuir waves are then understood by placing side by side such slabs where nearby ones have electrons in phase opposition.[23] Similarly, drift waves can be intuitively understood as the juxtaposition of the drift bumps described in Sect. II A of [78].

When teaching Landau damping, it is important to convey the physics behind it, in particular the fact that the damping of a Langmuir wave is due to a phase mixing of its constituting beam modes (van Kampen modes) and that these modes

[23] See for instance Sect. 14.2.1 of [52]. This sheds also a new light on the hydrodynamic or cold beam–plasma instability (Sect. 14.3.1 of [52]). Indeed a modulation with wavenumber k of the beam density generates a forcing of the plasma at pulsation $\omega = ku$, where u is the beam velocity, which feeds back on the beam density modulation. The response of a harmonic oscillator scales like $(\omega_\mathrm{p}^2 - \omega^2)^{-1}$. For $\omega \gg \omega_\mathrm{p}$ the electrons react weakly due to their inertia, which rules out a positive feedback for such ω's, and by continuity for $\omega > \omega_\mathrm{p}$. Then the plasma behaves like a classical dielectric, which screens the perturbing charge. As a result the unstable forcing must correspond to $\omega = ku \leq \omega_\mathrm{p}$ with a maximum for equality. This contrasts with the classical "negative energy" picture which rather suggests $\omega = -\omega_\mathrm{b} + ku$, where ω_b is the plasma frequency of the beam, and does not tell why the instability occurs rather for $ku \leq \omega_\mathrm{p}$ and why it is the strongest for $ku \simeq \omega_\mathrm{p}$. This forced harmonic oscillator picture works also for other reactive instabilities.

stay during the damping, as shown by the wave echo effect. The synchronization of particles with the wave brings the physical mechanism unifying Landau damping and growth (see Sect. 4.3.1). To the contrary one may wonder whether it is really important to teach the Landau calculation, especially when accounting that this calculation is not amenable to any intuitive interpretation.[24] Teaching Landau's calculation makes sense as a second stage of the introduction to Landau damping for students who are meant to become theoreticians.

As a result, it may be important to teach certain phenomena in a way very different from their initial derivation. Because of the limited capabilities of models, the status of analytical calculations is different from more axiomatic fields of physics. Heuristically, they are a powerful tool to uncover new hidden physics, but pedagogically one may avoid them to start with and keep the physical ideas and the corresponding images.[25] Another incentive to diminish the amount of taught analytical derivations is the growing importance of numerical simulations to uncover new hidden aspects of the complexity of plasmas: their results should be taught too.

Teaching physics requires teaching experimental facts. If less time is dedicated to calculations,[26] more time may be devoted to experimental results and to their error bars. This may be done in various ways. In particular, students' attention is captured when the historical path leading to the present view about a given phenomenon is described. Generally this path includes iterates of the interaction between experimental and theoretical results. This exhibits the fascinating character of the scientific adventure and shows how important are experiments[27] ... and experimentalists! This provides a balance to the fascination of computer work already mentioned in Sect. 4.2.1.6.

Taking Advantage of Nonlinear Dynamics and Chaos

Plasma physics developed in a progressive way, and often textbooks are more the accretion of successive layers of knowledge than a presentation of its global reconstruction. In particular, nonlinear dynamics and chaos might provide a way to revisit and unify separated chapters,[28] e.g., turbulent and collisional transport,

---

[24]Moreover this calculation gives no clue to the plasma behavior in the actual nonlinear regime where damping is a manifestation of stability of an infinite-dimensional Hamiltonian system (see Sect. 4.3.1).

[25]Using the blackboard is an efficient way to avoid "runaway lectures," especially when calculations are presented.

[26]Naturally this must be done without going up to a superficial presentation of the phenomena. Calculations are a way to anchor memory and to train students, especially at an undergraduate level.

[27]In particular the development of new diagnostics to touch other parts of the "elephant."

[28]This is all the more justified, since plasma physicists contributed a lot to the development of these topics.

the calculation of magnetic field lines,[29] or the introduction of fluid and of Vlasov equations. The statistical mechanics of systems with long-range interactions may bring useful complementary... and surprising views[30] [21, 32]. So does the theory of fluid turbulence.[31]

In particular it would be interesting to revisit with modern nonlinear dynamics and chaos what is usually called *collisional transport in plasmas*. The simplest instance of this transport deals with a uniform non-magnetized plasma. It is traditionally described by considering the motion of a test particle due to the Coulomb force of particles within the Debye sphere (radius λ_D) around it. Within this sphere two scales are important: the typical interparticle distance $d = n^{-1/3}$ and the classical distance of minimum approach $\lambda_{ma} = e^2/(4\pi\varepsilon_0 k_B T)$ where ε_0 is the vacuum permittivity, k_B is the Boltzmann constant, T is the temperature, n is the density, and e is the electron charge. These scales verify $\lambda_D \gg d \gg \lambda_{ma}$. The particles away from the test particles at a distance much larger than d are not felt individually, but act through their mean field. To the contrary, a particle at a distance much smaller than d is felt individually by the test particle and its Coulomb field dominates over that of all other particles. It is then natural to think of the interaction between these two particles as a two-body Rutherford collision. The effect of particles at a distance of the order of d can be described neither by a mean-field description nor by the two-body Rutherford picture: the test particle experiences Coulomb forces with the same order of magnitude due to several such particles, but not many.

Historically two groups at UC Berkeley's Radiation Laboratory derived at almost the same time a Fokker–Planck equation describing "collisions" in non-magnetized plasmas and quoted each other results in their respective papers: one by Gasiorowicz, Neuman and Riddell [71] and a year later one by Rosenbluth, MacDonald and Judd [108]. The first group of authors dealt with the mean-field part of the interaction by using perturbation theory in electric field amplitude. The second group of authors used the Rutherford picture. Each theory has a difficulty in describing the scales of the order of d. The mean-field approach cannot describe the graininess of these scales, and the Rutherford picture cannot describe the simultaneous "collisions" with several particles. Even for scales smaller than d, the Rutherford

[29] Unfortunately, the beauty and the flexibility of the derivation of the Hamiltonian description of magnetic field lines by a stationary action principle [28] have been largely overlooked. It was formulated in a simple way in [40, 102] showed a corresponding equivalence of canonical transformations and of changes of gauge.

[30] For instance the existence of negative -specific heat in a magnetically self-confined plasma torus [82]. The saturation of the cold and water-bag beam–plasma instability can be computed analytically by using Hamiltonian Eq. (4.1) introduced in Sect. 4.3 with a single wave (cold: [68]; water bag: [8, 9]). The mean-field derivation of Vlasov equation was already mentioned in Sect. 4.2.1.2.

[31] There are strong analogies with plasma turbulence, as exemplified by the Charney–Hasegawa–Mima model, but also strong differences since plasma turbulence is seldom fully developed. Furthermore the word "intermittency" is used with quite different meanings in the two fields.

collision is modified due to the fluctuating electric field of the other particles in the Debye sphere [71]. Using the more relevant description for scales smaller than d [108] and the one for the larger scales [71], the corresponding contributions to transport turn out to be of the same order of magnitude. Furthermore, if one accepts to cross the "validity border" d, and one performs the final integration on the whole range of scales $[\lambda_{\text{ma}}, \lambda_{\text{D}}]$ for either theory, the two results are found to agree [71, 108].

Because of gas dynamics, plasma physicists were led to think of the interactions of particles in kinetic unmagnetized plasmas within a Debye sphere as collisions. However, even though the Rosenbluth et al. paper provides the same result as Gasiorowicz et al.'s when both are applied to all scales within the Debye sphere, the Rutherford collision image is only correct for scales much smaller than d. Rigorously speaking one should not speak about collisional transport, but about "short range induced transport", "unscreened Coulomb interaction induced transport," or so.

The Gasiorowicz et al. approach has the merit to make a calculation of transport coefficients starting with the genuine N-body dynamics using explicit assumptions and avoiding the ad hoc truncation of integrals at the Debye length. However, within the same approximations, a more elegant derivation of the same Fokker–Planck equation describing "collisions" in non-magnetized plasmas is provided by taking the limit "infinite number of particles in the Debye sphere" of the Balescu–Lenard equation (see Sect. 8.4 of [7] and Sects. 7.3 and 7.4 of [76]). With a single calculation, this derivation provides both the dynamic friction and the diffusion coefficient. As Gasiorowicz et al.'s approach, it also avoids the ad hoc truncation of integrals at the Debye length. It requires the plasma to be stable, which is a serious caveat for the applicability of the traditional Fokker–Planck equation to magnetized plasmas which are cluttered with instabilities.[32]

However the Balescu–Lenard approach still has an intrinsic shortcoming. Indeed, due to short range interactions, particle dynamics is chaotic in reality (this is implicit in Rosenbluth et al.'s theory), and one is facing the calculation of transport coefficients for a chaotic motion. The Balescu–Lenard approach makes a perturbation calculation which is not a priori justified for chaotic dynamics, even for scales larger than d. Therefore, students should be warned to be cautious, since, for the motion of a charged particle in a spectrum of longitudinal waves, a perturbation calculation yields the quasilinear estimate for the diffusion coefficient, while a super quasilinear regime, a synergetic effect in chaos, is found to exist in this chaotic dynamics for intermediate resonance overlap (see Sect. 4.3.4). There diffusion becomes quasilinear for strong resonance overlap, but not because the perturbation calculation becomes valid again (see Sect. 4.3.4).

As a result, as yet, there is no correct calculation of the contribution of scales about d to short-range induced transport ("collisional transport"). This issue would

[32] The impact of instabilities was recently addressed in [4,5] by taking into account the spontaneous emission of waves by particles which induces a corresponding drag on top of the "collisional" one.

be worth more theoretical investigation: how good are the classical and neoclassical theories of transport? This suggests plasma physics courses to have a part devoted to dynamics with the successive introduction of Hamiltonian chaos, of the transport due to short-range interactions ("collisions"), and of turbulent transport. When dealing with "collisions," one might start with the true chaotic dynamics and exhibit the different nature of the interaction for the scales smaller and larger than d. Then one could introduce the corresponding approaches with appropriate caveats: (1) the approximate perturbative approach à la Balescu–Lenard, with a recall of Gasiorowicz et al.'s work; (2) the two-body approach of Rosenbluth et al. Finally one might point out that the matching of the two theories is still an open issue: as yet only two parts of the "elephant" have been touched.

4.3 Describing Plasma Dynamics with Finite-Dimensional Hamiltonian Systems

The main results of this section are summarized in the last three paragraphs of the extended summary in Appendix 1 of Sect. 4.5.

Due to its late development, plasma physics did not try to address its problems by a first principle approach, but borrowed many concepts and tools from other fields of physics like the kinetic theory of gases and fluid mechanics. In particular, in order to deal with kinetic aspects, people naturally looked for a description in terms of a velocity distribution function and therefore for some substitute of the Boltzmann equation, and this was the Vlasov equation. This equation was the starting point of most of the kinetic treatments of plasmas, and the Vlasovian description is a must of any plasma textbook. However, as recalled in Sect. 4.2.1.2, this equation is not justified for many timescales where it is used, and some of the calculations it enables to do are far from intuitive (e.g., Landau damping).

However, in a system where the transport due to short-range interactions ("collisions") is weak, it is natural to think about plasma dynamics by working directly with classical mechanics and taking into account that the collective field dominates over the graininess field. Though natural, this did not occur spontaneously before the Vlasovian approach ran into a major difficulty: the description of the nonlinear evolution of the weak warm beam–plasma instability or bump-on-tail instability. In the following, we only consider a one-dimensional plasma with electrons moving in a neutralizing uniform ion background.[33]

At the time where the Vlasovian approach ran into the difficulty of describing the nonlinear regime of the bump-on-tail instability, the theory of chaos for finite

[33]This is a tremendous simplification with respect to the physics of many actual plasmas. In particular density fluctuations may bring dramatic changes in the dynamics of Langmuir waves by Anderson localization [36, 59], by a transfer of particle momentum over an increased range of velocities [48], and by nonlinear decay and scattering processes [121].

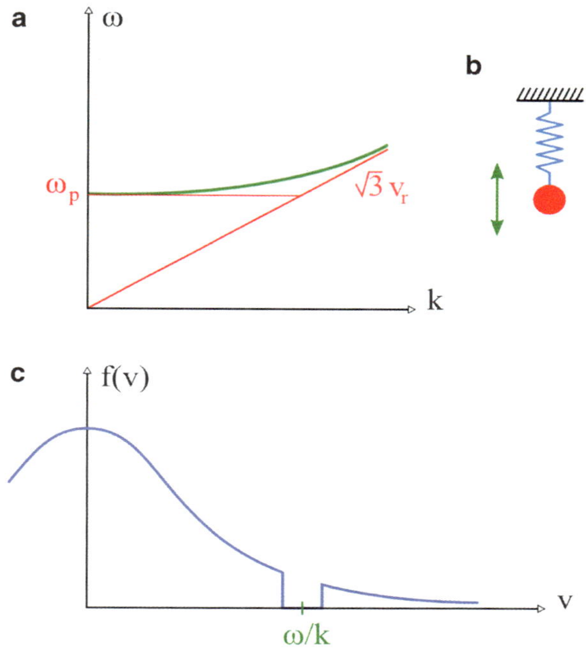

Fig. 4.1 Langmuir wave without resonant particles. (**a**) Bohm–Gross dispersion relation, (**b**) equivalent harmonic oscillator, and (**c**) electron velocity distribution function with a gap at the wave velocity

number of degrees of freedom Hamiltonian systems had been developing in the plasma physics community for more than a decade, and this was an incentive to tackle the weak warm beam–plasma instability by generalizing [51, 115] a model originally introduced for the numerical simulation of the cold beam–plasma instability [96,98]. There the beam was described as a set of particles while the wave was present as a harmonic oscillator. A Langmuir wave with a phase velocity ω/k where there are no resonant particles, as shown in Fig. 4.1c, verifies the Bohm–Gross dispersion relation[34] shown in Fig. 4.1a and is equivalent to a harmonic oscillator (Fig. 4.1b). If one considers a wave–particle interaction occurring in a finite range of velocities $[v_{\min}, v_{\max}]$, then it is sufficient to include in the Hamiltonian the waves with phase velocities in this interval, which defines their number M (Fig. 4.2). This finally yields the self-consistent Hamiltonian

[34]This relation makes sense, since we consider low-amplitude waves with phase velocities much above the thermal speed. If these conditions are not satisfied, the issue is a lot more involved [11–13].

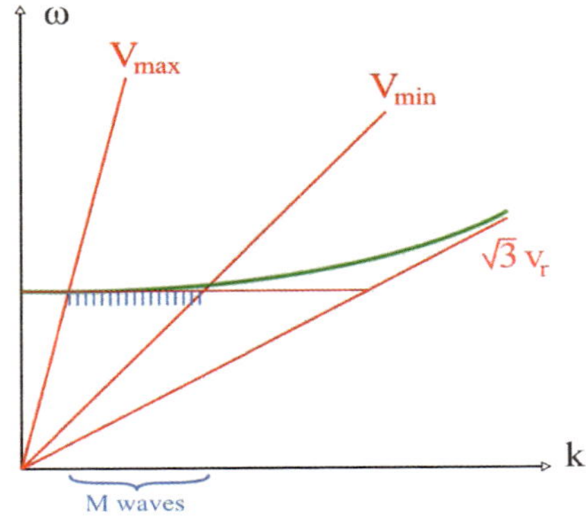

Fig. 4.2 Diagram showing Bohm–Gross dispersion relation with the velocity interval $v_{\min}, v_{\max}$ and a comb corresponding to the M waves with phase velocities in this interval

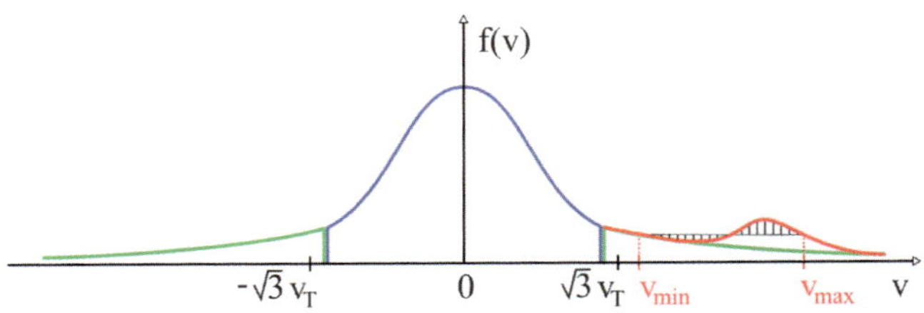

Fig. 4.3 Velocity distribution cut in three pieces: a nonresonant central part in *blue*, and left and right resonant parts in **green** for the case of a thermal plasma and in *red* for that of a bump-on-tail; the corresponding plateau in shown in *black*

$$\begin{aligned} H_{\mathrm{sc}} = & \sum_{r=1}^{N} \frac{p_r^2}{2} + \sum_{j=1}^{M} \omega_{j0} I_j \\ & - \varepsilon \sum_{r=1}^{N} \sum_{j=1}^{M} k_j^{-1} \beta_j \sqrt{2I_j} \cos(k_j x_r - \theta_j), \end{aligned} \tag{4.1}$$

where $\varepsilon = \omega_{\mathrm{p}}[2m\eta/N]^{1/2}$ is the coupling parameter and $\beta_j = [\partial\varepsilon_{\mathrm{d}}(k_j, \omega_{j0})/\partial\omega]^{-1/2}$, with ω_{p} the plasma frequency, m the mass of particles, η the ratio of the tail to the bulk density, $\varepsilon_{\mathrm{d}}(k, \omega)$ the bulk dielectric function, and k_j and ω_{j0} the wavenumber and pulsation of wave j. The conjugate variables for H_{sc} are (p_r, x_r) for the particles and (I_j, θ_j) for the waves. On top of the total energy $E_{\mathrm{sc}} = H_{\mathrm{sc}}$, the total momentum $P_{\mathrm{sc}} = \sum_{r=1}^{N} p_r + \sum_{j=1}^{M} k_j I_j$ is conserved.

This model was derived from the N-body description of the beam–plasma system [1]. More recently, this was done again in a heuristic way (see Sect. 2.1 of [46]) and in a rigorous one by a series of controlled approximations (see the remaining of chapter 2 of [46]), which enables replacing the many particles of the bulk by their collective vibrations. So, in Fig. 4.3, the blue central part of the distribution is no longer present as particle degrees of freedom; if one is interested in the evolution of the red bump, one may incorporate the left green wing into the bulk too. As shown in the next subsections, this approach helped into the investigation of the nonlinear evolution of the weak warm beam–plasma instability.[35] However its first contribution was to provide a rigorous mechanical understanding of Landau effect. It also provided a new insight into the transition from Landau damping to damping with trapping when the amplitude of a Langmuir wave is increased [97]: it turns out to be a second-order phase transition [66], a phenomenon which is hidden in the full N-body description of the plasma when the same Gibbsian approach is used. It is worth noting that a self-consistent Hamiltonian description is also powerful for the description of wave–particle interaction for waves in magnetized plasmas for which the Larmor precession plays an important role (see [83] and references therein).

4.3.1 Recovering Vlasovian Linear Theory with a Mechanical Understanding

Before applying this model to the saturation of the beam–plasma instability, it was necessary to make sure that it included the physics of Vlasovian linear theory. Therefore, one had to address the linear theory of the perturbation of a spatially uniform velocity distribution function by small waves. In order to stay in the spirit of classical mechanics, this unperturbed state should correspond to a single mechanical system and not to an ensemble of systems. This is naturally obtained by considering the unperturbed plasma as made up of a series of monokinetic beams and each beam as an array of equidistributed particles. If the waves have a vanishing amplitude, this state is invariant in time. Then perturbation theory is performed by using mere Fourier series and leads to a Floquet problem in $2(N+M)$ dimensions. In contrast with the simplest Floquet problem, the Mathieu equation, surprisingly this problem can be solved explicitly!

The solution includes the Landau instability [86] as an eigenmode if the distribution function has a positive slope. If the slope is negative, it does not provide Landau damping as an eigenmode, but only a series of beam modes. In agreement with van Kampen's theory [81], Landau damping is recovered as a result of the phase mixing of the latter. It must be stressed that in Hamiltonian mechanics, in agreement with time reversibility and with Liouville theorem, a damped eigenmode comes with an unstable one having the opposite exponentiation rate. Therefore Landau damping cannot be a damped eigenmode, since it would come together with an unstable one which would be seen with probability 1. Moreover, in Vlasovian theory, Landau damping is not an eigenmode, but a time-asymptotic damped solution obtained by analytic continuation. Furthermore, in Vlasovian theory, it is not sure a priori that the van Kampen phase-mixing solution actually exists and is not destroyed

[35]By taking advantage of the intuition developed by this approach it is possible to derive a more pedestrian approach to wave–particle interaction [52].

by nonlinear effects related to finite, though small, amplitude of the beam modes. Proving this nonlinear stability [95], and thereby the actual existence of Landau damping, was a mathematical tour de force, the equivalent of a KAM theorem for continuous systems, and led Cédric Villani to be awarded the 2010 Fields medal. In the frame of the finite-dimensional Hamiltonian approach, this nonlinear stability is the mere result of KAM theorem itself.

As André Samain pointed out, if the distribution function has a positive slope, an unstable eigenmode and a damped eigenmode are not enough to recover the Vlasovian result. Indeed, a typical initial perturbation excites both modes with the same amplitude at $t = 0$, but the damped one dies out, which leaves only the unstable one with half of the Vlasovian amplitude. In reality, a typical initial perturbation excites also a wealth of beam modes. When their contribution is properly taken into account, Yves Elskens found (Sect. 3.8.3 of [46]) that an initial perturbation with amplitude 1 evolves in time according to the time-reversible expression $e^{\gamma_{jL}t} + e^{-\gamma_{jL}t} - e^{-\gamma_{jL}|t|}$: the beam modes act subtractively to compensate the damped eigenmode and to restitute the Vlasovian solution. This apparent intricacy corresponds to experimental reality. If Langmuir waves are excited by a grid in a magnetized plasma column, this is done by the excitation of the various "monokinetic beams" going through the grid. Landau damping results from the phase mixing of these excitations which do not die out, as proved by echo experiments [6]. If a weak warm beam goes through the grid together with the background plasma, the beam modes are excited too and contribute to the Langmuir wave amplitude. As a result of this analysis, the Vlasovian limit, though very powerful, turns out to be a quite singular limit for the linear theory of waves.

At this point, we made sure the finite-dimensional Hamiltonian approach recovers Vlasovian linear theory. However, the former approach comes with an important bonus: it brings the information of particle dynamics in parallel with the wave's. This is absent in the Vlasovian description and has two important consequences. First, because of its lack of intuitive contents, the reality of collisionless Landau damping was fully recognized only after its experimental observation in 1964 by Malmberg and Wharton [91], almost two decades after its prediction. Second, textbooks are forced to come up with complementary models to try and explain intuitively the way Landau effect works. The finite-dimensional Hamiltonian approach enables to assess these models which are not all correct, unfortunately (see Sect. 4.3.1 of [46], in particular the exercise therein as a caveat[36]). Better, it shows Landau damping and instability to result from the same synchronization mechanism of particles with waves.

In order to avoid repeating here the whole argument, we give a simple proof that particles released at $t = 0$ with a velocity u, and a *uniform initial spatial distribution*, have an average velocity which comes closer to the wave phase velocity over a bounded time. Let $\ddot{x} = \varepsilon \cos x$ be the equation of motion of the particle in the wave,

[36] In particular, though initially published with a caveat, the surfer model induces in the mind of the students the wrong feeling that trapping is involved in Landau effect.

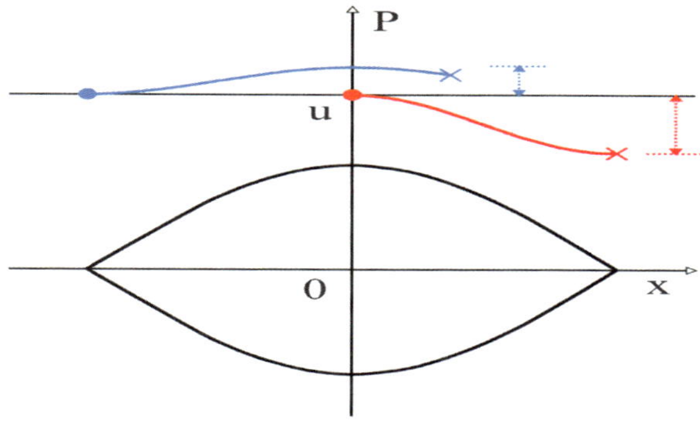

Fig. 4.4 Phase space plot displaying the average synchronization of two particles with a wave, one starting at the position of the X-point of the separatrix (*blue line*) and another one starting at the O-point of the trapping domain (*red line*)

expressed in the reference frame of the latter. Assume the unperturbed orbit to be $X_0(t) = x_0 + ut$. A perturbation calculation to second order in ε yields

$$\Delta u(t) = \varepsilon^2 \frac{\cos ut - 1 + \frac{1}{2} ut \sin(ut)}{u^3}. \tag{4.2}$$

$u\Delta u(t)$ is even in t and is negative from $t = 0$ up to $t = T \equiv 2\pi/|u|$, which means an average synchronization of the particles with the wave within this time interval whatever be the relative sign of their velocity to the wave. Since quantity $\Delta u(t)$ scales like $1/u^3$, the average synchronization is small for large $|u|$'s: it is a local effect in velocity. The effect is maximum for $|t| \simeq 3T/4$. For t small, Eq. (4.2) becomes

$$\Delta u(t) \equiv \langle \dot{x}(t) \rangle - u = -\frac{\varepsilon^2 ut^4}{24}, \tag{4.3}$$

to fourth order in t. We notice that the effect vanishes for small $|u|$'s. Therefore this effect is not related at all to trapping inside the wave troughs.

It can be intuitively understood as follows. Figure 4.4 displays a sketch of the phase space of particles moving in the presence of a wave. One particle is released at the position of the X-point of the separatrix (blue line) and another one starting at the O-point of the trapping domain (red line). The first one has an orbit further away from the separatrix than the second one. Therefore it is less modulated, which provides their average synchronization over the considered time duration. This average synchronization effect was proved to exist in an experiment with a traveling wave tube [37].

Due to this synchronization, particles change their momentum. Since wave–particle momentum is conserved by the self-consistent dynamics, the wave changes its momentum and thus its amplitude, in the opposite way. This brings the Landau effect [52].

The self-consistent calculation shows that the particles the most synchronized with the wave have a velocity about the growth rate γ in the present units. For such particles time T is about $1/\gamma$, the natural bound for the validity of a calculation with a wave of constant amplitude. The synchronization mechanism is the same for Landau damping and instability, which explains why the Landau effect is described by a single formula, though the physics of damping and instability display qualitative differences as far as the wave aspect is concerned. All this is hidden in the Vlasovian approach.

The Landau effect can also be recovered by a statistical approach ([61] and Sect. 4 of [46]). There the wave phase and amplitude evolutions are computed by perturbation theory in the coupling parameter ε of the self-consistent Hamiltonian. Together with the collective Landau effect, the calculation derives also the spontaneous emission of waves by particles. As a result, Landau damping turns out to be a relaxation mechanism driving waves to their thermal level.

4.3.2 Quasilinear Theory

In 1961, Romanov and Filippov [107] introduced the quasilinear equations which were made popular in 1962 by two papers published in the same issue of Nuclear Fusion [39, 118]. As indicated by their name, these equations were derived by considering the nonlinear dynamics of the beam–plasma instability as close to linear and more precisely by neglecting mode–mode coupling, except for its contribution to the evolution of the space-averaged velocity distribution function $\bar{f}(v,t)$. These equations are

$$\partial_t \bar{f} = \partial_v (D_{\mathrm{QL}}(v,t)\partial_v \bar{f}),$$

$$\partial_t \psi = 2\gamma_{\mathrm{L}}(v,t)\psi, \tag{4.4}$$

where $\gamma_{\mathrm{L}}(v,t) = \frac{\pi}{2}\eta k^{-2}\partial_v \bar{f}(v,t)$ and $D_{\mathrm{QL}}(v,t) = \pi\eta \frac{1}{k^2}\psi(v,t)$ are the instantaneous Landau growth rate and QL diffusion coefficient, while $\psi(v,t)$ is the power spectrum, a smooth function going through points $\psi(t,v_j) = k_j I_j(t)/(N\Delta v_j)$ with Δv_j the mismatch of phase velocity of wave j with its two neighbors.

These equations show that at time t an unstable Langmuir wave with phase velocity v grows with the Landau growth rate $\gamma_{\mathrm{L}}(v,t)$ computed with $\bar{f}(v,t)$ and that the instability saturates due to the diffusion of the velocities of particles, which levels out the bump on the tail of the distribution function and substitutes it with a plateau (see Fig. 4.3). They also predict the wave spectrum at saturation which is shown in the upper part of Fig. 4.5. Within experimental uncertainties, these predictions were confirmed by the first laboratory experiment looking at the bump-on-tail instability [106].

However, the perturbative approach used in the derivation of the quasilinear equations cannot be justified theoretically during the whole saturation of the instability. Indeed, waves scatter the particle positions with respect to their ballistic value. When the corresponding spreading of positions becomes on the order of the wavelength, the perturbative approach fails. The corresponding spreading time

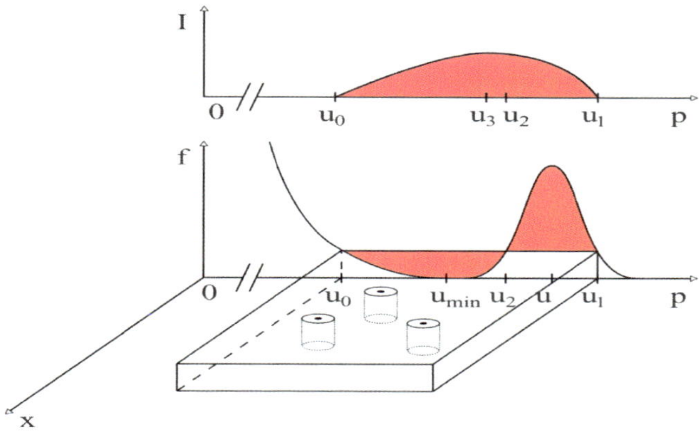

Fig. 4.5 Saturation of the bump-on-tail instability. *Upper part*: wave spectrum. *Lower part*: plateau both in velocity and in space. The *vertical cylinders* indicate groups of particles

τ_{spread} turns out to be the (Lyapunov) time of separation of nearby orbits in the chaos induced by the waves and also the typical trapping time of particles in the turbulent electrostatic potential. Obviously, leveling out the bump on the tail of the distribution function needs a time longer than the latter times. Therefore one might doubt at the validity of quasilinear equations to describe the saturation of the instability. In 1984, Laval and Pesme proposed a new Ansatz to substitute the quasilinear one and predicted that whenever $\gamma_{Landau}\tau_{spread} \ll 1$ both the wave growth rate and the velocity diffusion coefficient should be renormalized by a factor 2.2 [87]. This motivated Tsunoda, Doveil, and Malmberg to perform a new experiment with a traveling wave tube in order to decrease the noise due to the previous use of a magnetized plasma column [117]. It came with a surprising result: quasilinear predictions looked right, while quasilinear assumptions were completely wrong. Indeed no renormalization was measured, but mode–mode coupling was not negligible at all. Apparently one had "$A \Rightarrow B$" and "B right," but A wrong! This set the issue: would there be a rigorous way to derive the quasilinear equations?[37]

4.3.3 Dynamics When the Distribution Is a Plateau

First, it is important to notice that one does not need the quasilinear equations to be correct to prove the formation of the plateau. Indeed this formation comes from the chaos induced by the unstable Langmuir waves among the resonant particles, whatever be the precise description of the corresponding chaotic transport. When

[37] This academic issue has a broader relevance since the QL approximation is used everywhere in plasma physics.

the plateau forms in velocity, density becomes also almost uniform spatially in this range of velocities (see Fig. 4.5). Indeed, chaos tends at equidistributing particles all over the chaotic domain in phase–space. Actually, KAM tori, bounding the chaotic domain defined by a prescribed spectrum of waves, experience a sloshing motion due to the waves. This brings a small spatial modulation to the particle density which provides a source term for the Langmuir waves. However, if the plateau is broad, the evolution of the wave spectrum is slow, which brings only a small change to the previous simplistic picture of a uniform density (see Sect. 2.2 of [15]). Therefore, there is almost no density fluctuation to drive the wave evolution as defined by the self-consistent dynamics: the wave spectrum is frozen. Hence the particle dynamics is the one defined by a prescribed spectrum of waves. Clumps of particles may experience a strong turbulent trapping, but the distribution function is unaffected by this granular effect. As a result, self-consistency vanishes in the plateau regime if the plateau is broad enough, because particle transport only rearranges particles without changing the height of the distribution function itself within the plateau in phase–space, depriving waves from a source[38] (the little cylinders in Fig. 4.5 keep their height while moving). This is an instance where nonlinear effects increase the symmetry of the system and lead to a depletion of nonlinearity.[39]

This means that, when the plateau is formed, the diffusion coefficient $D(v)$ of particles with momentum v is the one found for the dynamics of particles in a prescribed spectrum of Langmuir waves. Let $D_{\mathrm{QL}}(v)$ be the quasilinear value of this coefficient. The next section discusses the possible values of D/D_{QL} in the resonance overlap regime.

4.3.4 Diffusion in a Given Spectrum of Waves

Quasilinear theory aims at describing the self-consistent evolution of waves and particles. One of its final coupled equations is a diffusion equation with a diffusion coefficient computed through perturbation theory. Since during saturation particle dynamics is chaotic in the beam velocity domain, one may wonder about the validity of such a formula, even if the wave spectrum is prescribed. When investigating this issue, some surprises were on the way!

---

[38]For a plateau with a finite width, the small remaining source brings a further evolution of the wave–particle system toward a Gibbsian state where the wave spectrum collapses toward small wavelengths together with the escape of initially resonant particles toward low bulk plasma thermal speeds [68]. This corresponds to a further step toward a new thermal equilibrium of the N-body system corresponding to the initial beam–plasma system. The description of the subsequent steps toward thermal equilibration requires to use a full N-body model.

[39]This phenomenon, also called depression of nonlinearity, was introduced in fluid mechanics [84] and was identified as a result of the emergence of long-lived vortices where the enstrophy cascade is inhibited. It also exists in systems with quadratic nonlinearities [20, 84].

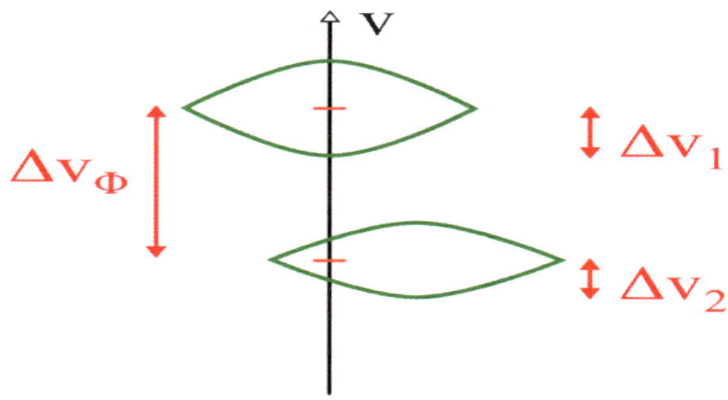

Fig. 4.6 Snapshot of the trapping domains of two nearby waves. They have half-widths Δv_i and a mismatch in velocity Δv_Φ

First, a single physical realization of the wave field acts on the particle velocity distribution to make it non-gaussian, which rules out diffusion: "chaotic" does not mean "stochastic." One needs an ensemble of "enough" independent realizations to make it gaussian [10]. A simple way to do this is for waves to have mutually independent random phases.

The motion of a particle in a discrete set of waves with random phases involves several times. First, a discretization time $\tau_{\text{discr}} = (k\Delta v)^{-1}$ where k is the typical wavenumber and Δv is the typical mismatch of nearby phase velocities. Second, the already defined spreading time τ_{spread}. Third, the autocorrelation time $\tau_{\text{ac}} = (k\Delta u)^{-1}$ where Δu is the full range of phase velocities. In the weak warm beam case, Δu is the width of the beam distribution function and τ_{ac} is the smallest of all three times.[40] The Chirikov overlap parameter [30] between two nearby waves is $s = (\Delta v_1 + \Delta v_2)/\Delta v_\Phi$ where Δv_Φ is the mismatch of their phase velocities and Δv_i's are the width in velocity of the trapping domain of wave i which scales like the square root of the wave potential[41] (see Fig. 4.6).

Whatever be the overlap of nearby resonances, perturbation theory is correct over a time τ_{spread}. Therefore, initially particle dynamics looks diffusive and the diffusion coefficient takes on the quasilinear value. A small value of Chirikov overlap parameter s is equivalent to $\tau_{\text{discr}} \ll \tau_{\text{spread}}$. Then at $t \simeq \tau_{\text{discr}}$ the particles

[40] In the opposite limit when $\tau_{\text{spread}}/\tau_{\text{ac}}$ is small, the time evolution of the waves is slow with respect to the trapping motion in the instantaneous wave potential. Then chaotic dynamics may be described in an adiabatic way with the picture of a slowly pulsating separatrix [44, 45] (see also Sect. 5.5 of [46] and Sect. 14.5.2 of [52]). In this limit, for the case of the motion in two waves, the resonance overlap defined hereafter is large.

[41] This criterion is a very useful rule of thumb which works, also experimentally [38], provided the two trapping domains are not too dissimilar. In particular, $\Delta v_1/\Delta v_2$ should not be too far from 1. Otherwise, one of the waves is a small perturbation for the other one, and the threshold of large scale chaos is a lot larger than 1 (see [50, 54] for more information). A more accurate way to understand the transition to large scale chaos is provided by a renormalization transformation [50, 54] (see also Sect. 5.4 of [46] and Sect. 14.5.4 of [52]). However Chirikov criterion can also be used to check whether high dimensional dynamics is chaotic enough. More specifically parameter s may be used as an observable whose Gibbsian estimate tells Gibbsian calculus makes sense when it is larger than 1 [60].

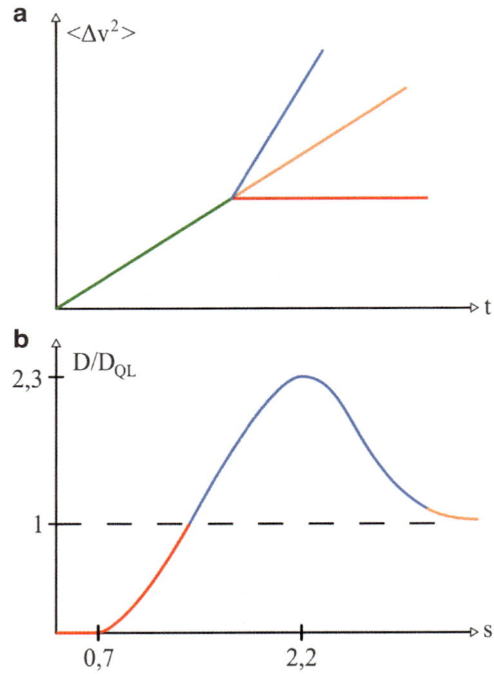

Fig. 4.7 Regimes of diffusion. (**a**) $\langle \Delta v^2 \rangle$ vs. time; initial quasilinear regime: *green line*; asymptotic saturation: *red line*; superquasilinear regime: *blue line*; time-asymptotic quasilinear regime: *brown line*. (**b**) D/D_{QL} vs. s; same color code as in (a), except for the *red growing segment* that corresponds to the weakly chaotic regime

feel they are in a quasiperiodic force field and the spreading of their velocities saturates. If the wave potential is periodic both in time and space, this saturation is due to the presence of KAM tori. In any case, till τ_{discr} particles feel the force field as a white noise and experience a stochastic diffusion. Figure 4.7a displays a cartoon of the variance $\langle \Delta v^2 \rangle$ of the velocities of particles all released with the same initial velocity in a prescribed spectrum of Langmuir waves. The stochastic diffusion corresponds to the green segment on the left and the saturation to the red segment on the right. As might be expected, chaos does not enter this picture.

When chaos becomes dominant, i.e., when $\tau_{\mathrm{discr}} \gg \tau_{\mathrm{spread}}$, numerical calculations revealed [29] that after a time $\tau_s \sim \tau_{\mathrm{spread}}$, $\langle \Delta v^2(t) \rangle$ grows with a slope in between the quasilinear one and 2.3 times this value[42] (in the range bounded by the brown and blue curves in Fig. 4.7a).

Figure 4.7b summarizes in a sketchy way the various regimes as to the value of the diffusion coefficient D measured over many τ_{spread}'s. For small values of s, this "time-asymptotic" value vanishes because of the saturation of $\langle \Delta v^2 \rangle$

[42]The necessity to go beyond τ_{spread} to see the chaotic diffusion is a caveat for the numerical measurement of a chaotic diffusion coefficient. This minimum time comes from the locality in velocity of wave–particle interaction [10, 46]. Indeed it can be shown that at a given moment the waves making particle dynamics chaotic have a phase velocity within $\Delta v \sim 1/(k\tau_{\mathrm{spread}})$ from the particle velocity. Those out of this range act perturbatively. If waves have random phases, after

after $\tau_{\rm discr}$. This corresponds to the horizontal red segment. When s grows above the chaotic threshold, D takes on positive values, but first below the quasilinear one (red growing curve in Fig. 4.7b). For intermediate values of s, D takes on superquasilinear values (blue curve in Fig. 4.7b). For large values of s, D takes on the quasilinear value[43] (brown curve in Fig. 4.7b).

This can be understood by considering the dynamics of a particle in a prescribed spectrum of waves defined by Hamiltonian

$$H(p,q,t) = \frac{p^2}{2} + A \sum_{m=\mu}^{M} \cos(mq - t + \varphi_m), \tag{4.5}$$

where the φ_m's are random variables, and $M \gg \mu \gg 1$. Let the particle have an initial velocity p_0 in between $1/m_0$ and $1/(m_0+1)$, with $M \gg m_0 \gg \mu$. We evaluate $\Delta p(t) = p(t) - p_0$ by integrating formally the equation of motion for p. For t small enough, the dependence of $\Delta q = q(t) - p_0 t - q_0$ on any two phases with all other phases fixed is weak. Then $\langle \Delta p(t) \rangle = 0$. We write $\langle \Delta p^2(t) \rangle = \Delta_0 + \Delta_+ + \Delta_-$, with

$$\Delta_j = -\varepsilon_j A^2 \int_0^t \int_0^t \sum_{m_1=\mu}^{M} \sum_{m_2=\mu}^{M} \frac{m_1 m_2}{2} \langle \cos[\Phi_{m_1}(t_1) + \varepsilon_j \Phi_{m_2}(t_2)] \rangle \mathrm{d}t_1 \mathrm{d}t_2, \tag{4.6}$$

where

$$\Phi_m(t) = m\Delta q(t) + \Omega_m t + m q_0 + \varphi_m, \tag{4.7}$$

where $\Omega_m = m p_0 - 1$, with $\varepsilon_\pm = \pm 1$ and $\varepsilon_0 = -1$, and under condition $m_1 \neq m_2$ for $j = -$ and condition $m_1 = m_2$ for $j = 0$. Let $t_- = t_1 - t_2$ and $t_+ = (t_1 + t_2)/2$.

For $t_- \ll \tau_{\rm spread}$, $\langle \exp[\mathrm{i}k_m(\Delta q(t_+ + t_-/2) - \Delta q(t_+ - t_-/2))] \rangle$ may be considered equal to 1. Therefore the support in t_- of the integrand in Δ_0 is on the order of $\tau_{\rm ac}$. Since $\tau_{\rm ac} \ll \tau_{\rm spread}$, the integration domain in t_- may be restricted to $|t_-| \leq \nu \tau_{\rm ac}$ where ν is a few units. In the limit where $\nu \tau_{\rm ac} \ll t \ll \tau_{\rm discr}$, we obtain

$$\begin{aligned} \Delta_0 &\simeq \sum_{m=\mu}^{M} \int_0^t 2D_{\rm QL}(p_0)\pi^{-1} \int_0^{\nu \tau_{\rm ac}} \langle \cos[\Omega_m t_-] \rangle p_0 \mathrm{d}t_- \mathrm{d}t_+ \\ &= 2D_{\rm QL}(p_0) \sum_{m=\mu}^{M} (\pi \Omega_m)^{-1} \langle \sin[\Omega_m \nu \tau_{\rm ac}] \rangle p_0 t = 2D_{\rm QL}(p_0) t, \end{aligned} \tag{4.8}$$

visiting several "resonance boxes" of width Δv, a particle feels as having been acted upon by a series of independent chaotic dynamics, which triggers a diffusive behavior. This decorrelation makes it possible to numerically measure the diffusion coefficient by following the dynamics either of a single particle for a series of random outcomes of the wave phases or of many particles for a single typical outcome of the phases. By extension this enables to reconcile the uniqueness of each realization of an N-body system with models invoking a probabilistic average over independent realizations.

[43] If the waves have random amplitudes A_m and phases φ_m such that $A_m \exp(\mathrm{i}\varphi_m)$ is a gaussian variable, then the superquasilinear bump does not exist, and $D/D_{\rm QL} \leq 1$ for all values of s, but still goes to 1 when s becomes large [42].

where the discrete sum over m was approximated by an integral and where $D_{\text{QL}}(p_0) = \pi(Am_0)^2/p_0$ is the quasilinear diffusion coefficient. $\Delta_{\pm}$ can be neglected since we assumed Δq to depend weakly on two phases with all other phases fixed. For small times, a particle feels a stochastic forcing due to many waves. Therefore its position has a weak dependence over any two random phases, which justifies the quasilinear estimate. If $s \gg 1$, it can be shown ([55], Sect. 6.8.2 of [15], and Appendix 2 of Sect. 4.6) that the position of a particle has a weak dependence over any two random phases over a time on the order of $\tau_{\text{QL}} = \tau_{\text{spread}} \ln(s)$. Therefore, if $s \gg 1$, the quasilinear estimate holds over a time $\tau_{\text{QL}} \gg \tau_{\text{spread}}$. Using this property, the estimate can be shown to be correct for all times ([55] and Appendix 2 of Sect. 4.6). This derivation of the quasilinear estimate in the $s \to \infty$ limit is not yet rigorous. However, by using probabilistic techniques, a rigorous proof can be obtained for the dynamics of particles in a set of waves with the same wavenumber and integer frequencies, if their electric field is gaussian [47] or just if their phases have enough randomness [43].

We have just shown that in the resonance overlap regime D/D_{QL} may cover a large range of values [29, 46]. In particular $D \simeq D_{\text{QL}}$ is obtained for random phases of the waves and strong resonance overlap [29, 43, 46, 47]. The plateau regime corresponds to $\gamma_{\text{L}} = 0$ and therefore to $\gamma_{\text{L}} \tau_{\text{spread}} = 0$. Since D/D_{QL} may cover a large range of values in this regime, $\gamma_{\text{L}} \tau_{\text{spread}} \ll 1$ does not imply per se any renormalization or non-renormalization of D/D_{QL} nor of γ/γ_{L} by wave–particle momentum conservation. This contradicts previous works using $\gamma_{\text{L}} \tau_{\text{spread}} \ll 1$ to try and prove the validity of quasilinear theory [46, 55, 88, 89] and the "turbulent trapping" Ansatz aiming at the contrary [87]. The value of D/D_{QL} in the plateau regime of the bump-on-tail instability depends on the kind of wave spectrum the beam–plasma system reaches during the saturation of the instability, and not only on condition $\gamma_{\text{L}} \tau_{\text{spread}} \ll 1$, as assumed by these works.

4.3.5 A Crucial Numerical Simulation

In order to find out the nature of the wave spectrum at saturation, numerical simulations were performed using a semi-Lagrangian code for the Vlasov-wave model [15]. This model is the mean-field limit of the granular dynamics defined by the self-consistent Hamiltonian: waves are still present as M harmonic oscillators, but particles are described by a continuous distribution function.

The simulations were benchmarked in various ways. In particular, they recovered that the wave spectrum is almost frozen when the plateau is formed. They were repeated for a large number of random realizations of the initial wave phases for a fixed initial spectrum of amplitudes. As shown by previous simulations, the final wave spectrum was found to be quite jaggy and not smooth as that predicted by QL theory. For each of the realizations, one computed the spreading of the velocities of test particles when acted upon by the final set of waves. The first four even moments of this spreading were compared with those of the solution to the quasilinear Fokker–Planck equation for velocity diffusion, using the velocity-dependent diffusion constant D_{QL} computed with the final wave spectrum. The agreement was found to be excellent: the plateau verified the predictions of QL theory. However, as found in previous numerical simulations and experimentally, mode–mode coupling was found to be very strong during the saturation.

At this point, the validity of QL predictions while QL assumptions are wrong sounded still like a mystery. However, the simulations brought an unexpected clue to elucidate it: the variation of the phase of a given wave with time was found to be almost non-fluctuating with the random realizations of the initial wave phases [16]. Therefore the simulations showed that the randomness of the final wave phases was a mere consequence of that of initial phases. As a result, the self-consistent dynamics was shown to display an important ingredient for the validity of a quasilinear diffusion coefficient for the dynamics in a prescribed spectrum.

4.3.6 New Analytical Calculations

The just mentioned almost non-fluctuating variation of the phase suggested to revisit the past analytical calculations of the wave phase and amplitude average evolutions. As mentioned at the end of Sect. 4.3.1, they were performed by averaging over the initial particle positions. The new numerical result suggested to perform instead an average over the initial wave phases, which is compatible with a nonuniform particle density. Furthermore, the previous calculations used a perturbative approach which made sense in the linear regime, but which might be unjustified for the chaotic regime of the instability. This was an incentive to use the Picard iteration technique which is the central tool to prove the existence and uniqueness of solutions to differential equations in the so-called Picard's existence theorem, Picard–Lindelöf theorem, or Cauchy–Lipschitz theorem.[44] The iteration turns out to be analytically tractable three times when starting from the ballistic solution. It can be shown analytically that the third-order Picard iterate is able to describe the separation between trapped and passing orbits of a nonlinear pendulum. Furthermore numerical calculations (Elskens, 2011, "private communication") indicate that for the chaotic motion of particles in a prescribed set of waves, such an iterated solution is already fairly good over the τ_{spread} timescale which is crucial for chaos to build up. However the accuracy of the third-order Picard iterate needs further assessment.

This work in progress already brings the following results. First, the modification of the average wave frequency due to the coupling with particles is exactly the principal part correction to the wave frequency provided by the Vlasovian calculation

[44]This iteration technique is very convenient to alleviate the algebra of many perturbation calculations. This is the case for the perturbation calculation of the dynamics defined by Hamiltonian Eq. (4.5) in the wave amplitude A. In particular for a single wave, which provides Eq. (4.2).

of the dispersion relation of Langmuir waves or by the equivalent calculations with the self-consistent Hamiltonian [46]. However, the latter calculations deal with a spatially uniform distribution of particles, while the present one holds whatever the spatial inhomogeneity of the distribution of tail particles, but requires an average over the phases of the Langmuir waves. Second, an estimate of phase fluctuations shows they scale like $\eta^{1/2}$, which makes them negligible, as shown by simulations. Therefore, if initial phases are random, they stay random for all times: there is no need for the traditional random phase approximation.

Third, assuming the wave spectrum of any realization to be smooth when averaged over a width in phase velocity on the order of $(k\tau_{\text{spread}})^{-1}$, where k is a typical wavenumber, the evolution of a wave amplitude A_j is given by

$$\frac{\mathrm{d}\langle|A_j|^2\rangle}{\mathrm{d}t} = 2\gamma_{j\mathrm{L}}\langle|A_j|^2\rangle + S_{\text{spont}j} + S_{\text{inhom}j}, \tag{4.9}$$

where γ_{L} is the Landau growth rate defined together with Eq. (4.4), $S_{\text{spont}} \sim \bar{f}(v_{\text{phase}})/N$, where $\bar{f}$ now is the space-averaged coarse-grained velocity distribution function of the tail particles, v_{phase} is the phase velocity of the wave, and

$$S_{\text{inhom}j} = \frac{N^2\varepsilon_j^4}{k_j^2}\int_0^t \mathrm{d}t' \int\int \mathrm{d}p\mathrm{d}p' \mathrm{e}^{\mathrm{i}[\Omega_j(p')t' - \Omega_j(p)t]}\langle\tilde{f}(-k_j,p',t')\tilde{f}(k_j,p,t)\rangle + \text{c.c.}, \tag{4.10}$$

where $\tilde{f}$ is the Fourier transform of the coarse-grained velocity distribution function, $\Omega_j(p) = k_j p - \omega_{j0}$, and t is on the order of τ_{spread}. Equation (4.9) displays successively the contribution to the wave amplitude evolution of Landau growth or damping, of spontaneous emission, and of the emission of spatial inhomogeneities (turbulent eddies). Because of the $1/N$ factor, spontaneous emission vanishes when $N \to \infty$, since plasma graininess becomes negligible. To the contrary, the contribution of inhomogeneities to wave emission does not vanish in this limit. Due to turbulent trapping, a gradient in the velocity distribution yields localized spatial gradients a quarter of trapping time later, but this exchange of slopes in space and velocity occurs in a fluctuating way. If $f(x,p)$ does not depend on x, S_{inhom} vanishes. This occurs in particular when the plateau forms at the end of the weak beam–plasma instability in the limit $N \to \infty$ [15]. If such an instability starts from a position-independent velocity distribution function, the $\tilde{f}$'s are only due to turbulent eddies. Then the size of $S_{\text{inhom}j}$ can be bounded by a quantity vanishing in the limit where the number of waves is large, i.e., for a continuous wave spectrum. Therefore, if these calculations make sense, in this limit the quasilinear equations might correctly describe the average behavior of the instability, even though a given realization be very far away from the average behavior.

4.4 Conclusion

The main messages of this chapter are summarized in the abstract and in Appendix 1 of Sect. 4.5 and are not recalled here. This short conclusive section is rather devoted to global remarks and prospects.

The part of this chapter devoted to wave–particle interaction in plasmas shows the description of collisionless plasmas as finite-dimensional Hamiltonian systems is relevant, simple and transparent for linear aspects, and powerful even for nonlinear and chaotic ones. It shows the irreversible evolution of a macroscopic system can be described by classical mechanics. Therefore, an old dream comes true, but is yet to be made rigorous. As a result, the finite-dimensional approach opens new avenues for the description of plasmas. However the mean-field description (Vlasov equation) stays a powerful tool, in particular for linear calculations, for exhibiting the metastable BGK modes, and for numerical simulations.

The description of complexity of plasmas is an intricate issue and even more so the future development of the corresponding methodology. Collective effects are important in plasma physics but also for its development: it requires collective efforts of plasma physicists. Therefore this chapter is more a way to start a brainstorming in the plasma community than a list of ready-to-use recipes. It would be very useful for this community to pay attention to the essence of its physics and of its practice. To this end it should review, retrace, and revisit past published material but also its past way of thinking, of interacting, and of meeting together. Such an attentive attitude is reminiscent of Cicero's quite philosophical proposal of the right way for the Roman citizens to be religious, linking it to elegance, diligence, and intelligence[45]: a nice inspiration for the work to come!

Acknowledgements I am indebted to Y. Camenen, L. Couëdel, F. Doveil, and Y. Elskens, for a thorough reading of a first version of this paper and for providing me with an extensive feedback. My thanks also go to D. Bonfiglio, S. Cappello, and F. Sattin who did the same for a second version. Y. Elskens also helped me a lot in improving the English. F. Baldovin, M. Bécoulet, D. Bénisti, N. Bian, A. Boozer, P. Diamond, M.-C. Firpo, M. Henneaux, T. Mendonça, B. Momo, K. Razumova, S. Ruffo, M. Valisa, and F. Zolla are thanked for very useful comments and new references. I thank D. Guyomarc'h for drawing all the figures. My thanks go to M. Farge who pointed out to me reference [110]. The topic of my talk at Chaos, Complexity and Transport 2011 was about the description of self-consistent wave–particle interaction with a finite-dimensional Hamiltonian described in Sect. 4.3. However, two seminars I gave later on in the north and south campuses of Marseilles were the occasion to start developing the ideas of Sect. 4.2, in kind of an

[45]"Qui autem omnia, quae ad cultum deorum pertinerent, diligenter retractarent et tamquam relegerent, sunt dicti religiosi ex relegendo, ut elegantes ex eligendo, ex diligendo diligentes, ex intellegendo intellegentes; his enim in verbis omnibus inest vis legendi eadem quae in religioso." Cicero, De Natura Deorum, 2, 28. English translation [31]: "Those on the other hand who carefully reviewed and so to speak retraced all the lore of ritual were called 'religious' from relegere (to retrace or re-read), like 'elegant' from eligere (to select), 'diligent' from diligere (to care for), 'intelligent' from 'intellegere' (to understand) ; for all these words contain the same sense of "picking out" (legere) that is present in 'religious'."

echo to Sect. 4.3. I thank the organizers of the conference for allowing me to extend the topic of my chapter beyond the original contents of my talk and to further develop my thoughts about plasma complexity and the way to tackle it.

4.5 Appendix 1: Extended Summary

The introduction recalls what are plasmas and provides a definition of complexity relevant to plasma physics. The chapter has two main parts. The first one is subjective and aims at favoring a brainstorming in the plasma community. It discusses the present theoretical description of plasmas, with a focus on hot weakly collisional plasmas. One of the purposes of this paper is to stop and to look backward to proceed better ahead. How do we work? How could our community improve its methodology? The **first part** of this chapter (Sect. 4.2) is made up of two subparts. The first one (Sect. 4.2.1) deals with the present status of this description, while the second one (Sect. 4.2.2) considers possible methodological improvements, some of them specific to plasma physics, but many may be of possible interest for other fields of science. The **second part** of this chapter (Sect. 4.3) is devoted to one instance where modern nonlinear dynamics and chaos helped revisiting and unifying the overall presentation of a paradigm of wave–particle interaction in plasmas.

Section 4.2.1, devoted to the present status of the description of plasma complexity, first recalls the path used for training students to this complexity. Then it recalls that most models used in plasma physics, even the Vlasov equation, have feet of clay, since they cannot be derived in an axiomatic way from first principles with conditions of validity suited to their actual applications. Each plasma physicist is shown to elaborate his own global view about plasma physics from many models which do not have any strict hierarchy. A principle of simplicity (Occam's razor principle) dominates the modeling activity. The validation of assumptions turns out to be more difficult for a complex system than for a simple one, because of the lack of information about it. In agreement with Popper's paradigm at any moment the description of plasma complexity is provisional. It results from a collective and somewhat unconscious process. This makes changing views more difficult. Numerical simulations are discussed as a complex tool to face complexity. However, the complexity they describe is still much smaller than in actual experiments, they often come without error bars on their predictions, the numerical coding of an analytical model often involves many uncontrolled approximations, and the role of intentionality is higher than in analytical calculations because of the choice of initial conditions and of parameters. Examples are provided at the various steps of this section.

Section 4.2.2, devoted to possible methodological improvements, motivates them by stating difficulties faced by plasma physicists, like information retrieval, the inflation of publications, and the growing importance of oriented programs. Working on complex systems is a hard task, but the present trend of scientific practice makes it even harder. Therefore plasma physicists would gain very much in any improvement of this practice and might be motivated into impelling a change. The proposals for improving the present situation go along the following lines: improving the ways papers are structured, improving the way scientific quality is assessed in the referral process, developing new databases, stimulating the scientific discussion of published results, diversifying the way results are made available, assessing more quality than quantity, and making available an incompressible time for creative thinking and non-purpose-oriented research. Some possible improvements for teaching are also indicated.

The suggested improvement to the structure of papers is the following: each paper, even letters, would have a "claim section" being a kind of executive summary. It would summarize the main results and their most relevant connection to previous literature. It would provide a clear information about the importance, the originality, the actual scientific contribution of the paper and about the "precedents, sources, and context of the reported work" as worded in the APS guidelines for professional conduct. Salient figures or formulas would be set there to support the claims. This procedure would improve the clarity of the papers by driving an author to state the essence of his results in a more accessible way and without having to care about the literary constraints of a normal text.

With this tool, the referral process might be improved by requiring referees to check the claims of the claim section and to motivate their possible disagreements with any of them. This procedure should make the referral process more scientifically rigorous, more ethical, and faster. Editors would benefit from a better refereeing process, which would avoid many authors' complaints, while accelerating the editorial process. Journals would benefit from the increased clarity of the contents of their published papers. The procedure might start with an experimental stage where the claim section would be optional for the authors, but not for the referees if the claim section is available.

The claim sections might be set by each scientific journals or publisher into a new dedicated database accessible through the Internet where cross-referenced papers would be hyperlinked. This would provide a new technique for data retrieval adapted to plasma complexity. It would ease the assessment of the state of the art of a given topic, with respect to what is available through present bibliographical databases.

One might consider broadening the way papers are commented in journals by adding a first friendlier step where a direct contact with the authors would lead to publishing a common short corrective communication, naturally linked to the original claim section of the original paper. A classical comment to the journal would be sent only if the authors could not agree about a common view.

As to possible improvements for teaching, student should be made conscious about the limited capabilities of models. One may avoid teaching many calculations to start with, but keep the physical ideas and the corresponding images. Nonlinear dynamics and chaos might provide a way to revisit and unify separated chapters, e.g., turbulent and "collisional transport," the calculation of magnetic field lines, or the introduction of fluid and of Vlasov equations.

The **second part** of this chapter (Sect. 4.3) is more specialized, and is a scientific presentation of a theoretical approach avoiding several shortcomings of the Vlasovian approach. It deals with Langmuir wave–electron interaction in one-dimensional plasmas. This topic is tackled by describing plasma dynamics with a finite-dimensional Hamiltonian system coupling N particles with M waves, the self-consistent Hamiltonian. This enables recovering Vlasovian linear theory with a mechanical understanding. In particular, the reason why Landau damping cannot be an eigenmode is shown to be rooted deeply in Hamiltonian mechanics. This damping is recovered as an analogue of van Kampen phase-mixing effect. This phase mixing in turn plays an essential role in the calculation of Landau instability. The self-consistent dynamics reveals that both Landau damping and instability result from the same synchronization mechanism of particles with waves.

The quasilinear description of the weak warm beam, or bump-on-tail, instability is then recalled, together with the apparent paradox that its predictions look correct while its assumptions are proved to be wrong. A recent analytical result shows that self-consistency vanishes when the plateau forms in the tail distribution function: the wave spectrum is frozen. This leads to consider the dynamics of particles in a frozen spectrum of waves with random phases. It involves a fundamental time-scale, the spreading time τ_{spread} after which the positions of particles are spread by a typical wavelength of the waves with respect to their ballistic values. Till a time at most τ_{spread}, particles feel the global force due to the waves as a stochastic force, and their velocities diffuse in a quasilinear way. If their dynamics is not chaotic, they eventually feel the quasiperiodic nature of the force, and diffusion stops. If their dynamics is chaotic, after a time τ_{spread}, they experience a chaotic diffusion that may be superquasilinear by a factor 2.3, but which becomes quasilinear in the limit of strong resonance overlap. The latter result is understood as a consequence of the weak dependence of the particle dynamics over any two phases over a time much larger than τ_{spread}.

Then is recalled a recent numerical simulation of the bump-on-tail instability aiming at checking whether diffusion is quasilinear when the plateau sets in and proving that it is indeed. It also brought the unexpected result that the variation of the phase of a given wave with time is almost not fluctuating for random realizations of the initial wave phases. This was an incentive to undertake new analytical calculations of the average behavior of the self-consistent dynamics when the initial wave phases are random. Using Picard iteration technique, they show that the modification of the average wave frequency due to the coupling with particles is exactly the principal part correction to the wave frequency provided by the Vlasovian calculation of the dispersion relation of Langmuir waves or by the equivalent calculations with the self-consistent Hamiltonian [46]. However the latter calculations deal with a spatially uniform distribution of particles, while the present one holds whatever the spatial inhomogeneity of the distribution of tail particles. An estimate of phase fluctuations shows they are negligible, confirming simulation results. The evolution of the wave amplitude involves the Landau effect and spontaneous emission, as already found for a spatially uniform distribution of particles and a "spontaneous emission" of spatial inhomogeneities.

4.6 Appendix 2: First Example of a Claim Section

Here is proposed a claim section for paper [55] quoted in Sect. 4.3.4. The title of the paper is "Proof of quasilinear equations in the chaotic regime of the weak warm beam instability," and its abstract is "The diffusion coefficient is proved rigorously to take on the quasilinear value for the chaotic motion of an electron in a prescribed set of strongly overlapping Langmuir waves with random phases. Natural approximations show this result to extend to the self-consistent chaotic motion of many particles in a set of many Langmuir waves corresponding to the weak warm beam instability. The weak influence of any particle on any wave and vice-versa is an essential ingredient of the derivation. Wave–particle momentum conservation implies the Landau growth rate to be related to the quasilinear diffusion coefficient." A possible claim section follows.

One considers the one-dimensional chaotic motion of an electron in a prescribed set of $M \gg 1$ strongly overlapping Langmuir waves with random phases and a regular enough spectrum. Let k be the typical wavenumber of a wave, q_0 and p_0 be the initial particle position and velocity, and $q(t)$ its position at time t. Let $\Delta q(t) = q(t) - q_0 - p_0 t$.

Claim 1: The variation of $k\Delta q(t)$ with any two phases stays small with respect to 2π over a time on the order of $\tau_{\text{QL}} = \tau_{\text{spread}} \ln(s)$, where $s \gg 1$ is the typical resonance overlap parameter of two nearby waves in the spectrum, and $\tau_{\text{spread}} = 4(k^2 D_{\text{QL}})^{-1/3}$, with D_{QL} the typical value of the quasilinear diffusion coefficient.

Claim 2: This implies the particle diffuses in a quasilinear way up to a time on the order of τ_{QL}.

Claim 3: The latter property implies the particle diffuses in the same way for larger times.

Claim 4: The same property holds for the self-consistent dynamics defined by Hamiltonian Eq. (4.1) provided the wave spectrum be regular enough too.

Most relevant connection to previous literature: [10, 29, 39, 87, 118].

4.7 Appendix 3: Second Example of a Claim Section

Here is proposed a claim section for paper [58] quoted in Sect. 4.2.1.4. The title of the paper is "Calculation of transport coefficient profiles in modulation experiments as an inverse problem," and its abstract is "The calculation of transport profiles from experimental measurements belongs in the category of inverse problems which are known to come with issues of ill-conditioning or singularity. A reformulation of the calculation, the matricial approach, is proposed for periodically modulated experiments, within the context of the standard advection-diffusion model where these issues are related to the vanishing of the determinant of a 2×2 matrix. This sheds light on the accuracy of calculations with transport codes, and provides a path for a more precise assessment of the profiles and of the related uncertainty." A possible claim section follows.

One applies the advection–diffusion model for the quantity $\zeta(r,t)$

$$\begin{aligned} \partial_t \zeta &= -\nabla \cdot \Gamma(\zeta) + S \\ \Gamma &= -\chi \nabla \zeta + V \zeta \end{aligned} \tag{4.11}$$

to modulation experiments. One considers cylindrical symmetry and a purely sinusoidal forcing term S with pulsation ω.

Claim 1: Decomposing the signal ζ into a real amplitude and phase, $\zeta = Ae^{i\phi}$, and S as $S = S_r + iS_i$ yields

$$\begin{aligned} &\mathbf{M} \cdot \mathbf{Y} = \mathbf{G} \\ &\mathbf{Y} = \begin{pmatrix} \chi \\ V \end{pmatrix}, \quad \mathbf{M} = \begin{bmatrix} -A' \cos\phi + A\phi' \sin\phi & A\cos\phi \\ -A' \sin\phi - A\phi' \cos\phi & A\sin\phi \end{bmatrix}, \\ &\mathbf{G} = \begin{bmatrix} \frac{1}{r}\int_0^r dz z (S_r(z) - \omega A(z) \sin\phi(z)) \\ \frac{1}{r}\int_0^r dz z (S_i(z) + \omega A(z) \cos\phi(z)) \end{bmatrix}, \end{aligned} \tag{4.12}$$

where the primes stand for differentiation with respect to r and where all quantities in the l.h.s. of the first equation are computed at radius r.

Claim 2: On this basis, and with a controllable smoothing of the experimental data, the profile of transport coefficients is computed by inverting matrix $\mathbf{M}(r)$ at each measurement point.

Claim 3: This method enables a precise estimate of the uncertainty on the transport coefficients from that on the measurements at each measurement point.

Claim 4: The smaller the uncertainty on the estimate of the derivatives of A and ϕ, the larger the precision in the reconstruction of transport profiles.

Claim 5: At a given r , the smaller the absolute value of an eigenvalue, the larger the uncertainty of $\mathbf{Y}(r)$ along the corresponding eigenvector of matrix $\mathbf{M}(r)$ for a given uncertainty on measured data for all radii.

Claim 6: This method is lighter computationally than classical transport codes.

Claim 7: The reconstruction radius by radius enables to see how different the uncertainties are over $\mathbf{Y}(r)$ as a function of r.

Claim 8: This uncertainty is larger in the regions where sources or sinks are present.

Claim 9: In contrast with transport codes, this method requires a single boundary condition only.

Most relevant connection to previous literature: [111].

References

1. M. Antoni, Y. Elskens, D.F. Escande, Phys. Plasmas **5**, 841 (1998)
2. N. Arcis, D.F. Escande, M. Ottaviani, Phys. Lett. A **347**, 241 (2005)
3. N. Arcis, D.F. Escande, M. Ottaviani, Phys. Plasmas **13**, 052305 (2006)
4. S.D. Baalrud, J.D. Callen, C.C. Hegna, Phys. Plasmas **15**, 092111 (2008)
5. S.D. Baalrud, J.D. Callen, C.C. Hegna, Phys. Plasmas **17**, 055704 (2010)
6. D.R. Baker, N.R. Ahern, A.Y. Wong, Phys. Rev. Lett., **20**, 318 (1968)
7. R. Balecu, *Statistical Dynamics Matter Out of Equilibrium* (World scientific, Singapore, 1993)
8. J. Barré et al., Phys. Rev. E **69**, 045501 (2004)
9. J. Barré et al., J. Stat. Phys. **119**, 677 (2005)
10. D. Bénisti, D.F. Escande, Phys. Plasmas **4**, 1576 (1997)
11. D. Bénisti, L. Gremillet, Phys. Plasmas **14**, 042304 (2007)
12. D. Bénisti, D.J. Strozzi, L. Gremillet, Phys. Plasmas **15**, 030701 (2008)
13. D. Bénisti, O. Morice, L. Gremillet, D.J. Strozzi, Transport Theory Statist. Phys. **40**, 185 (2011)
14. I.B. Bernstein, J.M. Greene, M.D. Kruskal, Phys. Rev. **108**, 546 (1957)
15. N. Besse, Y. Elskens, D.F. Escande, P. Bertrand, Plasma Phys. Control. Fusion **53**, 025012 (2011)
16. N. Besse, Y. Elskens, D.F. Escande, P. Bertrand, in *Proceedings of 38th EPS Conference on Controlled Fusion and Plasma Physics*, Strasbourg, 2011, P2.009
17. D.D. Biskamp, *Nonlinear Magnetohydrodynamics* (Cambridge University Press, Cambridge, 1993)
18. D. Bonfiglio, S. Cappello, D.F. Escande, Phys. Rev. Lett. **94**, 145001 (2005)
19. D. Bonfiglio, D.F. Escande, P. Zanca, S. Cappello, Nucl. Fusion **51**, 063016 (2011)
20. W.J. Bos, R. Rubinstein, L. Fang, http://hal.archives-ouvertes.fr/hal-00605446/en/
21. A. Campa, T. Dauxois, S. Ruffo, Phys. Rep. **480**, 57 (2009)
22. S. Cappello, D. Biskamp, *Proceedings of International Conference on Plasma Physics*, vol. 1, Nagoya, 1996, p. 854
23. S. Cappello, D.F. Escande, Phys. Rev. Lett. **85**, 3838 (2000)
24. S. Cappello, R. Paccagnella, in *Proceedings of Workshop on Theory of Fusion Plasmas*, ed. by E. Sindoni (Compositori, Bologna, 1990), p. 595
25. S. Cappello, R. Paccagnella, Phys. Fluids **B4**, 611 (1992)
26. S. Cappello et al., Nucl. Fusion **51**, 103012 (2011)
27. S. Cappello et al., *Theory of Fusion Plasmas*. AIP Conference Proceedings, vol. 1069 (2008), p. 27
28. J.R. Cary, R. Littlejohn, Ann. Phys. **151**, 1 (1983)
29. J.R. Cary, D.F. Escande, A.D. Verga, Phys. Rev. Lett. **65**, 3132 (1990)
30. B.V. Chirikov, Phys. Rep. **52**, 263 (1979)
31. M.T. Cicero, *De Natura Deorum* (De Natura Deorum Academica with an English Translation by H. Rackham). (Harvard University press, Cambridge, 1967), p. 193, http://ia600302.us.archive.org/27/items/denaturadeorumac00ciceuoft/denaturadeorumac00ciceuoft.pdf
32. T. Dauxois, S. Ruffo, L.F. Cugliandolo (ed.), *Long-Range Interacting Systems* (Oxford University Press, Oxford, 2010)
33. L. de Broglie, *Revolution in Physics* (Routledge and Kegan Paul, London, 1954)
34. D. del-Castillo-Negrete, in *Turbulent Transport in Fusion Plasmas: Proceedings of the First ITER Summer School*, ed. by S. Benkadda. AIP Conference Proceedings, vol. 1013 (American Institute of Physics, Meville, 2008), p. 207
35. D. del-Castillo-Negrete, Nonlinear process. Geophys. **17**, 795 (2010)
36. F. Doveil, Y. Vosluisant, S.I. Tsunoda, Phys. Rev. Lett. **69**, 2074 (1992)
37. F. Doveil, D.F. Escande, A. Macor, Phys. Rev. Lett. **94**, 085003 (2005)
38. F. Doveil et al., Phys. Plasmas **12**, 010702 (2005)

39. W.E. Drummond, D. Pines, Nucl. Fusion Suppl. **3**, 1049 (1962)
40. K. Elsässer, Plasma Phys. Control. Fusion **28**, 1743 (1986)
41. Y. Elskens, ESAIM Proc., ed. by F. Coquel, S. Cordier, **10**, 211 (2001), http://www.emath.fr/Maths/Proc
42. Y. Elskens, Commun. Nonlinear Sci. Numer. Simul. **15**, 10 (2010)
43. Y. Elskens, *J. Stat. Phys.* **148**, 591 (2012)
44. Y. Elskens, D.F. Escande, Nonlinearity **4**, 615 (1991)
45. Y. Elskens, D.F. Escande, Physica D **62**, 66 (1993)
46. Y. Elskens, D.F. Escande, *Microscopic Dynamics of Plasmas and Chaos* (IoP, Bristol, 2003)
47. Y. Elskens, E. Pardoux, Ann. Appl. Prob. **20**, 2022 (2010)
48. D.F. Escande, Phys. Rev. Lett. **35**, 995 (1975)
49. D.F. Escande, J.P.M. Schmitt, Phys. Fluids **19**, 1757 (1976).
50. D.F. Escande, Phys. Rep. **121**, 165 (1985)
51. D.F. Escande, in *Large Scale Structures in Nonlinear Physics*, ed. by J.D. Fournier P.L. Sulem. Lecture Notes in Physics, vol. 392 (Springer, New york, 1991), p. 73
52. D.F. Escande, in *Long-Range Interacting Systems*, ed. by T. Dauxois, S. Ruffo, L.F. Cugliandolo (Oxford University Press, Oxford, 2010), p. 469
53. D.F. Escande, *Proceedings of International Symposium on Waves, Coherent, Structures and Turbulence in Plasmas*. AIP Conference Proceedings, vol. 1308, 2010, p. 85
54. D.F. Escande, F. Doveil, J. Stat. Phys. **26**, 257 (1981)
55. D.F. Escande, Y. Elskens, Phys. Lett. A **302**, 110 (2002)
56. D.F. Escande, M. Ottaviani, Phys. Lett. A **323**, 278 (2004)
57. D.F. Escande, F. Sattin, Phys. Rev. Lett. **99**, 185005 (2007)
58. D.F. Escande, F. Sattin, Phys. Rev. Lett. **108**, 125007 (2012)
59. D.F. Escande, B. Souillard, Phys. Rev. Lett. **52**, 1296 (1984)
60. D.F. Escande, H. Kantz, R. Livi, S. Ruffo, J. Stat. Phys. **76**, 605 (1994)
61. D.F. Escande, S. Zekri, Y. Elskens, Phys. Plasmas **3**, 3534 (1996)
62. D.F. Escande, P. Martin, S. Ortolani et al., Phys. Rev. Lett. **85**, 1662 (2000)
63. D.F. Escande, R. Paccagnella et al., Phys. Rev. Lett. **85**, 3169 (2000)
64. W. Ettoumi, M.-C. Firpo, Phys. Rev. E **84**, 030103 (2011)
65. J.M. Finn, R.A. Nebel, C.C. Bathke, Phys. Fluids **B4**, 1262 (1992)
66. M.-C. Firpo, Y. Elskens, Phys. Rev. Lett. **84**, 3318 (2000)
67. M.-C. Firpo et al., Phys. Rev. E **64**, 026407 (2001)
68. M.-C. Firpo, F. Leyvraz, G. Attuel, Phys. Plasmas **13**, 122302 (2006)
69. J.P. Freidberg, *Ideal Magneto-Hydro-Dynamics* (Plenum Press, New York, 1987)
70. B. Gallet et al., Phys. Rev. Lett. **108**, 144501 (2012)
71. S. Gasiorowicz, M. Neuman, R.J. Riddell, Phys. Rev. **101**, 922 (1956)
72. M. Gobbin et al., Phys. Rev. Lett. **106**, 025001 (2011)
73. M. Greenwald, Plasma Phys. Control Fusion **44**, R27 (2002)
74. R.J. Hastie, F. Militello, F. Porcelli, Phys. Rev. Lett. **95**, 065001 (2005)
75. R.D. Hazeltine, J.D. Meiss, *Plasma Confinement* (Dover Press, Mineola, 2003)
76. R.D. Hazeltine, F.L. Waelbroeck, *The Framework of Plasma Physics* (Westview Press, Boulder, 2004)
77. T.C. Hender et al., Nucl. Fusion **47**, S128 (2007)
78. W. Horton, Rev. Mod. Phys. **71**, 735 (1999)
79. K. Ikeda et al., Nucl. Fusion **47**, S1 (2007)
80. B.B. Kadomtsev, *Tokamak Plasma: A Complex Physical System* (IOP, Bristol, 1992), p. 3
81. N.G. van Kampen, Physica **21**, 949 (1955)
82. M.K.H. Kiessling, T. Neukirch, Proc. Natl. Acad. Sci. **100**, 1510 (2003)
83. C. Krafft, A. Volokitin, A. Zaslavsky, Phys. Rev. E **82**, 066402 (2010)
84. R. Kraichnan, R. Panda, Phys. Fluids **31**, 2395 (1988)
85. T.S. Kuhn, *The Structure of Scientific Revolutions*, 1st edn, 1962. (University of Chicago Press, Chicago, 1996)

86. L.D. Landau, Zh. Eksp. Teor. Fiz. **16** 574 (1946); translation J. Phys. USSR **10**, 25 (1946); reprinted in *Collected Papers of Landau*, ed. by D. ter Haar (Pergamon, Oxford, 1965)
87. G. Laval, D. Pesme, Phys. Rev. Lett. **53**, 270 (1984)
88. Y.-M. Liang, P.H. Diamond, Comments Plasma Phys. Control. Fusion **15**, 139 (1993)
89. Y.-M. Liang, P.H. Diamond, Phys. Fluids **B5**, 4333 (1993)
90. R. Lorenzini et al., Nature Phys. **5**, 570 (2009)
91. J.H. Malmberg, C.B. Wharton, Phys. Rev. Lett. **13** 184 (1964)
92. F. Militello, F. Porcelli, Phys. Plasmas **11**, L13 (2004)
93. F. Militello, R.J. Hastie, F. Porcelli, Phys. Plasmas **13**, 112512 (2006)
94. R. Monchaux et al., Phys. Fluids **21**, 035108 (2009)
95. C. Mouhot, C. Villani, J. Math. Phys. **51**, 015204 (2010)
96. I.N. Onishchenko et al., ZhETF Pis. Red. **12**, 407 (1970); translation JETP Lett. **12**, 281 (1970)
97. T.M. O'Neil, Phys. Fluids **8**, 2255 (1965)
98. T.M. O'Neil, J.H. Winfrey, J.H. Malmberg, Phys. Fluids **4**, 1204 (1971)
99. S. Ortolani, D.D. Schnack, *Magnetohydrodynamics of Plasma Relaxation* (World Scientific, Singapore, 1993)
100. A.G. Peeters et al., Nucl. Fusion **51**, 094027 (2011)
101. F.W. Perkins et al., Nucl. Fusion **39**, 2137 (1999)
102. E. Piña, T. Ortiz, J. Phys. A: Math. Gen. **21**, 1293 (1988)
103. K.R. Popper, *The Logic of Scientific Discovery* (Routledge, London, 2002)
104. M.E. Puiatti et al., Phys. Plasmas **16**, 012505 (2009)
105. M.E. Puiatti et al., Nucl. Fusion **49**, 045012 (2009)
106. C. Roberson, K.W. Gentle, Phys. Fluids **14**, 2462 (1971)
107. Yu.A. Romanov, G.F. Filippov, Zh. Eksp. Theor. Phys. **40**, 123 (1961); translation Soviet Phys. JETP **13**, 87 (1961)
108. M.N. Rosenbluth, W.M. MacDonald, D.L. Judd, Phys. Rev. **107**, 1 (1957)
109. M.G. Rusbridge, Plasma Phys. Control Fusion **33**, 1381 (1991)
110. B. Russell, *The Scientific Outlook*, part I, chapter II. (George Allen and Unwin, London, 1931)
111. F. Ryter, R. Dux, P. Mantica, T. Tala, Plasma Phys. Control Fusion **52**, 124043 (2010)
112. P.B. Snyder et al., Nucl. Fusion **47**, 961 (2007)
113. H. Spohn, *Large Scale Dynamics of Interacting Particles* (Springer, Berlin, 1991)
114. J.B. Taylor, Phys. Rev. Lett. **33**, 1139 (1974)
115. J.L. Tennyson, J.D. Meiss, P.J. Morrison, Physica D **71**, 1 (1994)
116. P.W. Terry et al., Phys. Plasmas **15**, 062503 (2008)
117. S.I. Tsunoda, F. Doveil, J.H. Malmberg, Phys. Fluids **B3**, 2747 (1991)
118. A.A. Vedenov, E.P. Velikhov, R.Z. Sagdeev, Nucl. Fusion Suppl. **2**, 465 (1962)
119. J.A. Wesson, *Tokamaks* (Oxford University press, Oxford, 2004)
120. P.H. Yoon, T. Rhee, C.-M. Ryu, Phys. Rev. Lett. **95**, 215003 (2005)
121. L.F. Ziebell, P.H. Yoon, J. Pavan, R. Gaelzer, Astrophys. J. **727**, 16 (2011)
122. http://slow-science.org/
123. http://www.aps.org/policy/statements/02_2.cfm
124. http://public.lanl.gov/alp/plasma/people/alfven.html
125. http://www.iter.org/proj/iterhistory
126. http://www.ias.edu/about/mission-and-history
127. http://www.articleofthefuture.com/?utm_source=ESJ001&utm_campaign=&utm_content=&utm_medium=email&bid=UA81L4F:XXV56
128. http://www.quantiki.org/wiki/Quantiki:About

Chapter 5
First Principle Transport Modeling in Fusion Plasmas: Critical Issues for ITER

Yanick Sarazin

Abstract Tokamaks aim at confining hot plasmas by means of strong magnetic fields in view of reaching a net energy gain through fusion reactions. Plasma confinement turns out to be governed by small-scale instabilities which saturate nonlinearly and lead to turbulent fluctuations of a few percent. This paper recalls the basic equations for modeling such weakly collisional plasmas. It essentially relies on the kinetic, or more precisely the gyrokinetic, description, although some attempts are made to incorporate some of the kinetic properties, namely, wave-particle resonances, in fluid models by means of collisionless closures. Three main types of micro-instabilities are detailed and studied linearly, namely, drift waves, interchange, and bump-on-tail. Finally, some of the main critical issues in turbulence modeling are addressed: flux-driven versus gradient-driven models, the subsequent impact of mean profile relaxation on turbulent transport dynamics, and the role of large-scale flows, either at equilibrium or turbulence driven, on turbulence saturation and on the possible triggering of transport barriers. The significant progress in understanding and prediction of turbulent transport in tokamak plasmas thanks to first-principle simulations is highlighted.

5.1 Transport Issues in Controlled Fusion Devices

5.1.1 Magnetic Configuration and Main Plasma Parameters

Controlled magnetic fusion devices such as tokamaks or stellarators aim at harnessing fusion energy in view of electricity production by means of intense magnetic fields of several Teslas. When experiencing a fusion reaction, deuterium and

Y. Sarazin (✉)
CEA, IRFM, F-13108 Saint-Paul-Lez-Durance, France
e-mail: yanick.sarazin@cea.fr

X. Leoncini and M. Leonetti (eds.), *From Hamiltonian Chaos to Complex Systems: A Nonlinear Physics Approach*, Nonlinear Systems and Complexity 5,
DOI 10.1007/978-1-4614-6962-9_5,

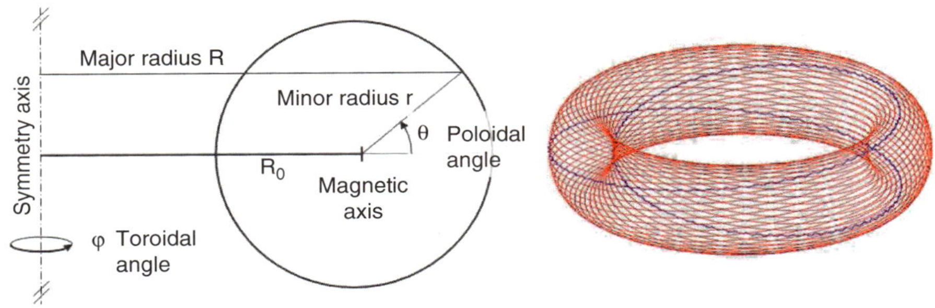

Fig. 5.1 *Left*: Elements of geometry in fusion toroidal devices. *Right*: Schematic magnetic configuration, showing field lines with $q = 2$

tritium nuclei produce a neutron at 14.06 MeV and a helium nucleus at 3.52 MeV. The associated reaction rate $\langle \sigma v \rangle_{DT}$ exhibits a sharp threshold in temperature: it drastically drops down below $T \approx 10\,\text{keV}$ and exhibits a maximum at a few tens of keV. At the envisaged working temperatures of order of 20 keV, fusion reactions are largely dominated by quantum physics, namely, tunnel effect.

The magnetic equilibrium in tokamaks is made of two components of the magnetic field. The toroidal magnetic field B_T is generated by N external poloidal coils, in which a current I of the order of one mega-Ampere circulates. From the Maxwell-Ampere equation, it follows that B_T decreases with the major radius R: $B_T = \mu_0 NI/2\pi R$. The poloidal field B_P is generated by the toroidal plasma current I_p. This latter current is inductively generated by varying the magnetic flux in the central solenoid, the plasma being the secondary of a transformer. The resulting magnetic field lines are helices which generate nested closed magnetic surfaces with a torus-like shape. The number of toroidal turns per poloidal turn defines the safety factor q, which usually ranges between 1 in the plasma core and a few units at the edge (Fig. 5.1).

Up to trapping processes due to electromagnetic mirror effects, ions and electrons move almost freely in the parallel direction, along the magnetic field lines. Their thermal speed is of the order of $v_{Ti} \approx 5.10^5\,\text{m.s}^{-1}$ for deuterium ions and $v_{Te} \approx 3.10^7\,\text{m.s}^{-1}$ for electrons. Conversely, should collisions and turbulence be negligible, their transverse motion would be limited to a few gyroradii ($\rho_i = m_i v_{\perp i}/eB \approx 3.10^{-3}\,\text{m}$ and $\rho_e \approx 5.10^{-5}\,\text{m}$) from their reference magnetic surface. The fast parallel motion leads to rapid homogenization of plasma characteristics on magnetic surfaces. This property allows one to define radial profiles of the physical quantities (density, pressure, current, *etc.*), which correspond to flux surface averages. The radial direction, transverse to the magnetic flux surfaces, defines the direction of the confinement.

Plasma densities in controlled fusion devices are necessarily low. An upper bound comes from the fact that the plasma beta $\beta = nT/(B^2/2\mu_0)$, the ratio of kinetic energy over magnetic energy, has to be smaller than unity for magnetic confinement to be effective. In practice, large-scale magnetohydrodynamical instabilities limit β to a few percent, leading to densities of the order of $n \approx 10^{20}\,\text{m}^{-3}$.

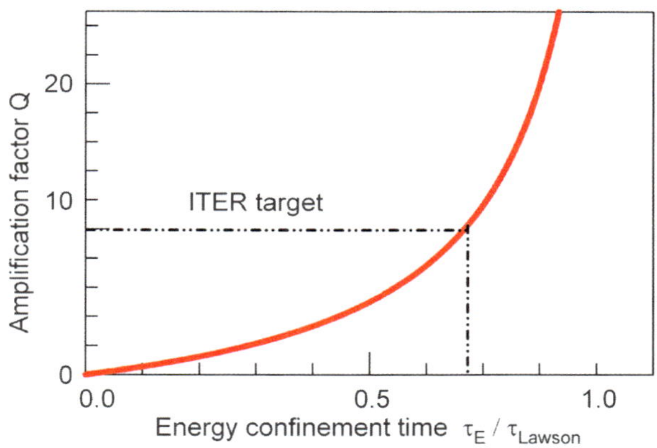

Fig. 5.2 Amplification factor Q as a function of the energy confinement time τ_E, normalized to the Lawson value

At such high temperatures and low densities, fusion plasmas are weakly collisional. Indeed, Coulomb collision frequencies scale like $\nu_{coll} \sim nT^{-3/2}$. It follows that the characteristic mean free path of both ion and electrons is of order of $\lambda_{mfp} = v_T/\nu_{coll} \approx 10\,\text{km}$, which translates into several hundreds of toroidal turns. In that respect, core fusion plasmas are almost collisionless.

5.1.2 Transport and Fusion Performance

In open systems such as controlled fusion plasmas, transport results from the existence of various sources of particle, momentum, and heat (at least) which drive the plasma out of thermodynamical equilibrium. Transport processes then allow the system to reaching statistical steady states, characterized by well-defined temporal means of density, temperature, *etc.* They allow heat and particles which are deposited in the core of the discharge to be expelled towards the "exterior," namely, the plasma-facing components [1].

Given the already mentioned constraints on fusion plasma densities and temperatures, heat transport largely governs the fusion performance in terms of quality factor Q. Indeed, the ratio of the fusion power over the additional power required to reach fusion reaction conditions $Q = P_{fus}/P_{add}$ strongly depends on the energy confinement time τ_E, as highlighted in Fig. 5.2:

$$Q \simeq \frac{k}{\frac{\tau_{Lawson}}{\tau_E} - 1} \tag{5.1}$$

where $\tau_{Lawson} \equiv \lambda knTV/P_{fus}$ stands for the Lawson time corresponding to self-sustained fusion, also called *ignition*. Here n, T, and V are the density, the temperature, and the volume of the plasma. k^{-1} is the fraction of P_{fus} which is directly reabsorbed by the plasma and λ some constant depending on the proportion of the main plasma species, namely, fusion ions and electrons. k and λ are of the order of 5 and 3, respectively, for deuterium-tritium homogeneous plasmas. If the confinement remains poor (small τ_E), Q increases linearly with τ_E. Conversely, close to the Lawson criterion $\tau_E \sim \tau_{Lawson}$, Q tends to infinity. In order to achieve $Q \approx 10$ in performing Iter plasmas, the confinement time should be of the order of a few seconds.

5.1.3 Transport and Turbulence

Although weak, binary collisions lead to some cross-field (radial) transport. The collisional diffusivity χ_{coll} scales like $\rho_s^2 \nu_{coll}$, s standing for the species. The proportionality factor is larger than one and depends on the collisionality regime, i.e., the ratio of ν_{coll} over some transit frequency. It is equal to $\varepsilon^{-3/2} q^2$ in the weakly collisional regime (so-called banana regime) and to q^2 in the "highly" collisional regime (so-called Pfirsch-Schlütter regime). Here, the small ε parameter is the ratio of the minor over the major radius $\varepsilon = r/R$. Core fusion plasmas are usually in the banana regime, for which ion diffusivity is of order of $10^{-2} - 10^{-1}\,\mathrm{m}^2.\mathrm{s}^{-1}$. It is one to two orders of magnitudes smaller than the experimentally measured transport coefficient, of order of a few square meters per second.

It turns out that heat transport is dominated by turbulence in core fusion plasmas. Small-amplitude fluctuations—of order of a few percent— develop due to primary instabilities (see Sect. 5.3). Especially, electric potential fluctuations lead to cross-field radial electric drift v_{Er} (see Sect. 5.2.2), which governs cross-field transport. The turbulent effective diffusivity can be approximated by $\langle v_{Er}^2 \rangle \tau_c$, where τ_c stands for turbulence correlation time and the brackets for statistical or time average. Experimental measurements or estimates of these quantities lead to turbulent diffusivity of order of $1\ m^2.s^{-1}$, in rough agreement with experimental values. Other evidences (such as the development of transport barriers in regions where turbulent fluctuations drop off) show that turbulence is indeed mainly responsible for cross-field transport in tokamak plasmas and ultimately governs the energy confinement time τ_E.

5.2 Turbulence Modeling: The Need for a kinetic Description

Predicting the turbulent transport level in fusion plasmas by means of first-principle simulations consists in solving self-consistently the dynamics of both particles and waves, namely, the fluctuations of the electromagnetic potentials $\delta\phi$ and $\delta\mathbf{A}$. The first part of the loop is well defined: Maxwell's equations relate the electromagnetic field to the charge and current densities of particles. As far as the plasma response to these fields is concerned, the two main levels of descriptions that can be distinguished (if one already excludes particle description) are:

- **Kinetic description** consists in solving Vlasov equation (or Fokker-Planck if collisions are retained). The number of degree of freedom is *a priori* 6 per plasma species, namely, electrons, main ions, and impurities. This number can be further reduced by considering the adiabatic limit (see Sect. 5.2.2). When averaging over the fast cyclotron timescales, the phase space is reduced down to 5 dimensions, with the result that the magnetic moment μ becomes a motion invariant. This corresponds to the so-called gyrokinetic theory.
- **Fluid description** relies on the moments of the distribution function, obtained by considering the velocity integral of f, weighted by the velocity to a certain power. This route is more tractable than the gyrokinetic approach, since the number of degree of freedom is 3 per plasma species. However, it suffers from critical drawbacks. First, it cannot properly account for the interactions between waves and particles, as long as they are resonant in the velocity space. This reveals critical for hot plasmas, where collisions remain negligible. Second, fluid equations hardly account for the various classes of particles, namely, trapped, passing, and suprathermal (e.g., helium ashes) particles. Third, some closure assumption has to be made so as to truncate the infinite hierarchy of fluid equations. Such a closure problem remains an open issue in core fusion plasmas, which are almost collisionless.

Given the large numerical resources required by the gyrokinetic approach, the fluid description still remains of valuable interest. Comparing gyrokinetic and fluid simulations of turbulent transport allows one to improve the present fluid closures. One can basically distinguish two classes of collisionless fluid closures:

- Those aiming at incorporating part of the linear or quasi-linear kinetic properties of the system in the fluid model. Such works were initiated by Hammett and Perkins in 1990 [2] and further developed and improved by several authors afterwards (see, e.g., [3–7]).
- Those pursuing another projection than that of the fluid moments. In this framework, the concept of water bags [8, 9] provides a powerful alternative. It consists in using the intrinsic property of the Vlasov equation, namely, that the distribution function f remains constant along the characteristics. The water bag model aims at bridging the gap between the collisionless kinetic description and the multi-fluid approach.

5.2.1 Collisionless Fluid Approaches "à la Hammett-Perkins"

In 1990, Hammett and Perkins proposed a way to account for part of the kinetic linear properties in the fluid closure [2]. In the following, their result is reformulated in a more general and systematic way (see also [10] for an alternative treatment based on the entropy production rate). The basic idea is to equal the kinetic and fluid linear response functions in the kinetic limit, namely, when the phase velocity is much smaller than the thermal velocity $\omega/\omega_{\parallel} \ll 1$, where $\omega_{\parallel} = k_{\parallel} v_T$ stands for the parallel transit frequency, with $k_{\parallel}$ the parallel wave vector. Such a limit encompasses the wave-particle resonant interactions which characterize the kinetic regime. The agreement can be rendered valid up to the $(\ell-1)^{th}$ order in the small parameter $\zeta \equiv \omega/|\omega_{\parallel}|$, where ℓ stands for the number of retained fluid moments. Here, the $(\ell+1)^{th}$ moment is assumed to depend linearly on all the lower-order moments.

So as to exemplify the method on a simple case, and following the pioneering work of Landau on the subject in 1946 [11], let's consider the collisionless dynamics of nonrelativistic electrons, embedded in a strong uniform magnetic field. The problem is assumed to be electrostatic, and the ions are at rest with a density n_0. The electron distribution function f and the electric potential ϕ then compose a self-consistent system governed by the Vlasov and Poisson equations:

$$\partial_t f + v\,\partial_x f + \partial_x \phi\,\partial_v f = 0 \tag{5.2}$$

$$\partial_x^2 \phi = \int_{-\infty}^{+\infty} f \mathrm{d}v - 1$$

Here, the radial position r along the magnetic field is normalized to the Debye length $x \equiv r/\lambda_D = r/(\varepsilon_0 T/n_e e^2)^{1/2}$, while time t is normalized to the inverse of the electron plasma frequency $t \to t\omega_{pe} = t\mathrm{v}_{Te}/\lambda_D$. Consistently, the velocity is normalized to the thermal velocity $\mathrm{v}_{Te} = (T_e/m_e)^{1/2}$: $v = \mathrm{v}/\mathrm{v}_{Te}$. Also, the distribution function is normalized to v_{Te}/n_e $(f \to f\mathrm{v}_{Te}/n_e)$ and the electric potential Φ to T_e/e: $\phi = e\Phi/T_e$. The problem is two-dimensional in phase space (x, v). We will focus on the linear characteristics of system 5.2.

In the linear regime, characterized by small-amplitude perturbations with respect to the equilibrium which will be chosen centered Maxwellian $f_{eq} = (2\pi)^{-1/2} \exp(-v^2/2)$ with $\phi_{eq} = 0$ (it can be easily generalized to a noncentered Maxwellian featuring nonvanishing mean velocity), the dispersion relation can be expressed as a function of the plasma dispersion function $Z(\zeta)$ [12]:

$$Z(\zeta) \equiv \int_{-\infty}^{+\infty} \frac{\mathrm{e}^{-v^2}}{v-\zeta} \frac{\mathrm{d}v}{\sqrt{\pi}} \tag{5.3}$$

This expression is valid for $\Im(\zeta) > 0$. The function is then analytically continued for $\Im(\zeta) \leq 0$. The linearized system then provides the relationship between modes of density and of electric potential:

$$\hat{n}_k = \left\{ 1 + \frac{\zeta}{\sqrt{2}} Z\left(\frac{\zeta}{\sqrt{2}}\right) \right\} \hat{\phi}_k \tag{5.4}$$

where $\zeta \equiv \omega/|k|$. Fourier decomposition has been used for density and potential perturbations: $(\tilde{n}, \tilde{\phi}) = \sum_{k,\omega} (\hat{n}_{k,\omega}, \hat{\phi}_{k,\omega}) \exp\{i(kx - \omega t)\}$. In the limit of small arguments $|\zeta| \ll 1$ (so-called kinetic regime), the response function then reads as follows:

$$\hat{R}_{cin} \equiv -\frac{\hat{n}_{k,\omega}}{\hat{\phi}_{k,\omega}} \simeq -1 - i\sqrt{\frac{\pi}{2}}\,\zeta + o(\zeta^2) \tag{5.5}$$

with $\zeta = \omega/|k|$. This limit, corresponding to a much smaller phase velocity of the mode than particle thermal velocity, keeps memory of the resonant nature of wave–particle interactions.

The analogous fluid system is:

$$\partial_t n + \partial_x(un) = 0 \tag{5.6}$$

$$\partial_t u + u\,\partial_x u + \frac{\partial_x p}{n} - \partial_x \phi = 0 \tag{5.7}$$

$$\partial_t p + \partial_x(up) + \partial_x q + 2p\,\partial_x u = 0 \tag{5.8}$$

where n, u, p, and q are density, flow velocity, pressure, and heat flux, respectively: $n \equiv \langle 1\rangle_f$, $nu \equiv \langle v\rangle_f$, $p \equiv \langle (v-u)^2\rangle_f$ and $q \equiv \langle (v-u)^3\rangle_f$, with $\langle \ldots\rangle_f \equiv \int_{-\infty}^{\infty} \ldots f \mathrm{d}v$. The system can then be linearized close to the fluid equilibrium consistent with the kinetic one. Closing the system consists in expressing each mode of the flux as a function of $\hat{n}_k$, $\hat{u}_k$, and $\hat{p}_k$: $\hat{q}_k = \hat{\alpha}_n \hat{n}_k + \hat{\alpha}_u \hat{u}_k + \hat{\alpha}_p \hat{p}_k$. So far, the operators $\hat{\alpha}_n$, $\hat{\alpha}_u$, and $\hat{\alpha}_p$ remain unconstrained, apart from the fact that they are independent of $\hat{n}_k$, $\hat{u}_k$, and $\hat{p}_k$, consistently with the linear framework. With these coefficients, the fluid response function takes the following form:

$$\hat{R}_{fl} = \frac{\hat{\alpha}_p - \varepsilon\,\zeta}{\hat{\alpha}_n + 3\varepsilon\,\zeta + \alpha_p\,\zeta^2 - \zeta^3} \tag{5.9}$$

with $\varepsilon \equiv \mathrm{sign}(k)$.

The closure proposed by Hammett and Perkins aims at matching the kinetic and fluid linear response functions in the kinetic limit, namely, Eqs. 5.4 and 5.9 when $\zeta \to 0$. The identification of each power of ζ, order by order, then yields the expression of the unknown operators. Further considering the special case $\hat{\alpha}_u = 0$ as suggested by Hammett and Perkins (although this additional constraint can be easily released), one obtains

$$\hat{\alpha}_n = -\hat{\alpha}_p = i\,\frac{2\sqrt{2}}{\sqrt{\pi}}\,\mathrm{sign}(k) \tag{5.10}$$

In the end, the heat flux turns out to depend on temperature only:

$$\hat{q}_{k,\omega} = -i\,\frac{2\sqrt{2}}{\sqrt{\pi}}\,\mathrm{sign}(k)\,\hat{T}_{k,\omega} \tag{5.11}$$

Such a closure, which depends on the sign of the wave vector, is actually nonlocal in space. Indeed, it can be shown to read as follows in configuration space (notice that equilibrium density does not appear explicitly because it enters the normalization):

$$q = -\left(\frac{2}{\pi}\right)^{3/2} \lim_{\varepsilon \to 0^+} \int_{\varepsilon}^{+\infty} \frac{T(x+x') - T(x-x')}{x'}\,\mathrm{d}x' \tag{5.12}$$

The last term bears some analogy with the gradient operator, therefore echoing the collisional closure which yields a conductive flux. Here however, x' is not to be taken in the vanishing limit: conversely, it varies from 0 to $+\infty$. The closure is therefore nonlocal in configuration space.

Still, fluid models hardly recover kinetic results in fusion plasmas. Linear instability thresholds can often be adjusted, but nonlinear results usually depart from the kinetic predictions. Fluid models tend to overestimate turbulent transport. One of the reasons seems to be the overdamping of large-scale self-generated flows (so-called zonal flows; see Sect. 5.4), which play a critical role in turbulence saturation [13]. Another reason could well be the large number of fluid moments sometimes observed to be required to account for the complexity of the distribution function observed in kinetic simulations [14].

5.2.2 Gyrokinetic Description

Strongly magnetized plasmas such as fusion ones are suitable to gyro-ordering. They are characterized by much smaller variations of the magnetic field, both in time and in space, as compared to the cyclotron motion of charged species (sometimes called the *adiabatic limit*). It corresponds to cases where $|\partial_t \log B| \ll \omega_c$ and $|\mathbf{v}.\boldsymbol{\nabla} \log B| \sim \rho_c |\nabla \log B| \ll 1$. In such cases, particles are said to be magnetized, in the sense that the cyclotron motion can be distinguished from secular drifts, which will be detailed hereafter. Should the observed processes occur at much slower time scale than ω_c^{-1}, it is then legitimate to proceed to a phase-space reduction by averaging out the fast cyclotron motion in Vlasov equation. The resulting equation is the gyrokinetic equation, involving the 5-dimensional distribution of gyro-particles $f_G(\mathbf{r}, v_\perp, v_\parallel, t)$. The system is then closed with the help of Maxwell's equations, which need being expressed as a function of f_G. A modern formulation of the problem can be found in reference [15]. We propose here a simplified derivation.

In this limit, particle motions can be decomposed in the cyclotron motion plus velocity drifts of the guiding center: $\mathbf{v} = \mathbf{v}_c + \mathbf{v}_G$. When gyroaveraged over the cyclotron motion, Newton's equation reads as follows:

$$m_s \frac{\mathrm{d}\mathbf{v}_G}{\mathrm{d}t} = e_s \left(\langle \mathbf{E} \rangle + \mathbf{v}_G \times \mathbf{B} \right) - \mu_s \boldsymbol{\nabla} B \tag{5.13}$$

where the bracket stands here for gyroaverage, namely, $\langle \ldots \rangle \equiv \oint \ldots \mathrm{d}\varphi_c / 2\pi$, with φ_c the cyclotron phase. Here, the magnetic field $\mathbf{B}$ is taken at the position of the gyro-center. The last term on the right-hand side represents the drag force exerted on the guiding center by the small inhomogeneity of the magnetic field at the cyclotron radius scale. $\mu_s = m_s v_\perp / e_s B$ is the magnetic moment. $\mathbf{v}_G$ represents the guiding-center velocity. It can be decomposed into parallel and transverse components:

$$\mathbf{v}_G \equiv v_{G\parallel} \mathbf{b} + \mathbf{v}_{G\perp}$$

with $\mathbf{b} = \mathbf{B}/B$. Equation 5.13 can be used to obtain the expressions of both $v_{G\parallel}$ and $\mathbf{v}_{G\perp}$.

After some manipulation, within the adiabatic limit, the transverse drift[1] can be shown to exhibit two components:

$$\mathbf{v}_{G\perp} = \frac{\langle \mathbf{E} \rangle \times \mathbf{B}}{B^2} + \frac{m_s v_{G\parallel}^2 + \mu_s B}{e_s B} \frac{\mathbf{B} \times \nabla B}{B^2} + \frac{m_s v_{G\parallel}^2}{e_s B^2} \mu_0 \mathbf{j}_\perp \tag{5.14}$$

The first term is the electric drift velocity $\mathbf{v}_E$. In the electrostatic limit, $\langle \mathbf{E} \rangle = -\nabla \langle \phi \rangle$, and $\mathbf{v}_E$ is equal to $\mathbf{v}_E = \mathbf{B} \times \nabla \langle \phi \rangle / B^2$. The two last terms, denoted $\mathbf{v}_{d,s}$ hereafter, are the magnetic drifts. They are made of the so-called *grad-B* and curvature drifts. They are of the order of $(T_s/e_s BR)$ for thermal particles. In tokamaks, these drifts are essentially along the vertical direction. They lead to vertical charge separation, ions and electrons drifting in opposite directions. By using the condition for magnetic equilibrium in tokamaks, stating that $\mathbf{j} \times \mathbf{B} = \nabla p$ (with $\mathbf{j}$ the plasma current and p its pressure), it can be shown that the last term on the right-hand side of Eq. 5.14 is smaller than the second one by the factor $\beta \equiv p/(B^2/2\mu_0)$, the ratio of kinetic to magnetic energy (μ_0 the permeability of free space).

As far as the parallel acceleration is concerned, it can be shown to exhibit the following expression [16]:

$$\frac{\mathrm{d}v_{G\parallel}}{\mathrm{d}t} = -\frac{1}{m_s} \left(\mathbf{b} + \frac{m_s v_{G\parallel}}{e_s B^2} \mu_0 \mathbf{j}_\perp \right) \cdot \nabla \Xi - v_{G\parallel} \frac{\mathbf{B} \times \nabla B}{B^3} \cdot \nabla \langle \phi \rangle \tag{5.15}$$

with $\nabla \Xi = \mu_s \nabla B + e_s \nabla \langle \phi \rangle$. The term depending on the transverse current $\mathbf{j}_\perp$ is smaller than the other ones by the ratio β. All others are *a priori* of the same order of magnitude. Up to small terms depending on the parallel plasma current (more precisely, in the limit $(m_s v_{G\parallel}/e_s B^2)\mu_0 j_\parallel \ll 1$), the gyrokinetic equation can be approximated by:

$$\partial_t f_G + \mathbf{v}_{G\perp}.\nabla f_G + \frac{\mathrm{d}v_{G\parallel}}{\mathrm{d}t} \partial_{v_{G\parallel}} f_G = 0 \tag{5.16}$$

where $\mathbf{v}_{G\perp}$ and $\mathrm{d}v_{G\parallel}/\mathrm{d}t$ are given by Eqs. 5.14 and 5.15, respectively.

Maxwell's equations can be reformulated as a function of f_G by noticing that f derives from f_G by the following relationship:

$$f_s(\mathbf{x},\mathbf{v},t) = f_{Gs}(\mathbf{x}_G,\mathbf{v}_G,t) + \frac{e_s}{B} \{ \phi(\mathbf{x},t) - \langle \phi(\mathbf{x}_G,\mathbf{v}_G,t) \rangle \} \, \partial_\mu f_{eq,s}(\mathbf{x}_G,\mathbf{v}_G) \tag{5.17}$$

[1] Notice that transverse drifts can also be derived within the fluid framework in the same adiabatic limit. At first order in the small ρ/R parameter, with R the curvature—and or the gradient— length of B, they read: $\mathbf{u}_\perp^{(1)} \equiv \mathbf{u}_E + \mathbf{u}_s^* = \frac{\mathbf{E}\times\mathbf{B}}{B^2} + \frac{\mathbf{B}\times\nabla p_s}{n_s e_s B^2}$. The first component, the *electric drift* $\mathbf{u}_E$, is also a particle drift. The latter one is not, since it depends on the pressure, which is a fluid quantity only. It is known as the *diamagnetic drift* $\mathbf{u}_s^*$. It is the same order of magnitude for ions and electrons. Since it depends on the charge of the species, it carries transverse current. The second-order fluid drift is the so-called polarization drift. It is often approximated as follows: $\mathbf{u}_\perp^{(2)} \equiv \mathbf{u}_{pol,s} = -\frac{m_s}{e_s B^2} \Big[\partial_t + (\mathbf{u}_E + \mathbf{u}_s^* + \mathbf{u}_\parallel).\nabla \Big] \nabla_\perp \phi$.

where $\mathbf{x}_G$ stands for the position of the gyro-center. $\mathbf{x}$ and $\mathbf{x}_G$ are related as follows: $\mathbf{x} = \mathbf{x}_G + \boldsymbol{\rho}_c$. For large wavelengths $k_\perp \rho_s \ll 1$, and assuming that the electron response is adiabatic, the quasi-neutrality condition reduces to

$$\frac{e}{T_e}(\phi - \langle\phi\rangle_{FS}) - \frac{1}{n_{eq}}\boldsymbol{\nabla}_\perp \cdot \left(\frac{m_i n_{eq}}{e_i B^2}\boldsymbol{\nabla}_\perp \phi\right) = \frac{1}{n_{eq}}\int \frac{2\pi B}{m_i}\,\mathrm{d}\mu \mathrm{d}v_{G\parallel}\langle f_G\rangle - 1 \quad (5.18)$$

with $\langle\phi\rangle_{FS}$ the flux surface average of ϕ.

Equations 5.16 and 5.18 form a closed system in the electrostatic limit. They allow one to model ion turbulent transport in tokamak plasmas from first-principle equations.

5.3 Main Micro-Instabilities in Fusion Plasmas

Two main instabilities can be identified in the core of fusion plasmas: drift-wave (DW) instability and interchange. Their mechanism is briefly explained in Sect. 5.3.1. Both occur in the adiabatic limit, where particles are magnetized and subject to drift velocities transverse to their fast parallel motion. The first one is essentially three-dimensional—since relying on the properties of the particle response to parallel perturbations—while the latter one already exists in two dimensions, when perturbations are constant along field lines. It turns out that the interchange instability is dominant in fusion plasmas, in that its growth rate usually exceeds the one of the DW instability. Although the associated turbulent transport will be illustrated by means of gyrokinetic models in Sects. 5.4, 5.3.2 analyzes some of their linear properties with fluid models for the sake of simplicity.

Another class of instabilities also plays a critical role, in that it involves fast particles. In particular, they could reveal potentially deleterious for the confinement of helium ashes in fusion reactors, hence preventing—or at least limiting—energy transfer between the main ions and alpha particles. Such instabilities belong to the family of the bump-on-tail instability, for which a simple model is discussed in Sect. 5.3.3.

5.3.1 Physical Understanding of Drift-Wave and Interchange Instabilities

5.3.1.1 Drift Waves

Both the origin of drift waves and of the associated instability are detailed in Fig. 5.3.

First consider an electric potential perturbation made of the plane wave $\exp i(k_y y - \omega t)$ drawn in Fig. 5.3a, in the homogeneous magnetic field $\mathbf{B} = B\mathbf{e}_z$. Due

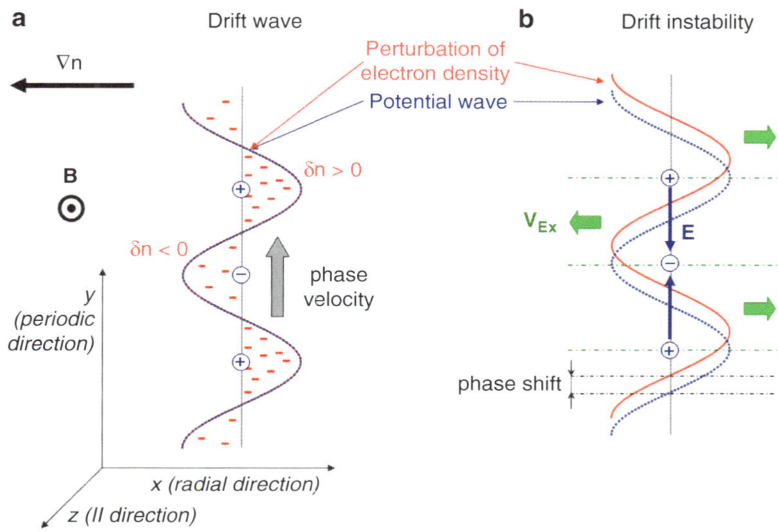

Fig. 5.3 Schematic view of the (**a**) drift wave and (**b**) drift-wave instability mechanisms

to their low inertia, electrons move rapidly along the magnetic field lines, such that they adjust quite instantaneously to any potential perturbation whose frequency ω is small with regard to their parallel dynamics, namely, $k_{\|} v_{T,e}$. Differently speaking, in the limit $\omega \gg k_{\|} v_{T,e}$, the dominant terms in the parallel electron fluid force balance equation

$$m_e n_e \frac{\mathrm{d}\mathbf{u}_e}{\mathrm{d}t} .\mathbf{b} = e n_e \nabla_{\|} \phi - \nabla_{\|} p_e + \frac{\nu_e m_e j_{\|}}{e} \tag{5.19}$$

are the electric field and the pressure gradient. The temperature gradient is usually small compared with the density gradient due to the fast electron thermal flow associated with $k_{\|} v_{T,e} \gg \omega$. In the isothermal thermal limit, the pressure gradient simply writes $\nabla_{\|} p_e = T_e \nabla_{\|} n_e$. Thus, much of the low-frequency drift-wave dynamics falls in the regime where the Boltzmann description of the electron response $n_e = n_0 \exp(e\phi/T_e)$ applies in the form:

$$\frac{\delta n_e}{n_{eq}} \simeq \frac{e\delta\phi}{T_e} \tag{5.20}$$

Such an electron response is sometimes called *adiabatic*. In this framework, electron density and electric potential perturbations are in phase.[2] The corresponding

[2]Notice that such a result intrinsically derives from the fast motion of the electrons in the *parallel direction* due to their small inertia. Therefore, only those modes which exhibit some structure in the parallel direction (i.e., such that $k_{\|} \neq 0$) are subject to an adiabatic response of the electrons.

sinusoidal electric field E_y goes from the super- ($\delta n_e > 0$) to the sub-density regions ($\delta n_e < 0$): its signs reverse at the extreme of the potential wave. As a result, the electric drift governs an inward (resp. outward) radial motion on half of the super-density (resp. sub-) lobe and an outward (resp. inward) in the other half. The net result is an oscillation of the wave with a phase velocity along the y-direction, directed upwards. A more careful treatment[3] shows that the drift wave frequency is actually of the order of the electron diamagnetic frequency: $\omega = \omega_e^* = -(k_y\rho_i)v_{T,i}\mathrm{d}(\log n_{eq})/\mathrm{d}r$.

5.3.1.2 Drift Wave Instability

Consider now the case when the electron density and potential perturbations are out of phase (Fig. 5.3b). The shape of the sinusoidal electric field is governed by the electric potential: the sign of E_y reverses at the extreme of the potential wave. The resulting radial component of the $E \times B$ drift is shown in Fig. 5.3b. Let's focus on the super-density region. Due to the nonvanishing phase shift between δn_e and $\delta\phi$, this region experiences a net outward motion on average. In this way, it turns out that the initial perturbation gets amplified. Applying the same reasoning to the sub-density region leads to a net inward motion of this region. It is important to notice that such an instability only develops for a negative phase shift (as shown in Fig. 5.3b), namely, for those modes k such that

$$\frac{\delta n_k}{n_{eq}} = (1 - i\delta_k)\frac{e\delta\phi_k}{T_e} \qquad \text{with } \delta_k > 0 \tag{5.21}$$

where k stands for the wave vector in the direction transverse both to the density gradient and to the magnetic field (y in the present case). Those modes with negative values of δ_k will be damped.

Density and electric potential perturbations can become out of phase due to various mechanisms. The two main ones are the plasma resistivity, which breaks up the assumption of an adiabatic response of the electrons, and wave-particle resonances. The first mechanism is detailed in Sect. 5.3.2, while the second requires a kinetic treatment.

5.3.1.3 Interchange Instability

The most deleterious micro-instabilities in core tokamak plasmas are of the so-called *interchange* type. Under certain circumstances, interchanging two flux tubes

---

[3]The ion density fluctuation δn_i comes from the continuity equation, namely, $\partial_t \delta n_i + u_{Er}\mathrm{d}n_{eq}/\mathrm{d}r = 0$, with $u_{Er} = -\partial_y\phi/B$. For the considered plane wave, this reads as follows: $-i\omega\delta n_i = i(k_y/B)(\mathrm{d}n_{eq}/\mathrm{d}r)\,\delta\phi$. The quasi-neutrality constraint $\delta n_i = \delta n_e$ then leads to the result. See also Sect. 5.3.2

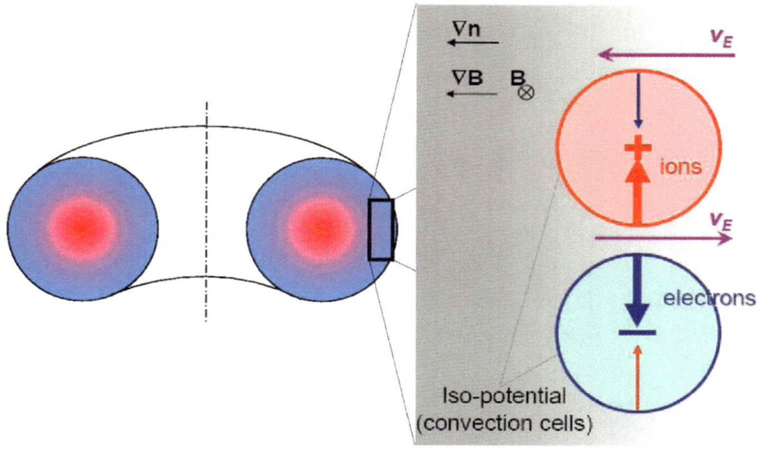

Fig. 5.4 Schematic view explaining the physical mechanism of the interchange instability in tokamaks

leads to a drop of energy and is therefore an unstable process. As we will see, the physics relies on both the inhomogeneity of the magnetic field and on the fact that the plasma departs from the thermodynamical equilibrium and exhibits large gradients. Figure 5.4 details the physics at work. Such an instability bears some similarities with the well-known Rayleigh–Bénard instability in neutral fluids.

Let us assume there exist small convection cells, *i.e.*, closed contour lines at constant electric potential, in the equatorial plane and on the low-field side of the tokamak. These fluctuations of the potential lead to local electric fields. In the case of magnetized species (see Sect. 5.2.2), the particles are subject to velocity drifts transverse to the magnetic field lines. In the configuration plotted in Fig. 5.4, the electric drift $\mathbf{v}_E$ goes from the left to the right in between the two cells, while the curvature and ∇B drifts are vertical, up (resp. down) for the ions (resp. electrons). Also, notice that the density gradient points to the left in this low-field side region. Consider the motion of particles located at the midplane. Due to the electric drift, both ions and electrons move to the right, into a less dense region. In addition, ions move vertically to the top, while electrons move down. Therefore, a large amount of ions goes to the already positive cell. Similarly, consider the motion of particles located at the top of the inset. This time, the electric drift goes from right to left, due to the inversion of the local electric field. Again, due to the vertical drift, a small amount of electrons is going towards the positive cell. The balance for the positive cell is clearly in favor of positive charges. The same reasoning would have led to a net increase of negative charges in the negative cell. As a result, the initially small convection cells are growing.

The same reasoning can be applied to the high-field side. In this case, you can convince yourself that the convective cells tend to die away: positive cells receive more negative charges than positive ones and the opposite for negative cells. This region is stable with regard to the interchange. However, remember that both regions are coupled. Indeed, since particles essentially move along the field lines, they experience stable and unstable regions. In this framework, the parallel current, which carries the electric charges from one cell parity to another, appears to be stabilizing. As a matter of fact, such an instability is all the more efficient since the parallel resistivity is large, *i.e.*, when the stable and unstable regions tend to be decoupled.

5.3.2 Simple Model for Drift-Wave and Interchange Instabilities

We develop here a fluid version of these instabilities, which proceeds from an extension of the well-known Hasegawa-Wakatani model [17, 18] when accounting for the inhomogeneity of the magnetic field.

Consider a plasma in a strong static magnetic field. Within the adiabatic limit, the transverse projection of the momentum balance equation can be replaced by the fluid drifts (see Sect. 5.2.2). For a single-charge species, quasi-neutrality simply reads $n_i = n_e = n$. The matter and charge balance equations then read:

$$\partial_t n + \boldsymbol{\nabla}.\{n\,(\mathbf{u}_E + \mathbf{u}_e^*)\} - \frac{\nabla_\parallel j_{\parallel e}}{e} = 0 \tag{5.22}$$

$$\boldsymbol{\nabla}.\left\{en\left(\mathbf{u}_i^* - \mathbf{u}_e^* + \mathbf{u}_{pol,i}\right)\right\} + \nabla_\parallel j_\parallel = 0 \tag{5.23}$$

with $j_\parallel = j_{\parallel i} + j_{\parallel e}$. Ions are further assumed to be cold, $T_i = 0$, such that $\mathbf{u}_i^* = 0$. Finally, electron temperature fluctuations are neglected: $T_e = Cst$.

Should $\mathbf{B}$ be uniform, the electric drift $\mathbf{u}_E$ would be divergence free. For the same reason, $\boldsymbol{\nabla}.(n\mathbf{u}_e^*)$ would be vanishing. In the presence of the inhomogeneous magnetic field illustrated in Fig. 5.5, both the electric drift $\mathbf{u}_E$ and the diamagnetic current $\mathbf{j}_e^* = -en\mathbf{u}_e^*$ are no longer divergence free:

$$\boldsymbol{\nabla}.\mathbf{u}_E = \boldsymbol{\nabla}\phi.\left(\boldsymbol{\nabla}\times\frac{\mathbf{B}}{B^2}\right) \tag{5.24}$$

$$\boldsymbol{\nabla}.\mathbf{j}_e^* = T_e\boldsymbol{\nabla}n.\left(\boldsymbol{\nabla}\times\frac{\mathbf{B}}{B^2}\right) \tag{5.25}$$

with $\boldsymbol{\nabla}\times(\mathbf{B}/B^2) = -(2/R_0B)\mathbf{e}_y$ in the considered configuration (Fig. 5.5). Since the term $n\boldsymbol{\nabla}.\mathbf{u}_E$ remains small with respect to the electric advection term $\mathbf{u}_E.\boldsymbol{\nabla}n$ (in the ratio $L_n/R \ll 1$), it can be neglected in the mass conservation, Eq. 5.22. Conversely, the latter term $\boldsymbol{\nabla}.\mathbf{j}_e^*$ has to be kept in the charge balance, Eq. 5.23, where it competes with the other terms.

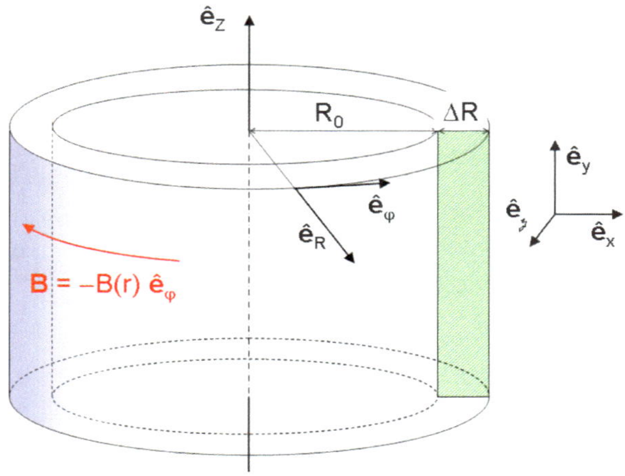

Fig. 5.5 Geometry for the Hasegawa-Wakatani model with an inhomogeneous magnetic field. We shall focus on the green area characterized by $\Delta R \ll R_0$

The divergence of the polarization current is more complex. Using the expression of the ion polarization drift $\mathbf{u}_{pol,i}$, it is given the following simplified form:

$$\boldsymbol{\nabla}.\left(en\mathbf{u}_{pol,i}\right) = -\boldsymbol{\nabla}.\left\{\frac{nm}{B^2}\left(\partial_t + \mathbf{u}_E.\boldsymbol{\nabla}\right)\boldsymbol{\nabla}_\perp\phi\right\} \approx -\frac{nm}{B^2}\left(\partial_t + \mathbf{u}_E.\boldsymbol{\nabla}\right)\nabla_\perp^2\phi$$

The system Eqs. 5.22–5.23 then write:

$$\partial_t n + \frac{1}{B}[\phi, n] = \frac{\nabla_\parallel j_{\parallel e}}{e} \tag{5.26}$$

$$\partial_t \nabla_\perp^2\phi + \frac{1}{B}[\phi, \nabla_\perp^2\phi] + \frac{2BT_e}{mR_0}\,\partial_y \log n = \frac{B^2}{nm}\nabla_\parallel j_\parallel \tag{5.27}$$

The Poisson brackets of two scalars f and g are defined by $[f,g] \equiv (\boldsymbol{\nabla} f \times \boldsymbol{\nabla} g).\mathbf{b} = \partial_x f \partial_y g - \partial_y f \partial_x g$, with (x,y) the cartesian coordinates in the plane transverse to $\mathbf{b} \equiv \mathbf{B}/B$.

The current derives from the parallel force balance on the electrons. When neglecting their inertia, namely, $nm_e \mathrm{d}u_{\parallel e}/\mathrm{d}t$, one obtains the generalized resistive Ohm's law: $-\nabla_\parallel p_e + en\nabla_\parallel\phi - nm_e\nu_{ei}\left(u_{\parallel e} - u_{\parallel i}\right) = 0$. Within the hypotheses of the model, one gets:

$$j_\parallel = \frac{T_e}{e\eta}\,\nabla_\parallel\left(\log n - \frac{e\phi}{T_e}\right)$$

where the resistivity is given by $\eta \equiv m_e \nu_{ei}/e^2 n$. Furthermore, due to their low inertia, the electrons carry most of the parallel current, such that $j_\parallel \approx j_{\parallel e}$.

Let us then introduce the following dimensionless variables:

$$\tau = \omega_c t \ ; \ (X,Y) = (x,y)/\rho_s$$

$$N = \log \frac{n}{n_0} \ ; \ \Phi = \frac{e\phi}{T_e}$$

with n_0 the constant density such that $n_0 \approx n_{eq}$. Time is normalized to the cyclotron frequency $\omega_c = eB/m$ and distances to the Larmor radius $\rho_s = mc_s/eB$, with $c_s^2 = T_e/m$ the sound speed. One then decomposes density into equilibrium $N_{eq}(X)$ and fluctuations $N(X,Y,\tau)$, with $\mathrm{d}N_{eq}/\mathrm{d}X = -1/L_N$. In this framework, Eqs. 5.26–5.27 lead to the extended Hasegawa-Wakatani model:

$$\partial_\tau N + [\Phi, N] + \frac{\partial_Y \Phi}{L_N} = C(\Phi - N) \tag{5.28}$$

$$\partial_\tau \nabla_\perp^2 \Phi + [\Phi, \nabla_\perp^2 \Phi] + g\partial_Y N = C(\Phi - N) \tag{5.29}$$

where $C \equiv (k_\parallel \rho_s)^2 B/en_{eq}\eta$ and $g \equiv 2\rho_s/R_0$. Here, $k_\parallel$ stands for the parallel operator: $\nabla_\parallel^2 \to -k_\parallel^2$. Notice that the phase shift between the density N and the electric potential Φ is all the larger since C is small, *i.e.*, since the plasma resistivity η is large. As will become clear hereafter, this phase shift is at the origin of the DW instability.

The dispersion relation can be easily derived:

$$\omega^2 + iC(1+k_\perp^{-2})\omega + i\frac{Ck_Y}{k_\perp^2}\left(g - \frac{1}{L_N}\right) + \frac{g}{L_N}\frac{k_Y^2}{k_\perp^2} = 0$$

Linear solutions are then:

$$\omega_\pm = -i\frac{C}{2}(1+k_\perp^{-2}) \pm \frac{i}{2}\left\{C^2(1+k_\perp^{-2})^2 + i\frac{4Ck_Y}{k_\perp^2}\left(g - \frac{1}{L_N}\right) + 4\frac{g}{L_N}\frac{k_Y^2}{k_\perp^2}\right\}^{1/2}$$

The main linear properties of the system are plotted in Fig. 5.6. Two limiting cases are especially instructive.

- In the limit $C = 0$, the solutions read $\omega_\pm = \pm i\frac{2k_Y}{k_\perp}(g/L_N)^{1/2}$. This corresponds to the interchange instability, the two driving terms being the magnetic field curvature g and the density gradient length L_N. Besides, density and potential fluctuations are in quadrature, such that the transport is maximum.
- The other limit $C \to +\infty$ is instructive as well. In this case, density and potential fluctuations are in phase: $N \approx \Phi$. Also, the growth rate turns out to vanish asymptotically like C^{-1}: $\max\{\mathrm{Im}(\omega)\} = \frac{k_Y^2}{Ck_\perp^2(1+k_\perp^{-2})}\left\{\frac{g}{L_N} + \frac{1}{16k_\perp^2(1+k_\perp^{-2})^2}\left(g - \frac{1}{L_N}\right)^2\right\}$. Such a result is consistent with the mechanism of the drift-wave instability: it only develops if density and electric potential fluctuations are out of phase.

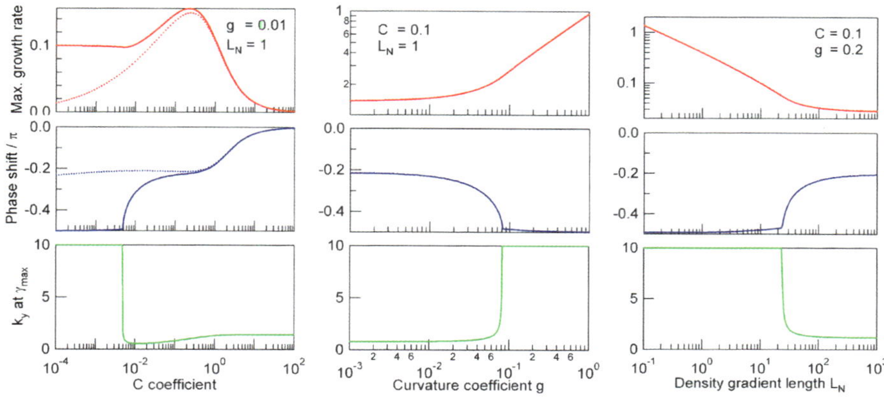

Fig. 5.6 Maximum growth rate γ_{max} (*top*), phase shift at γ_{max} (*middle*), and wave number k_Y at γ_{max} (*bottom*) when varying various control parameters in the system Eqs. 5.28–5.29. Transition from the drift-wave instability to the interchange instability is most evident on the phase shift, which is close to $-\pi/2$ in the latter case. *Dotted curves* (*right*) refer to the original Hasegawa-Wakatani case, with $g = 0$

5.3.3 Bump-on-Tail Instability

The so-called "bump-on-tail" instability develops when the equilibrium distribution function presents a positive slope with respect to velocity in the tail. In particular, it can occur when a low-density plasma beam of finite mean velocity interacts with the bulk plasma, at rest. Obviously, such an instability requires the kinetic description, for phase-space resonances reveal essential for its development. In tokamak, such a type of instability can be excited by fast particles emerging from specific heating schemes such as neutral beam injection (see, e.g., [19] and references therein) or alpha particles.

In the following, the bump-on-tail instability is studied by means of a simple model. Let us consider the system described by Eq. 5.2.

$$\partial_t f + v\partial_x f + \partial_x \phi \partial_v f = 0 \tag{5.30}$$

$$\partial_x^2 \phi = \int_{-\infty}^{+\infty} f \mathrm{d}v - 1 \tag{5.31}$$

Let us consider an equilibrium distribution function made of 2 Maxwellians $f_{eq} = f_1 + f_2$ (cf. Fig. 5.7): the bulk plasma particles f_1, at rest, of density $n_1 = (1-\varepsilon)$ and small temperature $T_1 = \varepsilon^2$, and a hot beam f_2, of small density $n_2 = \varepsilon$ and temperature $T_2 = 1$, with ε a small positive parameter $0 < \varepsilon \ll 1$:

$$f_1 = \frac{1-\varepsilon}{\sqrt{2\pi}\,\varepsilon}\, \mathrm{e}^{-v^2/2\varepsilon^2} \tag{5.32}$$

$$f_2 = \frac{\varepsilon}{\sqrt{2\pi}}\, \mathrm{e}^{-(v-v_0)^2/2} \tag{5.33}$$

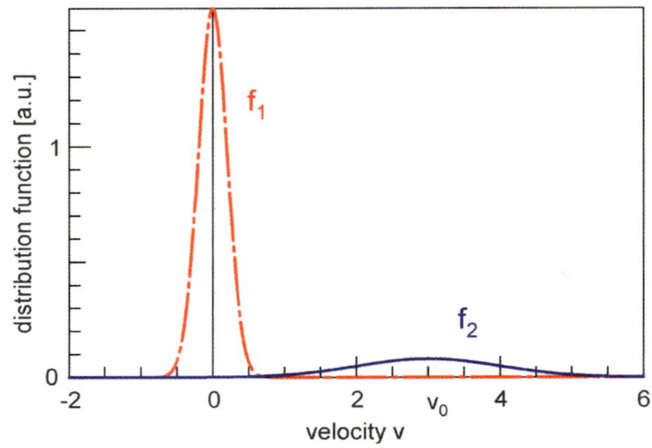

Fig. 5.7 Equilibrium distribution function f_{eq}. It is made of 2 Maxwellians: $f_{eq} = f_1 + f_2$

v_0 is the fluid velocity of the beam, assumed constant. The equilibrium electric potential is vanishing: $\phi_{eq} = 0$.

Let us consider small-amplitude fluctuations around the equilibrium given by Eqs. 5.32–5.33: $\tilde{\phi}(x,t) \ll 1$ and $f = f_{eq}(v) + \tilde{f}(x,v,t)$, with $\tilde{f} \ll f_{eq}$. Let us decompose fluctuations onto their Fourier components: $(\tilde{n}, \tilde{\phi}) = \sum_{k,\omega}(\hat{\phi}_{k,\omega}, \hat{n}_{k,\omega})$ $\exp\{i(kx - \omega t)\}$. Noticing that $\partial_v f_{eq} = -(v/\varepsilon^2)\, f_1 - (v - v_0)\, f_2$ in this case, the dispersion relation reads as follows, as long as ω remains real:

$$\mathscr{D}(k,\omega) = \mathscr{D}_r(k,\omega) + i\,\mathscr{D}_i(k,\omega) = 0 \tag{5.34}$$

$$\mathscr{D}_r(k,\omega) = k^2 - \mathrm{P}\!\int_{-\infty}^{+\infty} \frac{k}{\omega - kv} \left\{\frac{v}{\varepsilon^2}\, f_1 + (v - v_0)\, f_2\right\} \mathrm{d}v \tag{5.35}$$

$$\mathscr{D}_i(k,\omega) = -\pi k \int_{-\infty}^{+\infty} \left\{\frac{v}{\varepsilon^2}\, f_1 + (v - v_0)\, f_2\right\} \delta(\omega - kv)\, \mathrm{d}v \tag{5.36}$$

where $\mathrm{P}\!\int$ denotes the principal part. We will show that pure (i.e., neither damped nor excited) waves exist below the linear threshold only, while certain wave numbers become unstable above the linear threshold.

5.3.3.1 Linear Threshold

The linear threshold of this kinetic instability, characterized by a real frequency ω, fulfills the following set of equations:

$$\mathscr{D}_r(k,\omega_{r*}) = 0$$

$$\mathscr{D}_i(k,\omega_{r*}) = 0$$

At lowest order in $\varepsilon \ll 1$, canceling the imaginary part of the dispersion relation leads to:

$$\omega_{r*} = k v_0 \tag{5.37}$$

where ω_{r*} stands for the (real) frequency at the threshold.

Canceling the real part, $\mathscr{D}_r(k, \omega_{r*}) = 0$, requires some care. First notice that it can be rewritten as follows:

$$\mathscr{D}_r(k, \omega_{r*}) = k^2 + \varepsilon + \frac{1-\varepsilon}{\varepsilon^2} + \frac{v_0}{\varepsilon^2} \mathrm{P}\int_{-\infty}^{+\infty} \frac{f_1}{v - v_0}\, \mathrm{d}v = 0 \tag{5.38}$$

In the limit $\varepsilon \ll v_0$, f_1 almost vanishes at the pole (for $v = v_0$), so that the integrand can be Taylor expanded in the fluid hydrodynamical limit, i.e., for $v \ll v_0$. In this limit, the relation, Eq. 5.38, reads as follows:

$$\mathscr{D}_r(k, \omega_{r*}) \approx k^2 + \varepsilon - \frac{1-\varepsilon}{v_0^2} = 0 \tag{5.39}$$

The critical state, corresponding to the linear stability threshold, is then achieved for those values of the velocity v_0 solutions of Eq. 5.39:

$$v_{0*}^{\pm}(k) = \pm \left(\frac{1-\varepsilon}{k^2 + \varepsilon} \right)^{1/2} \tag{5.40}$$

Obviously, the case $v_0 = 0$ is stable.[4] It follows that the *bump-on-tail instability* develops at a given wave vector k if and only if $|v_0| > |v_{0*}(k)|$.

Alternatively, given v_0, the unstable branch is made of those wave vectors k which are larger than the critical one $|k| > k_*$, with $k_* = [1 - \varepsilon(1 + v_0^2)]^{1/2}/|v_0| \approx |v_0|^{-1}$.

5.3.3.2 Growth Rate in the Vicinity of the Linear Threshold

Above the linear threshold, the solutions ω of the dispersion relation, Eq. 5.34, exhibit a positive imaginary part. Close to the threshold, solutions are characterized by $\omega = \omega_r + i\gamma$, with $0 < \gamma \ll |\omega_r|$ and ω_r and γ real.

Taylor expanding Eq. 5.34 with respect to the small parameter γ/ω_r leads to the new approximate dispersion relation:

$$\mathscr{D}(k, \omega) \approx \mathscr{D}_r(k, \omega_r) - \gamma \partial_\omega \mathscr{D}_i(k, \omega_r) + i\{\mathscr{D}_i(k, \omega_r) + \gamma \partial_\omega \mathscr{D}_r(k, \omega_r)\} = 0 \tag{5.41}$$

The second term on the right-hand side can be neglected since it is second order ($\gamma/\omega_r \ll 1$ and $\mathscr{D}_i \ll \mathscr{D}_r$ since $\mathscr{D}_i$ only accounts for the resonance condition). It follows that, at leading order,

---

[4]Indeed, it corresponds to 2 centered Maxwellians, for which Landau damping only is expected.

$$\mathscr{D}_r(k,\omega_r) \approx 0 \tag{5.42}$$

$$\gamma = -\frac{\mathscr{D}_i(k,\omega_r)}{\partial_\omega \mathscr{D}_r(k,\omega_r)} \tag{5.43}$$

The real frequency ω_r is given by the implicit equation 5.42, where $\mathscr{D}_r$ is given by Eq. 5.35.

It can be anticipated that the phase velocity ω_r/k of the unstable wave is in between the thermal velocity ε of particles from f_1 and the mean fluid velocity v_0 of particles from f_2: $\varepsilon \ll |\omega_r/k| \leq |v_0|$. This remark will reveal useful when calculating the two integrals of $\mathscr{D}_r$. The integral with f_1 can be calculated in the hydrodynamical limit $|\omega_r| \gg |kv|$. In this framework, at lowest order of the Taylor expansion, one obtains:

$$\mathsf{P}\int_{-\infty}^{+\infty} \frac{kv}{\omega - kv}\, \frac{f_1}{\varepsilon^2}\, \mathrm{d}v \approx (1-\varepsilon)\, \frac{k^2}{\omega_r^2} \tag{5.44}$$

The integral with f_2 can be reformulated as follows:

$$\mathsf{P}\int_{-\infty}^{+\infty} \frac{k(v-v_0)}{\omega - kv}\, f_2\, \mathrm{d}v = -\varepsilon + \varepsilon(\omega_r - kv_0)\mathsf{P}\int_{-\infty}^{+\infty} \frac{\mathrm{e}^{-u^2/2}}{(\omega_r - kv_0) - ku}\, \frac{\mathrm{d}u}{\sqrt{2\pi}} \tag{5.45}$$

It has to be calculated in the kinetic limit $(\omega_r - kv_0) \ll ku$. In this case, it yields:

$$\mathsf{P}\int_{-\infty}^{+\infty} \frac{k(v-v_0)}{\omega - kv}\, f_2\, \mathrm{d}v \approx -\varepsilon + \varepsilon(\omega_r - kv_0)^2 \tag{5.46}$$

Injecting the two previous expressions in the one of $\mathscr{D}_r$ (Eq. 5.35), one obtains at 0th order in ε:

$$\mathscr{D}_r(k,\omega_r) \approx k^2 - \frac{k^2}{\omega_r^2} \tag{5.47}$$

Then, canceling $\mathscr{D}_r$ (cf. Eq. 5.42) leads to the expression of the real frequency (at 0th order in ε) just above the linear threshold:

$$\omega_r^{\pm} = \pm 1$$

As far as $\mathscr{D}_i(k,\omega_r)$ is concerned, it reads at leading order in ε:

$$\mathscr{D}_i(k,\omega_r) \approx \frac{-\varepsilon}{|k|}\sqrt{\frac{\pi}{2}}\,(kv_0 \mp 1)\, \exp\left\{-\frac{(kv_0 \mp 1)^2}{2k^2}\right\} \tag{5.48}$$

Using Eq. 5.43, one finally obtains the approximate expression of the linear growth rate:

$$\gamma \approx \frac{\varepsilon}{|k|^3}\sqrt{\frac{\pi}{8}}\,(kv_0 \mp 1)\, \exp\left\{-\frac{(kv_0 \mp 1)^2}{2k^2}\right\} \tag{5.49}$$

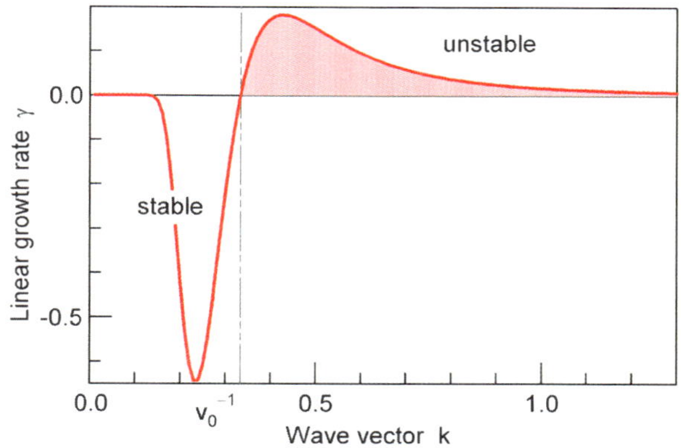

Fig. 5.8 Linear growth rate of the bump-on-tail instability, for the equilibrium distribution function plotted in Fig. 5.7 (Eqs. 5.32–5.33) as function of the wave vector k

It is plotted in Fig. 5.8. Notice that the growth rate is positive (bump-on-tail unstable) above $k_* \approx |v_0|^{-1}$, consistently with the calculation of the linear threshold. In true physical units, it reads $\gamma \approx \varepsilon\sqrt{\pi/8}\,(kv_0 \mp \omega_p)/|k\lambda_D|^3\ \exp\{-(kv_0 \mp \omega_p)^2/2k^2\lambda_D^2\}$.

5.4 Critical Issues in Turbulent Transport Modeling

5.4.1 Gradient-Versus Flux-Driven Models

Whatever the model used for turbulence simulations, two different classes can be distinguished: either *gradient-driven* or *flux-driven* models. They can be understood as follows. Let us decompose the distribution function in equilibrium and fluctuating parts: $f = f_{eq} + \tilde{f}$, where f_{eq} stands for the flux surface average for instance ($f_{eq} \equiv \langle f \rangle$). The Hamiltonian can be decomposed similarly as $H = H_{eq} + \tilde{h}$. Vlasov or gyrokinetic equations

$$\partial_t f - [H, f] = S \tag{5.50}$$

are then formally equivalent to the following system:

$$\partial_t f_{eq} - \langle [\tilde{h}, \tilde{f}] \rangle = S \tag{5.51}$$

$$\partial_t \tilde{f} - [H_{eq}, \tilde{f}] - [\tilde{h}, \tilde{f}] + \langle [\tilde{h}, \tilde{f}] \rangle = [\tilde{h}, f_{eq}] \tag{5.52}$$

Gradient-driven models would correspond to neglecting the time evolution of f_{eq}, Eq. 5.51, when studying the dynamics of the fluctuations. Notice that such an approach is equivalent to adjusting the source term in Eq. 5.51 so as to counterbalance the turbulent flux in real time and to prevent the subsequent mean profile relaxation:

$$\langle [\tilde{h}, \tilde{f}] \rangle + S = 0 \rightarrow \partial_t f_{eq} = 0$$

This approach assumes scale separation between equilibrium and fluctuations. The profile relaxation time is assumed to be governed by the energy confinement time $\tau_E \sim 1s$, much larger than the characteristic evolution time of fluctuations, of order of the inverse of the linear growth rate $\gamma^{-1} \sim 10^{-5}$. In this framework, turbulence roughly evolves in a frozen equilibrium, such that Eq. 5.52 is decoupled from Eq. 5.51: the back reaction of turbulent transport on the equilibrium profile is no longer accounted for.

However, it can be argued that this timescale separation assumption can actually break down: (i) First of all, the local gradient can evolve on much smaller time scales than τ_E and significantly impact on the excitation of unstable modes. (ii) Second, γ^{-1} is not necessarily the good estimate of the fluctuation characteristic time. Especially, their correlation time can significantly exceed γ^{-1}. (iii) Finally, tokamaks are open systems where fluxes (of particles, heat, and sometimes momentum) are prescribed, not gradients. The system then self-consistently finds the complex balance between gradients and turbulence level (and subsequent turbulent transport coefficient) to overcome the imposed driving fluxes.

Conversely, flux-driven systems allow mean profiles to fluctuate around time-averaged values, such that their time derivative is vanishing on (time) average only, on timescales of order of the energy confinement time:

$$\left\langle \langle [\tilde{h}, \tilde{f}] \rangle + S \right\rangle_{\tau_E} = 0 \rightarrow \left\langle \partial_t f_{eq} \right\rangle_{\tau_E} = 0$$

As a matter of fact, mean profile relaxation provides an efficient and natural way for turbulence to reach saturation. It competes with nonlinear mode–mode coupling, which is contained in the third term of Eq. 5.52, by which energy is transferred from linearly unstable to damped or stable modes. This latter saturation mechanism is the only one to be considered in gradient-driven simulations. As exemplified in Sect. 5.4.2, the self-consistent relaxation of equilibrium profiles can efficiently govern turbulence saturation via trapping mechanism.

5.4.2 *Profile Relaxation and Turbulence Trapping*

Let's consider the extended Hasegawa-Wakatani model Eqs. 5.28–5.29, in the simplified case $C = 0$. The equilibrium radial density gradient is denoted $\bar{n}'$, while $\tilde{n} = \sum \hat{n}_k(x) \mathrm{e}^{iky+\gamma t}$ stands for the density fluctuations. In the limit of weak fluctuations, the linearized system around an equilibrium at vanishing electric field reads for each Fourier mode:

$$\gamma \hat{n}_k = ik\bar{n}' \hat{\phi}_k$$

$$\hat{\phi}_k'' - \frac{gk^2}{\gamma^2} \frac{\bar{n}'}{\bar{n}} \hat{\phi}_k = k^2 \hat{\phi}_k$$

where $g > 0$ stands for the average curvature of the magnetic field line. By analogy with the Schrödinger equation, $V(x) \equiv (gk^2/\gamma^2)\bar{n}'/\bar{n}$ plays the role of the pseudo-potential, k^2 representing the "total energy." It is well known that the eigenmodes are radially localized in the local minima of the potential $V(x)$.

Indeed, let's consider the following equilibrium profile:

$$\bar{n} = n_0 \left[1 - \varepsilon \tanh \left(\frac{x - x_0}{\lambda_n} \right) \right]$$

with $\varepsilon \ll 1$. In this case, the pseudo-potential V is well centered at x_0 and of the form:

$$V(x) \approx \frac{-\varepsilon g k^2}{\gamma^2 \lambda_n} \cosh^{-2} \left(\frac{x - x_0}{\lambda_n} \right)$$

Looking for solutions localized close to x_0, such that $\rho \equiv (x - x_0)/\lambda_n \ll 1$, one finds that the radial modes satisfy the equation:

$$\frac{\partial^2 \hat{\phi}_k}{\partial \rho^2} + (k\lambda_n)^2 \left[\frac{\varepsilon g}{\gamma^2 \lambda_n} (1 - \rho^2) - 1 \right] \hat{\phi}_k = 0 \tag{5.53}$$

The solutions read $\hat{\phi}_k = \hat{\phi}_{k0} \exp\{\gamma t - \frac{1}{2}(\rho/\rho_0)^2\}$, where γ and ρ_0 are given by

$$\gamma^2 = \varepsilon g k^2 \rho_0^4 \lambda_n$$

$$\rho_0^2 = -\frac{1}{k^2 \lambda_n^2} \pm \frac{2}{k\lambda_n} \left[1 + \frac{1}{(2k\lambda_n)^2} \right]^{1/2}$$

In tokamaks, spatial scales are such that $k\rho_i < 1$ and $\lambda_n \gg \rho_i$, such that $k\lambda_n \gg 1$ is the general case. Within this limit, $\rho_0^2 \approx \pm 2/k\lambda_n$, and $\gamma \approx 2(\varepsilon g/\lambda_n)^{1/2}$. This latter expression highlights the two driving terms of the interchange instability, namely, the gradient length λ_n and the magnetic field curvature g. The resulting mode structure appears to be localized within the pseudo-potential well, i.e., close to x_0:

$$\hat{\phi}_k = \hat{\phi}_{k0} \exp \left\{ 2 \left(\frac{\varepsilon g}{\lambda_n} \right)^{1/2} t - \left(\frac{x - x_0}{\ell} \right)^2 \right\}$$

with $\ell = 2(\lambda_n/k)^{1/2}$. In the considered limit $\ell \ll \lambda_n$, Fig. 5.9, the mode remains localized in the large gradient region. One can further notice that it is more elongated along the periodic direction y than along x, since $k\ell \gg 1$. In the case of multiple regions of strong equilibrium gradient, the mode can even tunnel from one region to another.

Besides, the back reaction of the turbulent flux on the mean profile can lead to large-scale transport events, as predicted theoretically [20]. The sequence of events can be understood as follows. Let's consider the local steepening of the mean gradient (as expected close to the driving source), favoring the local increase of the

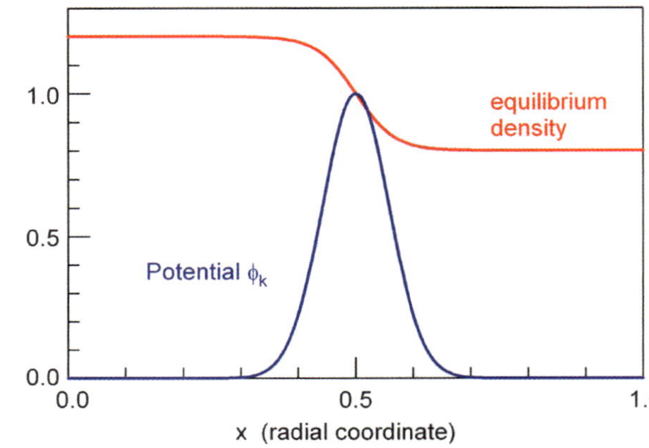

Fig. 5.9 The Fourier modes of the electric potential are trapped within the large density gradient region

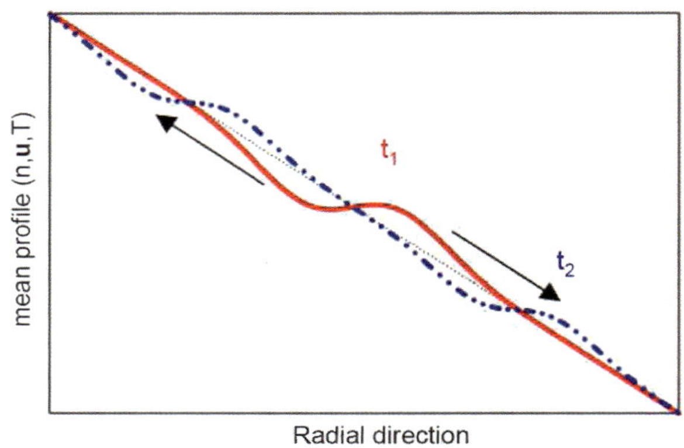

Fig. 5.10 Schematic mechanism of an avalanche event understood as a domino effect

fluctuations. The turbulent flux then increases locally, leading to the flattening of the profile at this location. As a consequence, the gradient increases in the close neighborhood (as illustrated in Fig. 5.10), as a result of the conservation of the transported quantity. The same process then propagates both uphill and downhill, similarly to fronts in sandpile avalanches. Note however that the turbulent flux is always outwards. Such a mechanism can be further amplified if kinds of "turbulence channels" preexist in the system. They would be the analogous to large-scale cyclones or anticyclones in atmospheric turbulence. They are known as "streamers" in tokamak plasmas and correspond to radially elongated convective cells. These streamers can either be coherent structures, i.e., such that their correlation time is much larger than their turnover time, or form transiently, via for instance percolation processes of smaller structures [21].

A good example of such a ballistic transport is provided by the simulation shown in Fig. 5.11, obtained with the global gyrokinetic code GYSELA, which solves the coupled set of gyrokinetic equation for ions and quasi-neutrality equation with adiabatic electrons in the flux-driven regime [22]. It models both drift-wave and interchange instabilities due to ions in tokamak plasmas (see Sects. 5.3.1–5.3.2).

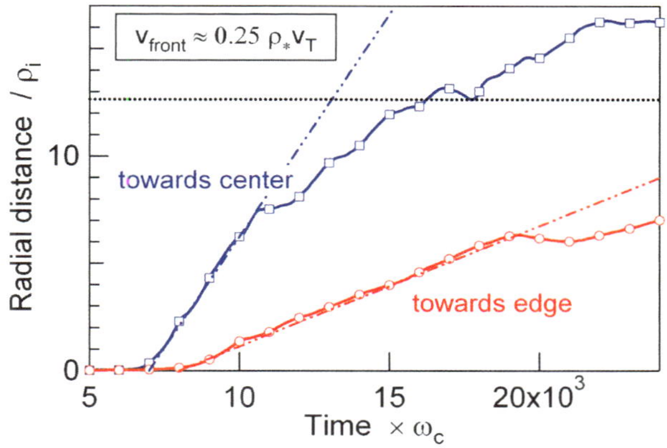

Fig. 5.11 Propagation radial distance – normalized to the ion Larmor radius – of the initial temperature gradient fronts as a function of time – normalized to the ion cyclotron frequency. The fronts invade both core (*blue*) and edge (*red*) initially stable regions. Simulations performed with the GYSELA code [22]

Here, the initial R/L_T profile, with R the major radius and L_T the temperature gradient length $|\nabla \log T| = L_T^{-1}$, which typically exhibits a double hyperbolic tangent radial shape, expands ballistically towards both radial ends of the simulation domain at a speed of the order of $v_{front} \approx 0.25\rho_* v_T$, where $\rho_* = \rho_i/a$ is the local normalized gyroradius and v_T the local ion temperature. Due to the larger temperature in the core ($\rho = 0.2$) than at the edge ($\rho = 0.8$), the front propagates faster in the core. Such a kind of dynamics persists in the saturated nonlinear regime, as exemplified in Fig. 5.12, which features the flux surface average of the radial turbulent heat flux as a function of radial coordinate and time. Propagating fronts are clearly visible, moving both outwards and inwards. In such regimes, gyrokinetic simulations have also found that the plasma can reach complex self-organization states, where large-scale sheared flows govern both the size and the radial localization of avalanche-like transport events [23].

5.4.3 *Large Scale Flows and Transport Barriers*

The crucial role of sheared flows was already identified a few decades ago, both for neutral fluids [24] and for magnetized plasmas. In this latter case, it is equivalent to sheared radial electric field, leading to the radially sheared poloidal component of the flux surface average $E \times B$ drift, namely, $\langle v_{E\theta} \rangle = \langle \partial_r \phi / B \rangle$. By shearing apart the turbulent convective cells, it tends to reduce their size, possibly reducing the turbulent transport. If the velocity shear is large enough, it can even lead to the complete suppression of turbulence. Whenever the heat source is maintained during

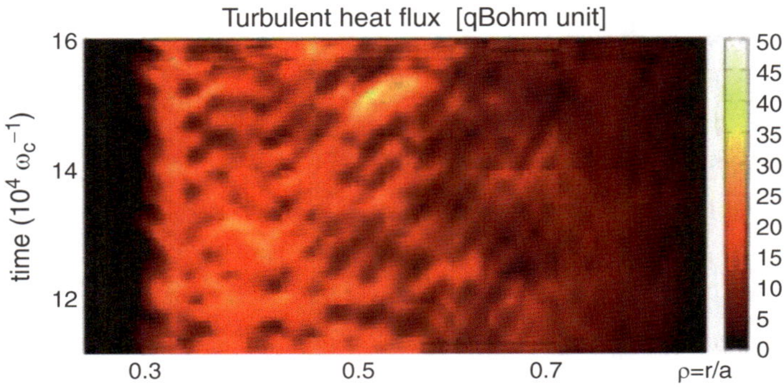

Fig. 5.12 Color plot of the turbulent heat flux in the two-dimensional space (r,t) in the saturated nonlinear regime of a GYSELA simulation [22]. A prescribed heat source term is located around $\rho = r/a = 0.25$

the transition, the system then locally develops large pressure gradient so as to expel the same amount of heat flux with reduced transport coefficient. Transport barriers are then characterized by steep pressure gradients. The threshold in velocity shear for triggering such transport barriers is still a matter of active research. Several theoretical predictions have been proposed, basically relying on the competition between the shearing rate $\gamma_E \sim \langle v_{E\theta} \rangle'$, where the prime denotes radial derivative, and the characteristic decorrelation rate of turbulence [25, 26], or alternatively the linear growth rate of the main instability [27].

More recently, the important role of the turbulence self-generated zonal flows was recognized, both experimentally [28], theoretically [29], and in nonlinear simulations [30]. These low-frequency flows correspond to turbulence-generated poloidal sheared flows. When present, they efficiently shear apart the turbulent convective cells, leading to less radially extended vortices, as illustrated in Fig. 5.13. The time evolution of zonal flows can be obtained by taking the flux surface average and velocity space integral of the gyroaveraged gyrokinetic equation, Eq. 5.16, with the use of quasi-neutrality Eq. 5.18. It then appears that

$$\frac{\partial \langle v_{E\theta} \rangle}{\partial t} = -\frac{1}{r^2} \frac{\partial}{\partial r} \left(r^2 \Pi_{r\theta} \right) \tag{5.54}$$

with $\Pi_{r\theta} \equiv \langle \tilde{v}_{Er} \tilde{v}_{E\theta} \rangle = -\sum_k (k_\theta / B^2)\ \mathrm{Im}\{\hat{\phi}_k^*\ \partial_r \hat{\phi}_k\}$ the (r,θ) component of the Reynolds stress tensor. These axisymmetric flows do not contribute to cross-field transport, while carrying a significant amount of turbulent energy. As such, they represent a wells where turbulent eddies can condense their energy, without any deleterious impact on the confinement properties of the discharge. They are all the more efficient since they are weakly damped (only by collisions in the linear regime [31, 32]), such that they actively contribute to the saturation of turbulence by shearing apart the vortices. They can even trigger transport barriers, as recently reported in numerical simulations [33].

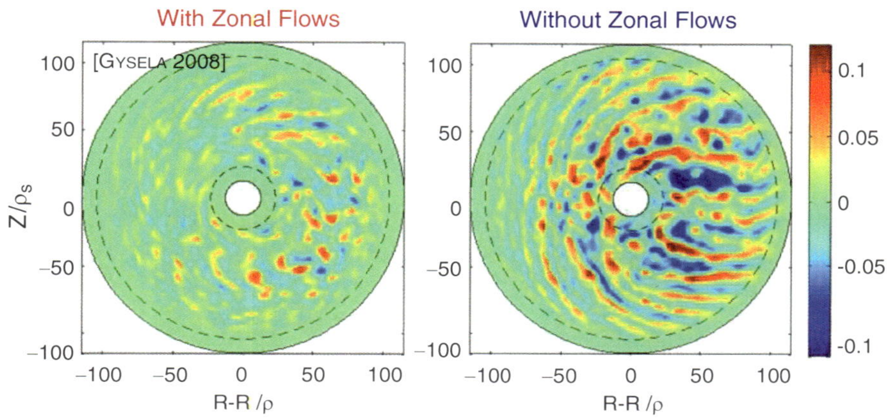

Fig. 5.13 Snapshot of the normalized electric potential fluctuations $e\delta\phi/T = e(\phi - \langle\phi\rangle)/T$ in the nonlinear regime of ion temperature gradient-driven turbulence (5D gyrokinetic code GYSELA). On the right, zonal flows have been artificially suppressed

The shear of the radial electric field is thought to be one of the key ingredients for the so-called L- to H-mode transition in tokamak plasmas (first discovered in the German ASDEX tokamak [34]), where the plasma spontaneously bifurcates from low- to high-confinement regimes when the heating power exceeds some threshold. During such a transition, turbulence is quenched and a transport barrier subsequently develops close to the last closed magnetic flux surface. Here, the radial electric field is observed to reverse sign [35]. The zonal flows could also play a role in triggering the transition. The common understanding is that the steep pressure gradient at the edge, associated to the transport barrier, then keeps the shear of E_r large enough to sustain the H-mode. Indeed, the radial force balance ensures the following relationship:

$$E_r = \frac{\nabla_r p}{en} - v_\theta B_\varphi + v_\varphi B_\theta \tag{5.55}$$

The toroidal velocity is expected to be small in the absence of injected torque, and the poloidal velocity is mainly governed by the neoclassical theory. It should be noticed that such a relation does not hold in the scrape-off layer, where the parallel boundary conditions on the target plates have to be considered. In this region of open magnetic surfaces, one rather expects $e\phi/T_e$ to be constant, leading to $E_r \sim -\nabla_r T_e/e$ in the opposite direction to $\nabla_r p$. The H-mode is further characterized by quasi-periodic relaxation events, during which the pressure pedestal collapses on a few hundreds of microseconds [36]. These so-called edge localized modes (ELMs) put an upper bound to the accessible pressure gradient, in good agreement with MHD stability criterion [37]. Various regimes are observed, mostly depending on the distance to the power threshold, in which their repetition frequency and the energy lost per ELM either increase or decrease with the injected heating power. Their frequency ranges from about ten to two hundred of Hertz. The ELM lost energy increases when frequency decreases and can reach 10 % of the stored diamagnetic energy in the plasma. Since leading to unacceptable power heat loads on the target plates for ITER, the search of control tools of such large-amplitude ELMs is currently a matter of active research.

5.5 Conclusion

Tokamak plasmas are open systems which are intrinsically out of thermodynamical equilibrium. The existence of large gradients can lead to several primary instabilities. They can be either reactive (of fluid type) or kinetic. Most of them saturate nonlinearly at scales of the order of a few ion Larmor radii, leading to turbulent fluctuations of a few percents in the plasma core. The resulting turbulent transport breaks the insulating property of the magnetic configuration, made of nested toroidal magnetic surfaces, hence reducing the fusion performance. The main types of encountered micro-instabilities are drift waves, interchange, and bump-on-tail. Understanding such mechanisms in view of their possible control is one of the challenges for ITER and beyond. Because of their small density and high temperature, fusion plasmas are weakly collisional. Therefore, they require a kinetic treatment, although attempts are made to account for some of the critical kinetic properties—especially wave-particle resonant interactions—in fluid models. In this framework, gyrokinetic models have been developed. They proceed from the phase-space reduction from 6 to 5 dimensions by averaging over the fast gyro-phase of the cyclotron motion. Self-consistent models then couple the gyrokinetic equation, involving the 5D distribution function of the gyro-centers, to Maxwell's equations. In practice, due to the small magnitude of magnetic fluctuations, the electrostatic limit is often considered, using quasi-neutrality in place of Maxwell-Gauss equation.

Although the set of relevant equations is essentially well posed, there remain a number of critical issues in turbulence modeling. First of all, flux-driven global full-f (i.e., accounting for the entire distribution function f, as opposed to δf models which evolve fluctuations only) models, where no scale separability assumption is assumed between fluctuations and mean quantities, reveal complex self-organization of turbulence. This rich dynamic results from the self-consistent interaction between turbulence, whose transport governs the mean profiles and mean gradients which drive the underlying instabilities. In particular, flux-driven simulations exhibit large-scale transport events, reminiscent of avalanches, which propagate almost ballistically on large radial distances, i.e., much larger than the turbulence Eulerian correlation length. A possible explanation is the turbulence trapping in local large-gradient regions, and the subsequent domino-like dynamics of steep gradients. Such avalanches were first predicted theoretically, then confirmed numerically, before being observed experimentally [38,39]. Secondly, full-f models self-consistently account for both equilibrium and turbulence-driven large-scale plasma flows. These flows are particularly important since they are known to contribute to turbulence saturation by tearing apart turbulence eddies. In some cases, they can even lead to the formation of transport barriers, which correspond to localized regions where turbulence is almost suppressed. One of these regimes actually constitutes the reference scenario in ITER.

In conclusion, the paper focusses on the contribution from first-principle modeling on the route towards fusion energy production. No need to say that numerous significant progresses have also been obtained recently in the fields of experimental observations, scenario developments, and technology. These aspects are not addressed in the present paper. It is fair to say that our understanding and prediction of turbulent transport in tokamak plasmas has gained a lot from first-principle nonlinear simulations. Recent critical achievements, especially regarding gyrokinetic models, have been possible thanks to the development of high performance computing resources. Still, flux-driven full-f global gyrokinetic simulations of ITER-like plasmas including both ions *and* electrons remain extremely challenging and certainly still out of reach of present-day supercomputers.

Acknowledgment It is my pleasure to acknowledge colleagues and friends who have most contributed to this paper through numerous enlightening discussions and common work on turbulence and transport for many years: X. Garbet and Ph. Ghendrih, P. Beyer, P.H. Diamond, G. Dif-Pradalier, and V. Grandgirard. Many thanks as well to the students J. Abiteboul, A. Strugarek, D. Zarzoso, and T. Cartier-Michaud. Last but not least, I wish to acknowledge C. Passeron for her precious support on numerical issues.

References

1. X. Garbet (Guest Editor), *Turbulent Transport in Fusion Magnetised Plasmas*, vol 6 (C.R. Physique, Amsterdam, 2006), 573–699
2. G.W. Hammett, F.W. Perkins, Phys. Rev. Lett. **64**, 3019 (1990)
3. M.A. Beer, Ph.D. thesis, Princeton University (1995)
4. B. Snyder, G.W. Hammett, W. Dorland, Phys. Plasmas **4**, 3974 (1997)
5. H. Sugama, T.-H. Watanabe, W. Horton, Phys. Plasmas **10**, 726 (2003)
6. T. Passot, P.L. Sulem, Phys. Plasmas **10**, 3906 (2003)
7. T. Chust, G. Belmont, Phys. Plasmas **13**, 012506 (2006)
8. P. Bertrand, M.R. Feix, Phys. Lett. **28**A, 68 (1968)
9. P. Morel, E. Gravier, N. Besse, R. Klein, A. Ghizzo, P. Bertrand, X. Garbet, P. Ghendrih, V. Grandgirard, Y. Sarazin, Phys. Plasmas **14**, 112109 (2007)
10. Y. Sarazin, G. Dif-Pradalier, D. Zarzoso, X. Garbet, Ph. Ghendrih, V. Grandgirard, Plasma Phys. Control. Fusion **51**, 115003 (2009)
11. L.D. Landau (1946), "On the vibrations of the electronic plasma", in *Collected Papers of L.D. Landau*, vol 61, ed. by D. Ter Haar (Pergamon Press, Oxford, 1965), p. 445
12. B.D. Fried, S.D. Conte, *The Plasma Dispersion Function* (Academic Press, New York NY, 1961)
13. A.M. Dimits et al., Phys. Plasmas **7**, 969 (2000)
14. Y. Sarazin, V. Grandgirard, G. Dif-Pradalier, E. Fleurence, X. Garbet, Ph Ghendrih, P. Bertrand, N. Besse, N. Crouseilles, E. Sonnendrücker, G. Latu, E. Violard, Plasma Phys. Control Fusion **48**, B179–B188 (2006)
15. A.J. Brizard, T.S. Hahm, Rev. Mod. Phys. **79**, 421 (2007)

16. V. Grandgirard ans Y. Sarazin, to appear in *Panoramas et Synthèses*, Société Mathématique de France (2013)
17. A. Hasegawa, M. Wakatani, Phys. Rev. Lett. **50**, 682 (1983)
18. M. Wakatani, A. Hasegawa, Phys. Fluids **27**, 611 (1984)
19. D. Zarzoso, X. Garbet, Y. Sarazin, R. Dumont, V. Grandgirard, Phys. Plasmas **19**, 022102 (2012)
20. P.H. Diamond, T.S. Hahm, Phys. Plasmas **2**, 3640 (1995)
21. Y. Sarazin, V. Grandgirard, J. Abiteboul, S. Allfrey, G. Dif-Pradalier, X. Garbet, Ph. Ghendrih, G. Latu, A. Strugarek, Nucl. Fusion **50**, 054004 (2010)
22. V. Grandgirard et al., Commun. Nonlinear Sci. Numer. Simulation **13**, 81–87 (2008)
23. G. Dif-Pradalier, P.H. Diamond, V. Grandgirard, Y. Sarazin, J. Abiteboul, X. Garbet, Ph. Ghendrih, A. Strugarek, S. Ku, C.S. Chang, Phys. Rev. E **82**, 025401(R) (2010)
24. L.N. Howard, R. Krishnamurti, J. Fluid Mech. **170**, 385–410 (1986)
25. H. Biglari, P. Diamond, P. Terry, Phys. Fluids B **2**, 1 (1990)
26. T.S. Hahm, K.H. Burrell, Phys. Plasmas **2**, 1648 (1995)
27. R.E Waltz, G.D. Kerbel, J. Milovich, Phys. Plasmas **1**, 2229 (1994)
28. A. Fujisawa, K. Itoh, H. Iguchi et al., Phys. Rev. Lett. **93**, 165002 (2004)
29. P.H. Diamond, M.N. Rosenbluth, F.L. Hinton et al., *Plasma Physics Control Nuclear Fusion Research* (IAEA, Vienna, 1998)
30. Z. Lin, T.S. Hahm, W.W. Lee, W.M. Tang, P.H. Diamond, Phys. Rev. Lett. **83**, 3645 (1999)
31. M.N. Rosenbluth, F.L. Hinton, Phys. Rev. Lett. **80**, 724 (1998)
32. F.L. Hinton, M.N. Rosenbluth, Plasma Phys. Controlled Fusion **41**, A653 (1999)
33. Y. Sarazin, V. Grandgirard, G. Dif-Pradalier et al., Phys. Plasmas **13**, 092307 (2006)
34. F. Wagner et al., Phys. Rev. Lett. **49**, 1408 (1982)
35. R. Moyer, K. Burrell, T. Carlstrom et al., Phys. Plasmas **2**, 2397 (1995)
36. H. Zohm, Plasma Phys. Control. Fusion **38**, 105 (1996)
37. J.W. Connor, Plasma Phys. Control. Fusion **40**, 191 (1998)
38. P.A. Politzer, Phys. Rev. Lett. **84**, 1192 (2000)
39. Y. Sarazin, M. Bécoulet, P. Beyer, X. Garbet, Ph. Ghendrih, T.C. Hender, E. Joffrin, X. Litaudon, P.J. Lomas, G.F. Matthews, V. Parail, G. Saibene, R. Sartori, Plasma Phys. Control. Fusion **44**, 2445 (2002)

Part III
From Kinetics to Fluids and Solids

Chapter 6
Turbulent Thermal Convection and Emergence of Isolated Large Single Vortices in Soap Bubbles

Hamid Kellay

Abstract Experiments using a novel thermal convection cell consisting of half a soap bubble heated at the equator to study turbulent thermal convection and the movement of isolated vortices are reviewed. The soap bubble, subject to stratification, develops thermal convection at its equator. A particular feature of this cell is the emergence of isolated vortices. These vortices resemble hurricanes or cyclones and similarities between these structures and their natural counterparts are found. This is brought forth through a study of the mean square displacement of these objects showing signs of superdiffusion. In addition to these features, the study of the statistical properties of the turbulence engendered in these soap bubbles shows a clear indication for the existence of the so-called Bolgiano–Obukhov scaling both for the temperature and the velocity fluctuations. A remarkable transition is uncovered: the temperature and the velocity structure functions show intermittency for small temperature gradients; this intermittency then disappears for large gradients.

6.1 Introduction

Turbulent thermal convection is ubiquitous in several natural settings such as the atmosphere or the inner core of planets and has attracted and continues to attract considerable attention from experimentalists and theorists [1, 2]. Experiments have demonstrated several robust features of this phenomenon such as the importance of thermal plumes and the onset of a large-scale circulation [3]. Several experiments use three-dimensional geometries, but recent experiments have demonstrated similar features in two dimensions [4–6]. These experiments use vertical, stably

H. Kellay (✉)
Université Bordeaux1, LOMA UMR 5798 du CNRS,
351 cours de la Libération, 33405 Talence, France
e-mail: hamid.kellay@u-bordeaux1.fr

X. Leoncini and M. Leonetti (eds.), *From Hamiltonian Chaos to Complex Systems: A Nonlinear Physics Approach*, Nonlinear Systems and Complexity 5,
DOI 10.1007/978-1-4614-6962-9_6,

stratified, soap films or soap bubbles [7, 8] as two-dimensional fluids. The flow occurs in the plane of the film and the velocity in the third dimension is strongly inhibited due to viscous dissipation. Soap films have now become good model systems to study two-dimensional hydrodynamics and turbulence [9] and these recent experiments extend their use to turbulent thermal convection. Interest in two-dimensional turbulence stems from the fact that atmospheric turbulence at large scales displays two-dimensional features due to the small thickness of the atmosphere [10]. According to some authors, this two dimensionality may have strong repercussions: The great red spot of Jupiter has been brought forth as a sign of the two-dimensional nature of atmospheric turbulence for example [11].

Besides its importance for the geophysical context, other fundamental issues arise. As for three-dimensional hydrodynamic turbulence [12–14], the statistical properties of temperature and velocity fluctuations in turbulent thermal convection, a state which can be reached for a high enough temperature difference between the bottom and the top of the container, can also be described by scaling laws [15,16]. While several experiments have been carried out to measure these statistical properties, a number of issues regarding the scaling properties remain unresolved [17]. In the two-dimensional version, which has been put forth recently using either vertical soap films or soap bubbles [5–7], a detailed examination of the statistical properties of the velocity fluctuations, the temperature fluctuations, and the density variations [5–8] showed that they indeed display scaling laws predicted by Bolgiano and Obukhov for stratified turbulence in the 1950s [15–17]. Such scaling laws have so far been elusive in three-dimensional experiments for reasons still debated today [17, 18].

This paper reviews the two-dimensional experiments carried out in a soap bubble and focuses both on the emergence of isolated vortices and on the statistical properties of the turbulent thermal convection produced. The paper is organized as follows: first we bring forth the emergence of these large-scale vortices and outline their specific properties, then we describe the main features of thermal convection and its statistical properties in this novel setup.

6.2 Isolated Vortices

A specific feature of recent experiments on soap bubbles subjected to thermal convection is the emergence and persistence of large isolated single vortices. The soap bubble (actually a half bubble is used) is heated at the equator giving rise to thermal convection. A prominent feature of this setup is the emergence of long-lived isolated vortices reminiscent of natural ones such as the red spot or hurricanes and cyclones. As we will see below, these vortices wander around the bubble randomly. The mean square displacement of these vortices varies as a power law in time. This scaling is different from the one expected for diffusive behavior and shows signs of superdiffusion. Surprisingly, analysis of the trajectories of natural hurricanes in the earth's atmosphere gives rise to a similar scaling law for their mean square displacement versus time. Thus, the properties of these isolated vortices in this novel experimental system mimic some features of the position fluctuations of natural hurricanes. This suggests that a small experimental setup such as the one used here may allow a careful study of such large-scale phenomena of importance for atmospheric science and for meteorology. A major difference between this two-dimensional setup and previously used cells to study thermal convection is the absence of lateral walls. We believe that this absence of walls is at the origin of the emergence of long-lived isolated vortices as opposed to a large-scale circulation in cells with lateral walls.

The setup consists of a hollow brass ring with an inlet and outlet for water circulation to thermostat the full apparatus at the desired temperature. This brass ring has a circular slot which can be filled with soap water. The middle part of the ring was covered with a Teflon disk. The half bubble was blown with a straw using the soap solution in the circular slot. The ring could be kept at the desired temperature to a precision of $\pm 0.1\,^{\circ}\mathrm{C}$ using the water circulation thermostat. The temperature of the solution (the soap solution is water at different concentrations c of detergent ranging from 0.2 % to 5 %) in contact with the ring and the temperature at the top of the bubble were measured using a needlelike thermistor. The temperature difference between the bottom and the top parts of the half bubble will be denoted ΔT which is our control parameter. The room was kept at a constant temperature of $17\,^{\circ}\mathrm{C}$. The difference in temperature ΔT can be changed in the range 5–45 °C. The ring has two concentric slots so we could vary the diameter of the half bubble which could be fixed to either 8 or 10 cm. Typical half bubbles with strong convective patterns are shown in Fig. 6.1. The patterns are filmed using a 3CCD camera.

Figure 6.1 shows typical half bubbles heated at the equator at different temperatures and illuminated with white light. Interference colors mark the surface of the bubble indicating variations in the thickness of the soap film. When no temperature gradient or a small gradient is present the thickness of the bubble decreases as the height increases giving rise to the horizontal bands seen in the photograph. The two-dimensional density of the soap film being ρh, where ρ is the density of soap water and h its thickness, the film is stably stratified with dense fluid at the bottom and lighter fluid at the top. When a sufficient temperature gradient is applied, the region near the equator is host to rising plumes just like in conventional thermal convection. This convection zone extends all around the equator and grows in height as the gradient increases as shown in Fig. 6.1. The upper part of the bubble is more quiescent than the zone near the equator. No particular difference appears at this stage with conventional thermal convection. A major difference is the absence of a large-scale circulation and the emergence and persistence of single isolated vortices in the upper part of the half bubble as displayed in Fig. 6.1d. Their presence is more frequent as the temperature difference increases.

These isolated vortices emerge randomly on the surface of the bubble, grow in size rapidly as illustrated in Fig. 6.2, and persist for relatively long times almost equivalent to the lifetime of the half bubble itself which could last several minutes. The occurrence of these vortices becomes more probable for higher ΔT while they are almost absent for small ΔT. At first sight, these isolated vortices move around

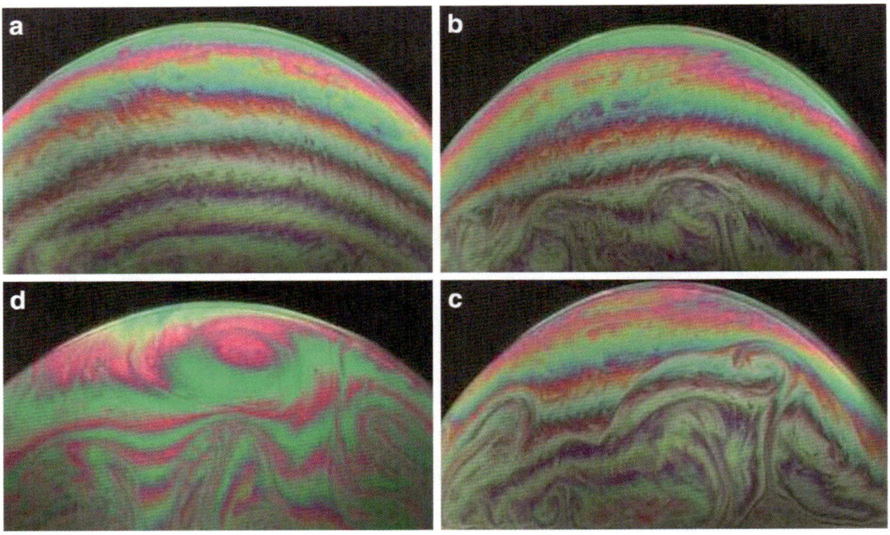

Fig. 6.1 Images of bubbles at different temperature gradients. ΔT increases from (**a**) to (**c**) with $\Delta T = 9, 17$, and $31\,^{\circ}$C, respectively. The convection zone grows in extent as ΔT increases. Plumes can be seen in this zone. (**d**) A bubble with a convection zone and an isolated vortex near the top for $\Delta T = 45\,^{\circ}$C

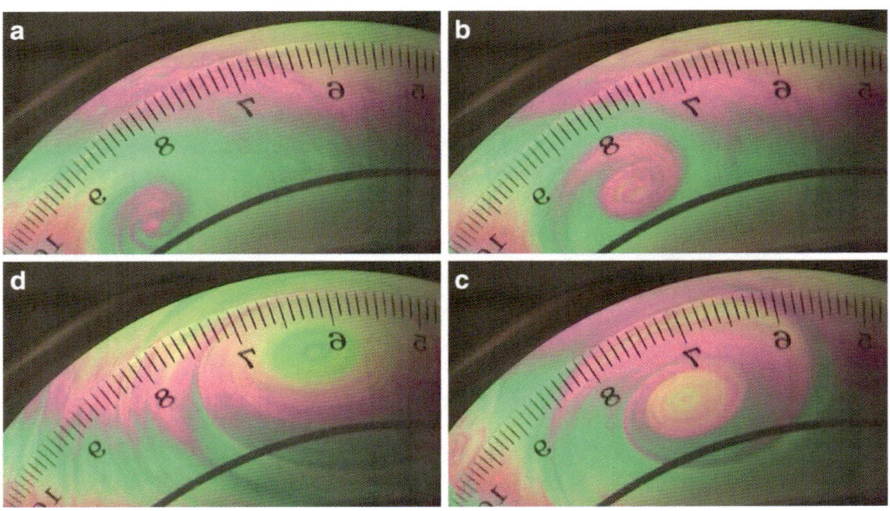

Fig. 6.2 Birth and growth of a single vortex. The time between successive images is 0.16 s, 0.44 s, and 0.76 s. The image of a transparent ruler (in cm) was projected on the bubble

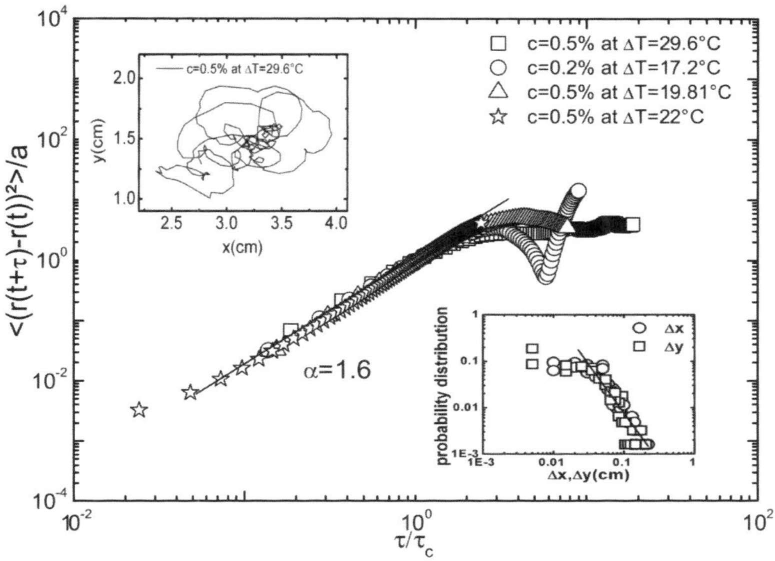

Fig. 6.3 Mean square displacement of the isolated vortices for different ΔT. *Upper inset*: a track of an isolated vortex. *Lower inset*: the pdf of the increment Δx and Δy for a fixed time interval Δt

the bubble randomly. A typical trajectory is shown in Fig. 6.3. These vortices move around the bubble with velocities near 1 cm/s. We analyze these trajectories by calculating the mean square displacement $< r^2(\tau) >=< (r(t+\tau)-r(t))^2 >$ for different time increments τ. This analysis shows that $< r^2(\tau) >\sim \tau^{\alpha}$ with $\alpha \sim 1.6$. This scaling law is to be contrasted with Brownian motion for which the scaling exponent is 1. Figure 6.3 shows this result for different temperature differences ΔT. Here we plot $\frac{<r^2(\tau)>}{<r^2(\tau_c)>}$ versus $\frac{\tau}{\tau c}$. The characteristic time τ_c is the correlation time obtained from the correlation function $< r(t+\tau)r(t) >$. The rescaling by τ_c and the corresponding mean squared displacement $< r^2(\tau_c) >$ collapses all of the data for different temperature differences ΔT. The scaling law observed is valid for more than a decade in timescales below τ_c. Above τ_c the mean square displacement seems to flatten with no systematic dependence. Despite the complexity of the problem, a single scaling law summarizes all of the data and strongly suggests that these vortices are superdiffusive, a behavior observed here for the first time as far as we know. Superdiffusion arises in the so-called Lévy flights [19, 20] for which the spatial steps for a fixed time increment are distributed according to a power law pdf$(\delta r) \sim r^{-\beta}$. The exponent of this power law is directly related to the exponent α by the relation $\beta = 1 + 2/\alpha$. A pdf of the displacements Δx and Δy in the horizontal and vertical directions is shown in the inset to Fig. 6.3. A small power law range can be observed here with an exponent $\beta = 2.2$ in good agreement with that obtained

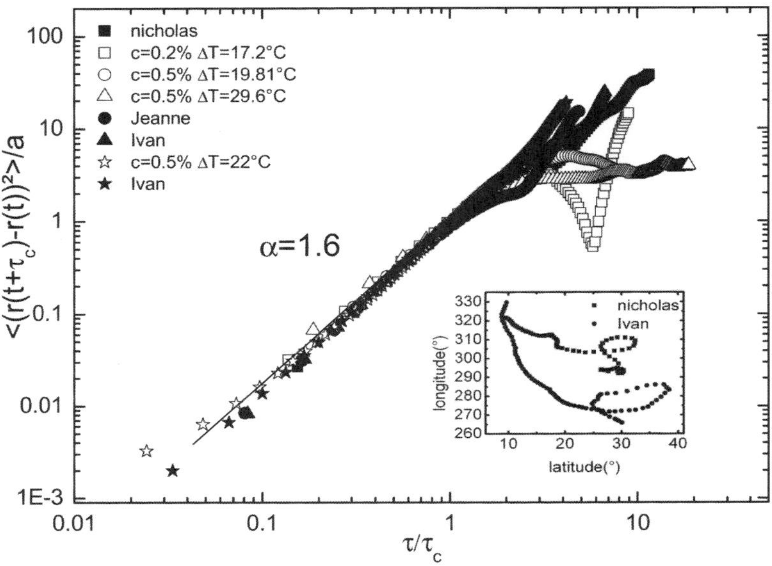

Fig. 6.4 Mean squared displacement for natural hurricanes plotted along with some of the data from the isolated vortices in this experiment. *Lower inset*: two hurricane tracks

from the mean square displacement as indicated by the solid line. This indicates that the movement of the vortices can be recast into the random walks known as Lévy flights.

Isolated vortices occur in natural settings as well. Because of the similarities between our isolated vortices and hurricanes or cyclones we have analyzed the trajectories of certain natural hurricanes along similar lines. It should be signalled here that natural hurricanes seem to travel along relatively well-defined mean trajectories for which the Coriolis force and the beta-effect play a central role. However, they do show fluctuations around this mean trajectory. An analysis of the mean square displacement of different hurricanes including Nicholas (2003), Jeanne (2004), and Ivan (2004) shows a very similar behavior as our isolated vortices. The hurricane trajectories were obtained from the National Hurricane Center web site and consist of either satellite observations or of radar data. Their trajectories are sampled every 6 h for satellite data and every 15 min for radar data. The analysis of these trajectories is summarized in Fig. 6.4 where the trajectories are displayed in the inset. Here, the data of Fig. 6.3 are replotted so as to illustrate the similarity. This plot shows that the hurricanes and our isolated vortices display very similar features especially the power law scaling at times smaller than τ_c. The scaling exponent turns out to be very close to the value extracted from our vortices, namely, $\alpha = 1.6$.

Superdiffusive behavior can be traced to a nontrivial interaction between the moving object and the medium. An example of entities that interact with the medium itself has been illustrated through a study of the superdiffusion of passive beads in a bath of self-propelling bacteria [21]. Another example concerns the movement of passive beads in a laminar rotating flow where vortices may trap and release the particles giving rise to superdiffusion [22]. The isolated vortices here are kicked by the turbulent agitation of the surrounding flow. These vortices, being part of the flow, must have an important reaction on the medium itself. In addition, the movement of vortices is sensitive to the sign of vorticity gradients [23] which in a turbulent medium may show a complicated spatial and temporal distribution and a nontrivial interaction with the moving vortex giving rise to a complex trajectory and dynamics.

6.3 Statistical Properties of the Temperature and Velocity Fields

As stated above, the turbulent convection produced in such a cell can be characterized with respect to the velocity and temperature fluctuations. And a question that needs to be answered is about the relevance of known scaling laws to this situation. The two-dimensional nature of the system allows for tests that are difficult to carry out in three dimensions. We therefore explored the temperature field in this unusual thermal convection cell: half a soap bubble heated from below (see Fig. 6.1) [7]. As mentioned above, this geometry has the advantage of avoiding the presence of side walls and therefore the presence of the large-scale circulation often observed when lateral walls are present. By focusing on the structure functions of the temperature field a transition from an intermittent to a non-intermittent behavior has been observed. The results show that the scaling of these functions switches regimes from the so-called Obukhov–Corrsin-like scaling [13, 14] with intermittency at low temperatures to Bolgiano–Obukhov-like scaling without intermittency at higher temperatures. Our results are unique and surprising since previous numerical work indicated the presence of strong intermittency for the temperature field [24, 25]. Intermittency in fluid turbulence is an important problem in hydrodynamics and our experiments bring to light how a simple system evolves from an intermittent to a non-intermittent state.

The setup is in a room kept at a constant temperature of 17 °C with a humidity rate of nearly 75 % near the bubble. The temperature gradient between the bottom and the top of the half bubble ΔT could be varied up to 55 °C. The temperature measurements used a calibrated 14 bit infrared camera (resolution 256×360) working in the spectral range 3.6–5 μm with a sensitivity of 20 mK and an adjustable exposure time set between 0.5 and 1 ms. Images of the same region (between 100 and 500 images at a rate of 50 or 100 frames/second) were recorded and a homemade program was used to calculate temperature differences across different scales r. Averaging over the area of interest and over several images allowed us to improve the statistics (between 1 and 2.5 million points were used) and calculate the high-order moments of these differences. The temperature field was recorded for periods of up to 10 s which is greater than the temperature correlation time (of order 0.1 s).

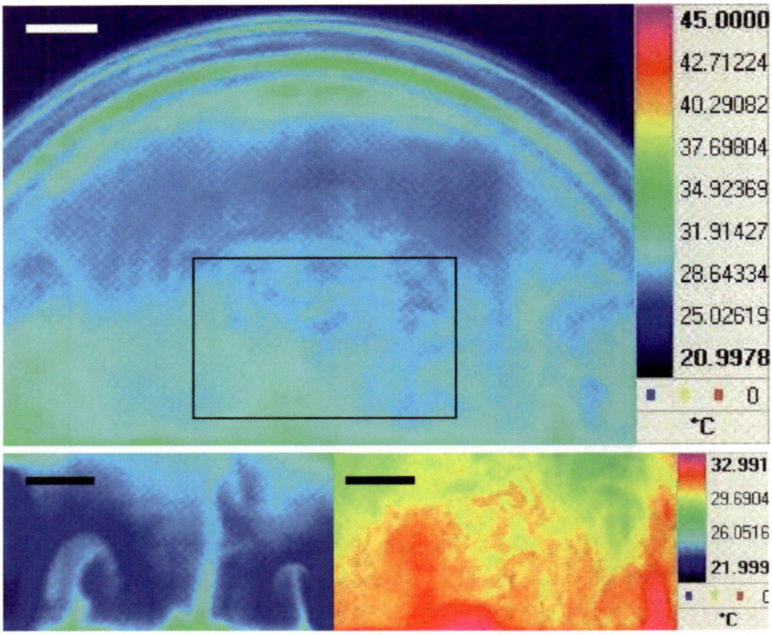

Fig. 6.5 Infrared images of the bubble (*top* $\Delta T = 50\,^{\circ}\mathrm{C}$) and a region near the *bottom*: $\Delta T = 21\,^{\circ}\mathrm{C}$ (*bottom left*) and $50\,^{\circ}\mathrm{C}$ (*bottom right*). The region delimited by a *rectangle* in the upper image indicates the area covered by the temperature and velocity measurements. The *brass ring* is located a few millimeters from the bottom of the images

The error in r, introduced by the curved geometry of the bubble, turned out to be less than a few percent over a 1 cm region. The effect of evaporation was estimated to be small and the lifetime of the bubble, which should decrease with increased evaporation, actually increases by a factor of about 4 when a temperature gradient is imposed indicating that convection is more important than both evaporation and draining by gravity.

Figure 6.5 shows a full view of the bubble as well as images obtained with the infrared camera in a region near the bottom of the half bubble where the thermal convection is strongest. One can easily identify thermal plumes rising from the bottom of the cell which are clearly visible for the low temperature gradient. The thermal convection becomes more intense as the temperature gradient increases and well-defined thermal plumes are difficult to discern. From such spatial images we extract the temperature difference $\delta T(r_x) = T(x + r_x) - T(x)$ and $\delta T(r_y) = T(y + r_y) - T(y)$ and calculate the nth moments as $< |\delta T(r_x)|^n >$ and $< \left|\delta T(r_y)\right|^n >$. Here x and y refer to the horizontal and vertical coordinates and the brackets refer to an average over space and time. The temperature structure functions are important quantities in the study of turbulence and different scaling relations have been proposed for their variation versus the scale r. In 3D turbulent flows, where

Kolmogorov-like scaling is believed to prevail for the low-order moments, Obukhov and Corrsin [13, 14] generalized the scaling arguments of Kolmogorov to a scalar field like the temperature and used both the energy dissipation rate ε and the scalar dissipation rate ε_θ to predict that the second-order structure functions should scale as $\varepsilon_\theta \varepsilon^{-1/3} r^{2/3}$. Similar scaling arguments can be used, as suggested by Bolgiano and Obukhov [15–17] for stably stratified turbulence, to the case of Rayleigh-Benard convection for which the fluid thermal expansion rate β, the gravity constant g, and the dissipation rate ε_θ fix the scaling relation of the second-order structure function of the temperature as $\varepsilon_\theta^{4/5}(\beta g)^{-2/5} r^{2/5}$ [17]. The nth order moments are expected to vary as a power law of the separation distance r with an exponent ζ_n^T of $n/5$ in the Bolgiano–Obukhov regime and $n/3$ for the Obukhov–Corrsin regime. To compare the experimental conditions here to their classical counterparts, we estimated the Rayleigh number ($Ra = \beta \Delta T g R^3/\nu\kappa$ where ν and κ are the kinematic viscosity and the thermal diffusivity of water) to be between 7×10^7 and 2×10^8 while the Reynolds number ($Re = V_{\text{mean}} R/\nu$ where V_{mean} is the characteristic horizontal velocity) is estimated to be about 3,000.

The temperature structure functions are displayed in Fig. 6.6a, b for two different ΔT: 21 °C and 50 °C. For the low ΔT, Fig. 6.6a, the temperature structure functions are roughly isotropic as the values of the differences for the two orthogonal spatial increments r_x and r_y are similar. These functions display power law scaling for spatial scales between roughly 1 and 10 mm as the compensated moments show. The scaling exponents vary in a nontrivial manner versus the order n of the moment. This exponent is in agreement with predictions of Obukhov and Corrsin [13, 14] for low n in Kolmogorov-like turbulence [12]. However, for higher moments, the exponents deviate from this prediction. The growth is nonlinear versus n which is the hallmark of intermittency. The relation between the higher-order moments and the low-order ones is nontrivial indicating that the functional shape of the probability distribution functions of the increments varies with r_x or r_y. This behavior is similar to that observed in three-dimensional experiments where Bolgiano-like scaling has not been observed so far; rather Obukhov–Corrsin scaling with deviations, just like passive scalar fields in three-dimensional hydrodynamic turbulence, is observed [17, 18, 26].

The high ΔT results are shown in Fig. 6.6b. While the structure functions show isotropy and power law scaling versus r, the variation of the exponents versus n turns out to be different from the previous results. Bolgiano–Obukhov-like scaling is observed in the range 1–10 mm as shown in Fig. 6.6b which displays the compensated moments as well as the scaling exponents extracted from such an analysis. Estimates of the Bolgiano length scale (above which such a scaling is believed to prevail) give $L_B \sim 1$ mm which is in good agreement with the range observed here and in previous experiments using vertical films [5, 6]. The surprising aspect is that a linear variation of the exponents versus n is observed. This linear variation indicates that intermittency is absent. This behavior has been observed for an imposed ΔT higher than about 35 °C.

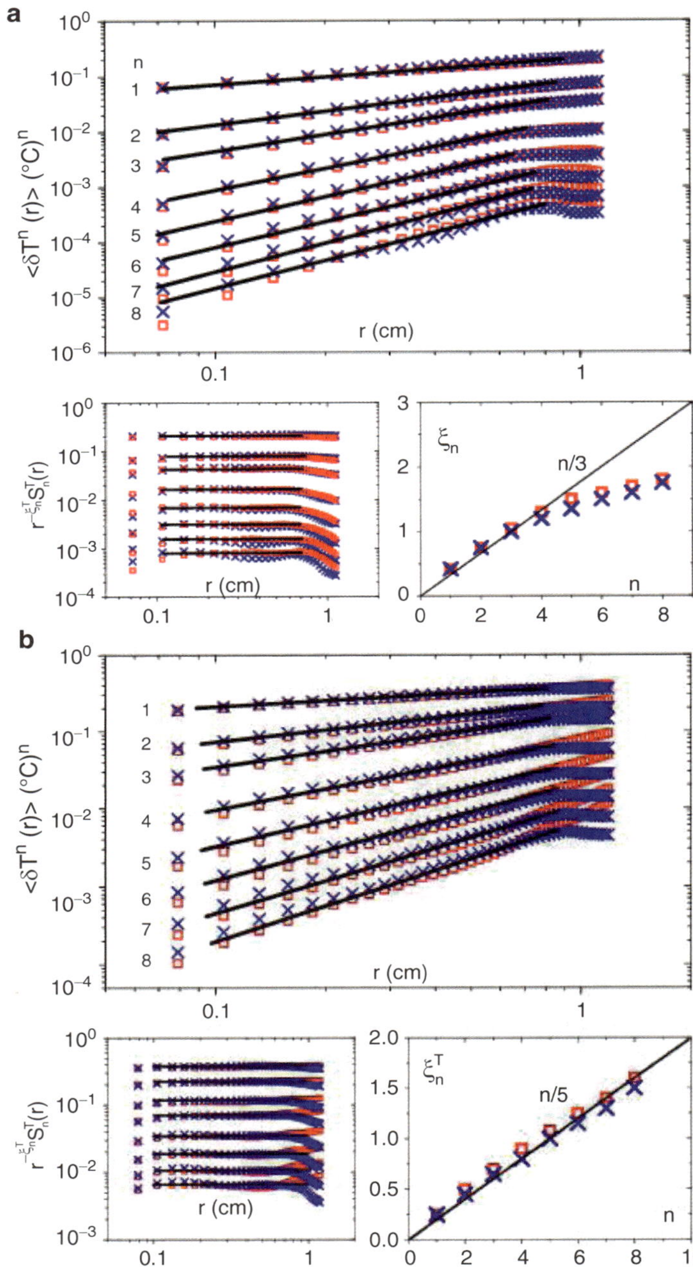

Fig. 6.6 Temperature structure functions for $\Delta T = 21\,°\mathrm{C}$ (**a**) $\Delta T = 50\,°\mathrm{C}$ (**b**). The *horizontal* (*squares*) and *vertical* (*crosses*) components are plotted up to order 8. The data is shifted by a multiplicative factor (x2 for n = 3 up to x64 for n = 8). Insets: compensated moments and scaling exponents

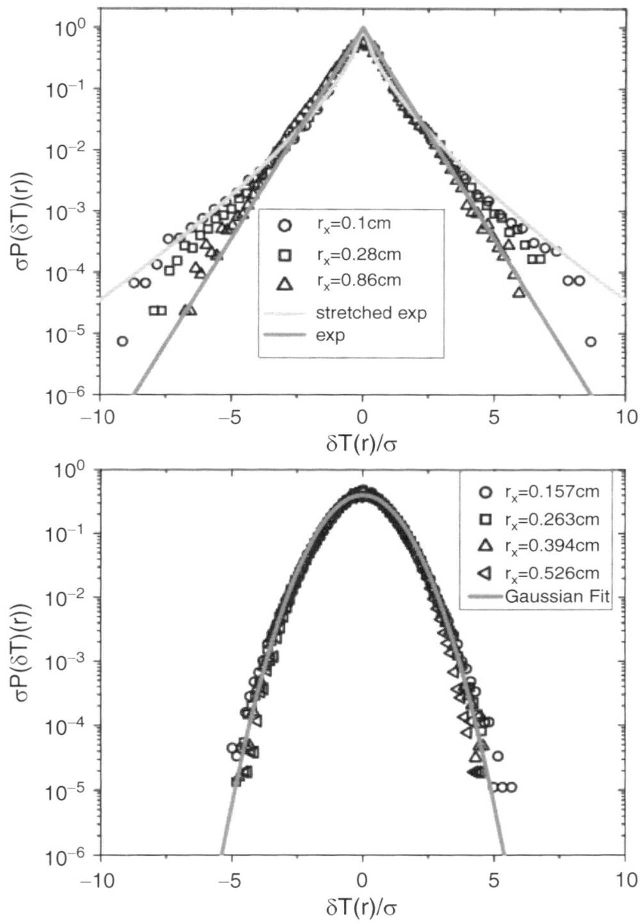

Fig. 6.7 Pdfs of temperature increments for $\Delta T = 21\,^{\circ}\mathrm{C}$ (*upper set*) $\Delta T = 50\,^{\circ}\mathrm{C}$ (*lower set*). The axes were rescaled by the standard deviation σ of the increments. *Solid lines*: fits using a stretched exponential function and an exponential function. A Gaussian function is used in the lower set of pdfs

Additional insight into this transition comes from an examination of the probability density functions (pdfs) of the temperature increments in the range of scales for which power laws are observed. These results are shown in Fig. 6.7. The horizontal axis has been rescaled by the standard deviation $\sigma(=\sqrt{<|\delta T(r)|^2>})$ of the temperature increment at the scale r while the vertical axis has been normalized in such a way that the integral of the function is unity. Note that for the small gradient, the pdfs start out as a stretched exponential with an exponent near 0.7 at the small-scale end of the scaling range (1 mm) and end up as an exponential for the large-scale end (9 mm). The pdfs evolve gradually as the scale increases from 1 to 10 mm, indicating a change of the functional shape of the pdf across the scales. For the high ΔT, the pdfs remain roughly Gaussian as the scale changes from 0.9 up to 8 mm. The normalization of the pdfs by σ collapses all the pdfs together indicating that they depend solely on the width of the distribution. An examination of the flatness of these pdfs shows that for the low ΔT, the flatness decreases from roughly 10 to 5 as the scale increases from 1 to 10 mm. On the other hand, the flatness for the high ΔT case remains roughly constant near a value of 2.8 which is not far from the flatness of a Gaussian distribution. These features are at the origin of the dependence of the scaling exponents versus n. In short, intermittency of the scaling exponents is associated with the gradual change of the functional shape of the probability density of temperature increments. On the other hand, the absence of intermittency is related to the Gaussian pdfs of the increments all through the scaling range.

To complement these observations, we measured the velocity fluctuations at a single point and constructed the one-dimensional velocity spectra as well as the horizontal velocity structure functions of order n as $< |\delta V(\delta t)|^n >$ where δt is a temporal increment. These measurements use a Laser Doppler Velocimeter and a soap solution seeded with 1 μm-sized polystyrene spheres. We record between 5×10^5 and 5×10^6 points over a total period of time of nearly 100 s which is greater than the velocity correlation time (0.2 s). The choice of the horizontal component of the velocity V is justified by the presence over sufficiently long periods of time of a mean flow in this direction, at the location of the measurements, allowing the use of the Taylor frozen turbulence hypothesis to convert δt to a scale r as $r = V_{\mathrm{mean}} \delta t$. This hypothesis is not tested here however. The velocity structure functions are expected to vary as power laws of r with an exponent $\zeta_n^V = 3n/5$ following similar arguments as for the temperature in the Bolgiano–Obukhov regime.

The properties of the convective zone share some similarities with those observed in experiments of convection in vertical soap films. A common feature is the scaling of the velocity field. The results of Zhang and Wu [6] show that in the turbulent regime obtained for high ΔT, the second-order structure function of the velocity differences scales as $<\delta v^2(r)>=<(v(r)-v(0))^2>\sim r$. This scaling is consistent with Bolgiano's prediction [15] for the energy density spectrum which reads: $E(k) \sim k^{-11/5}$. Our results for the velocity spectra are displayed in Fig. 6.8. The horizontal axis is frequency because the one-point measurements are time series of the velocity. The frequency axis can be converted to a wave number in the direction of the mean flow using Taylor's frozen turbulence assumption. The mean velocity here is horizontal so the wave number is in the horizontal direction and is given by $k_x = 2\pi f / V_{\mathrm{mean}}$. A histogram of the horizontal velocity fluctuations is shown in the inset to Fig. 6.8 showing a well-defined nonzero velocity. The scaling obtained from our data gives an exponent of -2.2 which is consistent with the findings of Zhang and Wu [6] and is in agreement with the Bolgiano scaling expected for the buoyancy subrange of turbulence in stably stratified fluids [15].

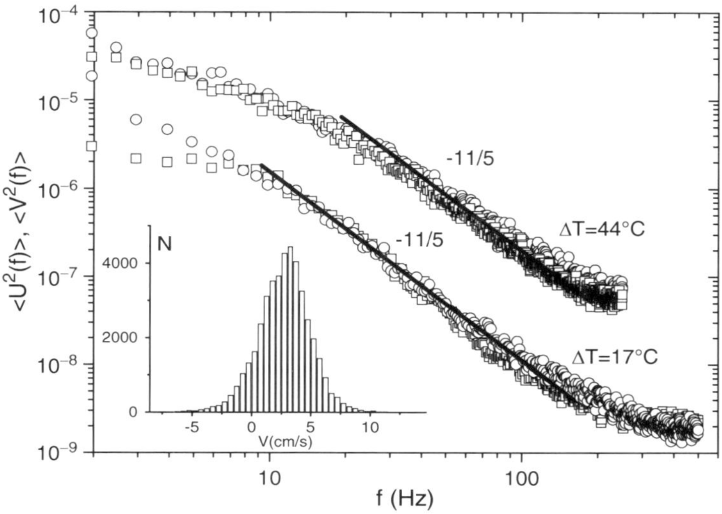

Fig. 6.8 Velocity spectra in the convection zone. *Circles* are for the vertical velocity while *squares* are for the horizontal velocity. The *solid line* is the Bolgiano prediction. Inset: a histogram of the horizontal velocity fluctuations for $\Delta T = 44\,^{\circ}\mathrm{C}$

The structure functions of the velocity differences are displayed in Fig. 6.9a, b for the low and high ΔT, respectively. Power laws are obtained for the different moments examined. The variation of the exponents, obtained from an analysis of the compensated moments, is nonlinear for the low ΔT case and linear for the higher one. The expected $3n/5$ variation is shown as a solid line. The scaling range in r, determined using the Taylor hypothesis, turns out to be similar to that obtained from the temperature structure functions. The variation of the exponents for the low ΔT case (as well as the flatness of the distributions which decreases from roughly 10 to nearly 3) is similar to that obtained by Zhang and Wu [6]. The agreement between our results and the spatial measurements of Zhang and Wu seems to validate our use of the Taylor hypothesis for this case. The low temperature gradient statistics therefore show that intermittency is observed for both the velocity and the temperature. For the high ΔT, the scaling exponents vary linearly with n and follow the $3n/5$ law in good agreement with the predictions of Bolgiano and Obukhov for the same range of scales r as the temperature. The flatness of the distributions in this case remains roughly constant near a value of 5. These results are therefore consistent with the temperature measurements and indicate an absence of intermittency for the velocity as well. The Taylor hypothesis in this case has not been tested however and the results of Zhang and Wu did not show an absence of intermittency in their rectangular cells for similar temperature gradients. Our results, even though consistent with the temperature measurements, would need additional confirmation. We noted by examining the pdfs of velocity differences that their

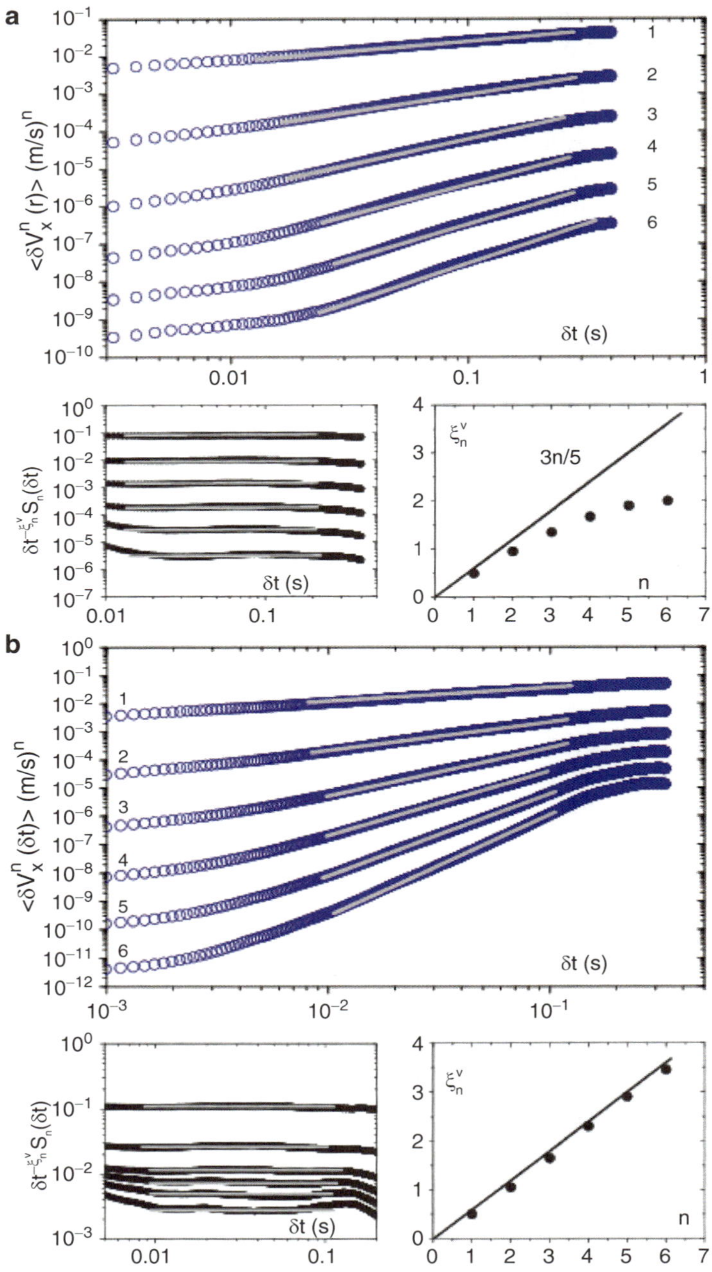

Fig. 6.9 ($< |\delta V(\delta t)|^n >$ for $\Delta T = 21\,°C$ (**a**) $\Delta T = 50\,°C$ (**b**). The mean horizontal velocity is 5 cm/s in (a) and 8 cm/s in (b). Insets: compensated moments and scaling exponents

evolution with ΔT is less convincing than that of the temperature: while a better collapse can be achieved for the high ΔT case, the pdfs are not Gaussian and show roughly exponential tails.

6.4 Conclusion

In conclusion, our novel quasi two-dimensional convection cell allows for a detailed study of the statistical properties of temperature fluctuations in turbulent thermal convection. These properties show that a transition from an intermittent state to a non-intermittent one occurs as the temperature gradient increases. Bolgiano–Obukhov-like scaling with no intermittency is recovered for the high-gradient case. Our results raise fundamental questions about the role of lateral walls and the ensuing large-scale circulation often observed in traditional convection cells as well as the role of thermal plumes in setting the properties of temperature fluctuations in turbulent thermal convection.

In addition, our experiments show that the curved nature of the bubble used allows for isolated vortices to emerge. The absence of walls is the most probable reason for the emergence of such structures. These isolated vortices resemble natural hurricanes for certain aspects. In particular, when the mean squared displacement of the eye is examined superdiffusion is recovered. This superdiffusion is probably indicative of Lévy flights and calls for further theoretical work on the movement of isolated vortices in a turbulent medium which is important for turbulence in general and for atmospheric and meteorological studies in particular.

Acknowledgements This work was carried out with my colleagues (M. Bessafi and Y. Amarouchene) and students (F. Seychelles, F. Ingremeau, T. Meuel, G. Prado) to whom I am very grateful. This work was supported by Grant "Cyclobulle" from the ANR.

References

1. L.P. Kadanoff, Phys. Today **54**, 34 (2001)
2. E.D. Siggia, Ann. Rev. Fluid Mech. **26**, 137 (1994)
3. G. Ahlers, Physics, **2**, 74 (2009)
4. B. Martin, X.L. Wu, Phys. Rev. Lett. **80**, 1892 (1998)
5. J. Zhang, X.L. Wu, K.Q. Xia, Phys. Rev. Lett. **94**, 174503 (2005)
6. J. Zhang, X.L. Wu, Phys. Rev. Lett. **94**, 234501 (2005)
7. F. Seychelles et al., Phys. Rev. Lett. **100**, 144501 (2008)
8. F. Seychelles, F. Ingremeau, H. Kellay, Phys. Rev. Lett. **105**, 264502 (2010)
9. H. Kellay, W.I. Goldburg, Rep. Prog. Phys. **65**, 1 (2002).
10. P. Morel, M. Larcheveque, J. Atmos. Sci. **31**, 2189 (1974)
11. P.S. Marcus, Nature **428**, 828 (2004)
12. A.N. Kolmogorov, Dokl. Akad. Nauk. SSSR **30**, 299 (1941)
13. S. Corrsin, J. Appl. Phys. **22**, 469 (1951)

14. A.M. Obukhov, Izv. Akad. Nauk SSSR Ser. Geog. Geofiz **13**, 58 (1949)
15. R. Bolgiano, J. Geophys. Research **64**, 2226 (1959)
16. A.M. Obukhov, Sov. Phys. Dokl. **4**, 61 (1959)
17. D. Lohse, K.-Q. Xia, Annu. Rev. Fluid Mech. **42**, 335 (2010)
18. C. Sun, Q. Zhou, K.-Q. Xia, Phys. Rev. Lett. **97**, 144504 (2006)
19. M.F. Shlesinger, G.M. Zaslavsky, U. Frisch (eds.), Levy Flights and Related Topics in Physics. Lecture Notes in Physics (Springer, Berlin, 1995)
20. J.P. Bouchaud, A. George, Phys. Rep. **195**, 127 (1990)
21. X.L. Wu, A. Libchaber, Phys. Rev. Lett. **84**, 3017 (2000)
22. T.H. Solomon, E.R. Weeks, H.L. Swinney, Phys. Rev. Lett. **71**, 3975 (1993)
23. D.A. Schecter, D.H.E. Dubin, Phys. Rev. Lett. **83**, 2191 (1999)
24. A. Celani, A. Mazzino, M. Vergassola, Phys. Fluids **13**, 2133 (2001)
25. A. Celani et al., Phys. Rev. Lett. **88**, 054503 (2002)
26. Z. Warhaft, Annu. Rev. Fluid Mech. **32**, 203 (2000)

Chapter 7
On the Occurrence of Elastic Singularities in Compressed Thin Sheets: Stress Focusing and Defocusing

Alain Pocheau

Abstract Compressing thin sheets usually yields the formation of singularities which focus curvature and stretch on points or lines. In particular, following the common experience of crumpled paper where a paper sheet is crushed in a paper ball, one might guess that elastic singularities should be the rule beyond some compression level. In contrast, we show here that, somewhat surprisingly, compressing a sheet between cylinders makes singularities spontaneously disappear at large compression. This "stress-defocusing" phenomenon is qualitatively explained from scale invariance and further linked to a criterion based on a balance between stretch and curvature energies on defocused states. This criterion is made quantitative using the scalings relevant to sheet elasticity and compared to experiment. These results are synthesized in a phase diagram completed with plastic transitions. They end up with a renewed vision of elastic singularities as a thermodynamic condensed phase where stress is focused, in competition with a regular diluted phase where stress is defocused. Different compression routes may be followed in this diagram by managing differently the two principal curvatures of a sheet, as experimentally achieved here. In practice, besides the famous Elastica and crumpled paper routes, this offers interesting alternatives for compressing a sheet with an amazing spontaneous regularization of geometry and stress that repels the occurrence of plastic damages.

A. Pocheau (✉)
IRPHE, Aix-Marseille Université, 49, rue F. Joliot Curie, B.P. 146,
13384 Marseille Cedex 13, France
e-mail: alain.pocheau@irphe.univ-mrs.fr

X. Leoncini and M. Leonetti (eds.), *From Hamiltonian Chaos to Complex Systems: A Nonlinear Physics Approach*, Nonlinear Systems and Complexity 5,
DOI 10.1007/978-1-4614-6962-9_7,

7.1 Introduction

Thin envelops, thin layers, or thin films stand as an efficient mean to separate domains, treat surfaces, or confine volumes. Examples include graphene sheets [1], epitaxial deposit at sub-micrometric scales [2], membranes at micrometric scales [3], packaging at sub-millimeter scales [4], metallurgical structures at millimeters scales and beyond, and geological layers at even larger scales [5], the scale meaning here the thickness of the object. Their common feature is to display a weak dimension, their thickness, in comparison to their length and width, according to which most of their properties can be recovered by treating them as 2D surfaces involving flexural effects. In many instances, however, these thin sheets undergo geometrical constraints that force them to fit into a reduced space. They then have to adapt their form to restrictive conditions, something they may do smoothly or sharply, i.e., with small or large curvature as compared to their inverse thickness. In the latter case, they then escape the 2D surface assumption, especially at the locations of large curvature where they show up surface singularities. The occurrence of these singularities is essential in various instances. In practice, they involve large elastic stresses and 3D interactions that make them escape the 2D modeling and possibly even the physical regime of the remainder of the sheet. In particular, the sheet properties are usually altered there regarding electronic properties, robustness, or even the elasticity regime, with a possible transition to plasticity at the core of singularities.

On a more general viewpoint, these singularities enable the elastic stress to relax in the remaining sheet parts: the bending stress for the so-called ridges [6–8] and the stretching stress for the so-called developable cones (d-cones) [9–12]. In this sense, they appear as inner degrees of freedom for adapting the geometric constraints imposed to compressed sheets. Doing so, they thus focus the sheet's stress on the singularity cores, a phenomenon called stress focusing [13].

Stress focusing is a particular example of the phenomenon of energy focusing which widely occurs in out-of-equilibrium systems beyond some distance to equilibrium. Some of its manifestations are vorticity concentration in turbulent fluids, rogue waves occurrence, shock wave formation in thermodynamic systems, dielectric breakdown in media submitted to electrostatic field, fracture in stressed solids, etc. (Fig. 7.1). In all these phenomena, the energy density which moved the system far from equilibrium spontaneously turned from a homogeneous distribution to a highly localized concentration. This phenomenon thus stands as an emblematic example of self-organization.

Stress focusing is all the more surprising that one might have naively guessed that energy seeks to spread over instead of concentrating on singular objects. In particular, this energy focusing goes against the equidistribution of energy and thus questions the statistical description of these systems. Moreover, the emergence of definite locations or structures (sometimes called "coherent structures") where energy is concentrated, largely governs the system behavior and its properties. This

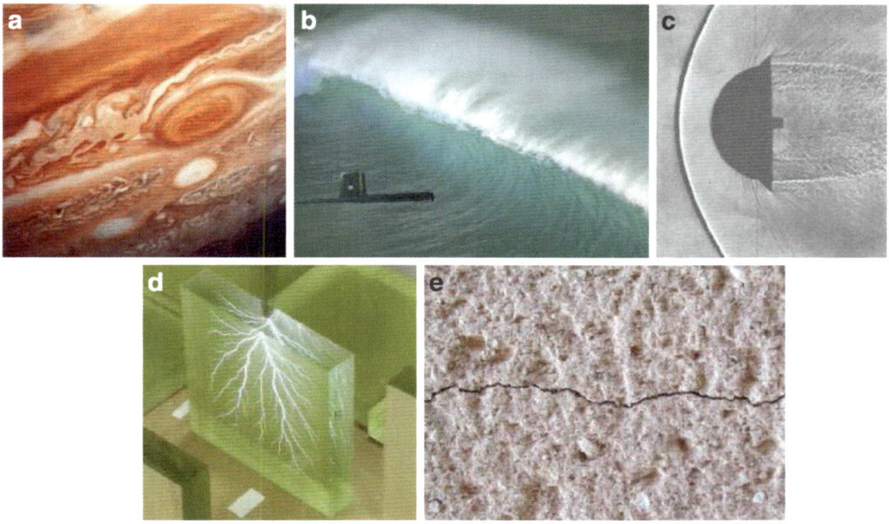

Fig. 7.1 Examples of energy focusing : (**a**) vorticity concentration in the Jupiter red spot taken by Voyager 2 (Credit NASA image), (**b**) rogue wave (Credit Toptenz.net) (**c**) shock wave produced by blunt bodies (Credit NASA image) (**d**) dielectric breakdown yielding so-called lichtenberg figures (Credit Theodore Gray as shown on http://www.capturedlightning.com) (**e**) fracture in solid concrete

is why vast efforts have been devoted in all the above systems to characterize the conditions for energy focusing as well as the resulting energetically dense structures and their implications.

Here, we address this issue in the context of sheet compaction where forms and stresses are governed by elasticity. Two major rules for self-organization are then in order. First, as there is no intrinsic scale in elasticity, scale invariance and scaling arguments apply. Second, as elasticity is non-dissipative in its elastic regime, energy landscapes can be used to infer the preferred states, including those involving elastic singularities. In particular, the occurrence of a stress-focused state may be understood as the fact that it became energetically preferred as compared to a stress-distributed state. Applying both these rules should then largely help elucidating stress focusing, but with possible surprises. In particular, the popular example of crumpled paper where the compaction of a sheet in a ball generates scars (Fig. 7.3-right) usually yields the common guess that singularities and stress focusing should irremediably persist when increasing compaction. On the opposite, we shall find here that, surprisingly, the two above rules deny this belief, in the sense that scalings imply that singular states should no longer be preferred at large compaction, if they previously were: stress should thus *defocus* at large compaction.

This phenomenon of stress defocusing implies that the ultimate state of a compressed sheet should be smooth and regular. Its actual existence will be evidenced on a dedicated experiment and the apparent paradox regarding the usual experience of crumpled paper will be clarified [14]. This will enable us to identify singularities as a thermodynamic condensed phase surrounded in phase space by the regular diluted phase corresponding to regular geometries and defocused stress. In particular, the persistence of singularities on some actual compression routes will be shown to refer to plasticity instead of elasticity. In this regard, the popular demonstration of paper crumpling by the hands will appear as a misleading example of (linear) elasticity since the singularities that form should disappear at large compression but actually do not because of plasticity only.

Altogether, this study will thus provide a modified vision of the nature of elastic singularities and of sheet adaptation to compression. In particular, on compression routes, singularities, instead of being the rule beyond some compression level, will actually appear as a transient state.

In the following, we first emphasize in Sect. 7.2 the relevance of an intermediate compression route between Elastica and crumpled paper to address singularity occurrence. We then recall in Sect. 7.3 some basics on linear elasticity, especially regarding the Gaussian curvature and its implications. We then report in Sect. 7.4 an experiment of compression between cylinders and the resulting evidence of stress defocusing. Energy arguments for stress focusing or defocusing are then addressed in Sect. 7.5 together with scaling arguments. They are applied in Sect. 7.6 to show the necessity of defocusing and derive a phase diagram for singularities, taking into account plasticity. This is followed by a conclusion on the implications of this study for the nature of singularities in elasticity.

7.2 On Singularity Occurrence in Sheet Elasticity: From Elastica to Crumpled Paper

On thin sheets, two kinds of stresses may be defined: one related to the stretch of a sheet viewed as a 2D surface and one related to the sheet's curvature [11, 13, 15]. As regards to stress focusing, it appears that the stretch stress is the dominant stress on singularities and on the non-singular states on which they appear. Accordingly, singularity formation actually corresponds to focusing that stretch on singularities, leaving in between unstretched but possibly curved domains.

Interestingly, stretch is generated by sheet deformations that involve a Gaussian curvature G, actually equal to the product of the principal curvatures c_1, c_2, at a point: $G = c_1 c_2$ (see Sect. 7.3.2). Usually, compacting sheets cannot avoid generating Gaussian curvatures, thus stretch, until provoking stress focusing in singularities, as on crumpled paper (Fig. 7.3-right). To improve our understanding of this phenomenon, the clue of this work is to notice that, as *two* principal curvatures are involved in the Gaussian curvature G, several physically different compacting routes may be explored depending on the correlations involved between them. In particular, three different routes worth being distinguished:

- Elastica: $G = 0$, $c_2 = 0$

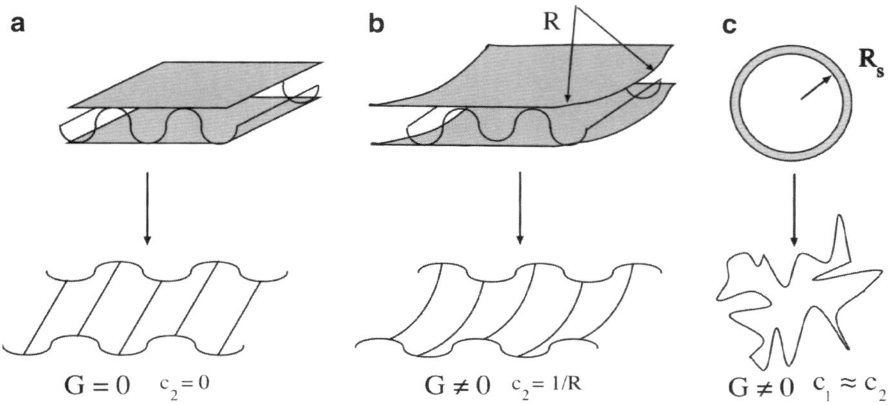

Fig. 7.2 Different compression routes regarding singularity occurrence: (**a**) Compression between flat plates. Fold axes are straight so that a principal curvature c_2 is forced to vanish: $c_2 = 0$, $G = 0$. The compression route shows no singularity. (**b**) Compression between cylinders. Fold axes are bent by the cylinders so that a principal curvature c_2 is forced to be that of the cylinders: $c_2 = 1/R$, $G \neq 0$. (**c**) Compression by a shrinking sphere of radius R_s. This corresponds to the crumpled paper configuration. Principal curvatures do not vanish and are about the same: $c_1 \equiv c_2 \equiv 1/R_s$, $G \neq 0$. This generates singularities

One may annihilate one curvature, simply by forbidding curvature on a direction (Fig. 7.2a). This is achieved in practice by compressing sheets in between parallel plates. Then, a family of parallel folds is generated by iterated bucklings, all parallel to one direction of the plates along which they are thus uncurved (Fig. 7.3a). One curvature, c_1, is thus provided by folds but the other, $c_2 = 0$, vanishes since it corresponds to the uncurved fold axis direction. Then $G = 0$ so that no stretch is generated and, therefore, no singularity at any compression level. The sheet is thus equivalent to a set of rods whose elastic evolutions are modeled by the so-called Euler's "Elastica" [16, 17].

- Isotropic compression and crumpled paper: $G \neq 0$, $c_1 \equiv c_2$.
  On the opposite, one may force the two principal curvatures not to vanish and to take statistically similar values (Fig. 7.2c). This is achieved in practice by compressing a sheet into a shrinking spherical domain, as when crumpling a paper with hands (Fig. 7.3-right). Then, because of isotropy, the two principal curvatures are both nonzero and statistically equivalent $c_1 \equiv c_2$, so the term "isotropic" compression. A Gaussian curvature is thus generated and increases with compaction until singularities occur.
- Anisotropic compression: $G \neq 0$, c_2 fixed
  Finally, a third route, actually intermediate between the two above opposite routes, may be designed by compressing sheets not between plates or a sphere but between cylinders (Fig. 7.2b). Taking the cylinders curvature radius R large compared to the gap between them, this looks locally similar to a compression in between parallel plates so that a family of parallel folds is expected. However,

Fig. 7.3 *Left*: buckling cascade on a sheet compressed between parallel plates [18]. *Right*: crumpled paper crushed in hands

the fixed cylinder curvature nevertheless bends their fold axes and this makes all the difference. The curvature c_2, instead of being zero as on the Elastica route, is fixed here to a nonzero value equal to the inverse cylinder curvature radius $c_2 = 1/R$. A nonzero Gaussian curvature is then generated yielding singularity formation beyond a compression level. However, in contrast with crumpled paper, the imposed curvature c_2 of the fold axes is kept constant here. It is thus decorrelated from the remaining fold curvature c_1, so the term "anisotropic" compression. Should this difference be relevant and yield a different sheet evolution?

The experiment reported in Sect. 7.4 will provide the answer. Interestingly, we note that the end result could be anticipated from scale invariance, but we postpone the explanation to Sect. 7.6.1.

7.3 Basics on Linear Elasticity of Sheets

The linear response of materials to deformation has been synthesized by Robert Hooke in the rule "ut tensio sic vis" which means that stresses and strains follow each other proportionally. The development of elasticity generalized this to a tensorial relationship between stress and strain whose possible forms can be simply grasped by considering the elastic energy [11, 13, 15].

We shall call in the remainder $\underline{\gamma}$ and $\underline{\sigma}$ the strain tensor and the stress tensor, (i, j) the indexes of the coordinates tangent to the sheet, and k the index of the normal coordinate to the sheet. The volumic density of elastic energy ε will then satisfy $\partial\varepsilon/\partial\gamma_{ij} = \sigma_{ij}$.

Reducing attention to sheets here enables one to decompose stresses into their average along the sheet depth, $\bar{\sigma}_{ij}$, and the complementary part σ'_{ij}: $\sigma_{ij} = \bar{\sigma}_{ij} + \sigma'_{ij}$.

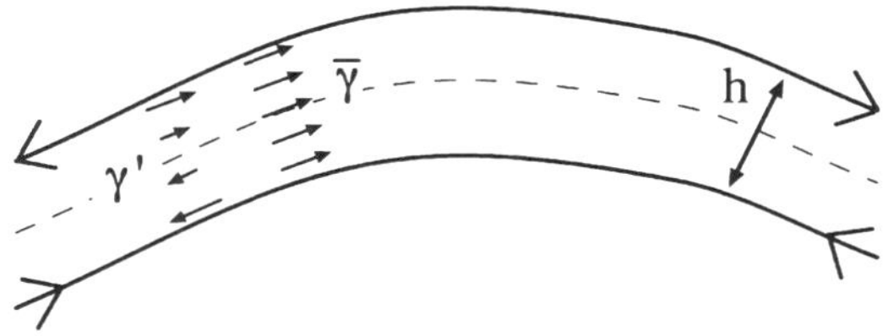

Fig. 7.4 Strain in a sheet made of the uniform stretching part $\bar{\gamma}$ and the flexural part γ' equal to the difference of strains with respect to the mid-height surface (*dotted line*)

The former stress corresponds to viewing the sheet as a superposition of identical 2D surfaces and the latter stress expresses the actual differences undergone by these surfaces depending on their position on the sheet depth: σ'_{ij} depends on x_k.

As 2D surfaces have zero thicknesses, their stresses only refer to stretching. However, following sheet curvature, the different surfaces which compose a sheet actually experience different strains since outer or inner surfaces stand at slightly different distances from the curvature centers. They thus involve some differences that are related to curvature. Within the thin sheet approximation, one can assume a linear variation of the complementary stresses σ'_{ij} with the normal component x_k: $\sigma'_{ij} \propto x_k$, the origin $x_k = 0$ being placed at the middle of the sheet thickness. Then the stresses σ_{ij} involve a mean part $\bar{\sigma}_{ij}$ independent of x_k and a complementary part σ'_{ij} linearly varying with x_k: $\sigma'_{ij} = x_k \tilde{\sigma}'_{ij}$. The same is true for the corresponding strains $\underline{\gamma} = \underline{\bar{\gamma}} + \underline{\gamma'}$ with $\underline{\bar{\gamma}}$ and $\underline{\gamma'}$, respectively, independent of and proportional to x_k: $\gamma'_{ij} = x_k \tilde{\gamma}'_{ij}$(Fig. 7.4).

By definition, the surfacic energy density e satisfies $\delta e = \int \sigma_{ij} \delta \gamma_{ij} dx_k$, the integral being taken between $\pm h/2$, h designing the sheet thickness. Integration thus yields $\delta e = \delta e_s + \delta e_b$ with $\delta e_s = h \bar{\sigma}_{ij} \delta \bar{\gamma}_{ij}$ and $\delta e_b = \frac{h^3}{12} \tilde{\sigma}'_{ij} \delta \tilde{\gamma}'_{ij}$. Here e_s denotes a stretching energy density and e_b a bending energy density. Interestingly, e_s is proportional to h but e_b is proportional to h^3.

On this basis, the objective remains to clarify the link between stress and strain. For this, a convenient way consists in using the elastic energy, thanks to its scalar nature. This, together with an emphasis on the role of the Gaussian curvature, will yield the expression of the link between sheet form and elastic stresses in equilibrium states in the form of the Föppl–von Kármán equations.

7.3.1 Sheet Elastic Energy

Following the linear relationship between stress and strain, the volumic elastic energy density ε is quadratically related to the strain tensor $\underline{\gamma}$. However, the energy density being scalar, it must be related to those parts of the strain tensor that are scalar and, because of global rotational invariance, isotropic. These constraints select two candidates only, the trace of the tensor square $\mathrm{Tr}(\underline{\gamma}^2)$ and the square of the tensor trace $[\mathrm{Tr}(\underline{\gamma})]^2$, yielding a simple relationship with coefficients λ and μ

called Lamé coefficients: $\varepsilon = \frac{1}{2}\lambda\,[\mathrm{Tr}(\underline{\gamma})]^2 + \mu\mathrm{Tr}(\underline{\gamma}^2)$. Two coefficients only are thus required to characterize an elastic medium. Formally, they are actually the analogous of the two viscosities required to characterize a viscous fluid.

Given the expression of the elastic energy density of a volumic material, one now wishes to apply it to the specific case of a thin sheet whose strain tensor $\underline{\gamma}$ can be decomposed into a thickness-uniform strain tensor $\bar{\underline{\gamma}}$ and a linearly thickness-dependent strain tensor $\underline{\gamma}'$: $\underline{\gamma}' = -x_k\underline{C}$. Here, the tensor $\underline{C}$ corresponds to the curvature tensor defined by $C_{i,j} = \mathbf{n}\partial^2\mathbf{r}/(\partial x_i \partial x_j)$. The dependence on x_k then states that the sheet surface that is the farthest from the center of curvature is stretched whereas that which is the nearest from this center is compressed, as compared to the mid-thickness surface $x_k = 0$ (Fig. 7.4).

Integration of the volumic energy density over the sheet thickness yields:

* No cross contribution between $\bar{\underline{\gamma}}$ and $\underline{\gamma}'$, i.e., between stretching and bending, for parity reason in x_k
* A surfacic stretching energy $e_\mathrm{s} = \frac{h}{2}[\lambda\,[\mathrm{Tr}(\bar{\underline{\gamma}})]^2 + \mu\mathrm{Tr}(\bar{\underline{\gamma}}^2)]$
* A surfacic bending energy $e_\mathrm{b} = \frac{h^3}{24}[\lambda\,[\mathrm{Tr}(\underline{C})]^2 + \mu\mathrm{Tr}(\underline{C}^2)]$

Here, both the strain tensors $\bar{\underline{\gamma}}$ and $\underline{C}$ only depend on the in-plane components x_i, x_j and are thus 2D. Interestingly, the trace of their square then satisfies $\mathrm{Tr}(T^2) = \mathrm{Tr}(T)^2 - 2\mathrm{Det}(T)$, as may be directly checked from the algebra of 2*2 matrices. As their trace and their determinant easily express as the sum and the product of their eigenvalues, it appears convenient to rewrite the surfacic energies in term of them:

* $e_\mathrm{s} = \frac{h}{2}\frac{E}{(1-\nu^2)}\{[\mathrm{Tr}(\bar{\underline{\gamma}})]^2 - 2(1-\nu)\mathrm{Det}(\bar{\underline{\gamma}})\}$.
* $e_\mathrm{b} = \frac{h^3}{24}\frac{E}{(1-\nu^2)}\{[\mathrm{Tr}(\underline{C})]^2 - 2(1-\nu)\mathrm{Det}(\underline{C})\}$ where $\nu = \lambda/2(\lambda+\mu)$ is the Poisson ratio and $E = \mu(3\lambda+2\mu)/(\lambda+\mu)$ the Young modulus.

Regarding the bending energy density, we note that $\mathrm{Tr}(\underline{C}) = c_1 + c_2$ corresponds to the sum of the principal curvature, c_1, c_2, and thus to twice the mean curvature $C = (c_1+c_2)/2$. On the other hand, $\mathrm{Det}(\underline{C})$ stands as the product of the principal curvatures, i.e., the Gaussian curvature G. The bending energy density thus also expresses as $e_\mathrm{b} = B[2C^2 - (1-\nu)G]$ where $B = \frac{h^3}{12}\,\frac{E}{(1-\nu^2)}$ is the bending modulus.

7.3.2 *Gaussian Curvature and Theorema Egregium*

Among the deformations that can be undergone by a sheet, it will appear relevant to determine the characteristics of those that induce no stretch, i.e., the isometric deformations. In 2D, they correspond to global translations and rotations. In 3D, they may also include curvature modes, i.e., bending, under conditions to clarify.

Quite generally, for a 3D volume, the evolution of its metrics may be deduced from that of elementary distances ds between nearby points M and $M + \mathbf{dM}$. Calling

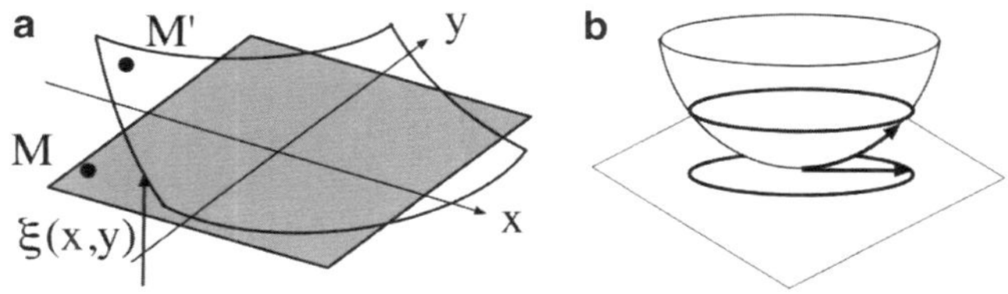

Fig. 7.5 (**a**) Gauss' Theorema Egregium. One considers a mapping $M \to M'$ from the plane (x,y) to the surface $z = \xi(x,y)$. For an isometric mapping to exist, the Gaussian curvature of the surface must be equal to that of the plane, i.e., zero. (**b**) An axisymmetric mapping from the plane to the paraboloïd that would preserve the perimeter of a circle would inevitably stretch its radius, as expected from the difference of Gaussian curvature between the plane ($G = 0$) and the paraboloïd ($G > 0$)

$\mathbf{u}(M)$ the displacement undergone by a point M, one gets $dM_i = dx_i$ in the rest state and $dM_i = dx_i + \partial u_i/\partial x_j dx_j$ in the stretched state, the space directions being indexed (i,j,k). The distance squared between nearby points $ds^2 = \mathbf{dM}^2$ then reads $ds^2 = dM_i dM_i = g_{ij} dx_i dx_j$ and thus $g_{ij} = \delta_{ij}$ in the rest state and $g_{ij} = \delta_{ij} + 2\gamma_{ij}$ in the stretched state, the strain tensor $\underline{\gamma}$ being $\gamma_{ij} = 1/2(\partial u_i/\partial x_j + \partial u_j/\partial x_i) + 1/2(\partial u_k/\partial x_i)(\partial u_k/\partial x_j)$.

Let us first start by determining to what conditions on a surface of cartesian equation $\xi(x,y)$ can an elementary isometric mapping exist between the plane $z = 0$ and this surface (Fig. 7.5a). Any mapping between the plane $(x,y,0)$ and the surface $(x',y',z' = \xi(x',y'))$ involves the displacement $(u,v,w) = [x'-x, y'-y, \xi(x',y')]$. Isometry imposes that the elementary length elements on the plane and on the mapped surface are the same: $ds^2 = ds'^2$ with $ds^2 = dx^2 + dy^2$ and $ds'^2 = dx'^2 + dy'^2 + dz'^2$. To express this constraint, let us notice that the latter expression writes $ds'^2 = dx^2(1+a) + dy^2(1+b) + 2c\,dxdy$ with $a = 2\partial_x u + (\partial_x \xi)^2$, $b = 2\partial_y v + (\partial_y \xi)^2$, $c = \partial_y u + \partial_x v + \partial_x \xi \partial_y \xi$. Isometry therefore imposes $a = b = c = 0$. This requirement may be transposed to a constraint on the sole surface $\xi(x,y)$ by noticing that the mapping (u,v) disappears from the combination $c - (a+b)/2 = 0$ which reads: $\partial_x^2 \xi \partial_y^2 \xi - (\partial_x \partial_y \xi)^2 = 0$.

A simple geometrical interpretation of this condition may be obtained by considering, up to a global rotation, the cartesian axes as the principal curvature axes of the surface, so that $\xi = \frac{1}{2}c_1 x^2 + \frac{1}{2}c_2 y^2 + \text{h.o.t.}$, c_1, c_2 denoting the principal curvatures and h.o.t. "higher order terms". The above criterion then reduces to $c_1 c_2 = 0$. More generally, the left-hand side of this criterion appears proportional to the determinant of the curvature tensor, $\text{Det}(\underline{C})$, and thus of the Gaussian curvature $G = c_1 c_2$. Accordingly, the condition for an isometric mapping to exist from a plane to a surface is that its Gaussian curvature is zero: $G = 0$.

One may straightforwardly generalize this constraint to isometric mappings from one surface, not necessarily a plane, to another, respectively, indexed 1 and 2. For this, one simply has to state that the evolution of their metrics from their tangent plane is the same: ${ds'_1}^2 = {ds'_2}^2$, this common length evolution being possibly not zero. The above algebra then shows that this turns back to the equality of their a, b, and c terms and thus of their combination $c-(a+b)/2$ or, equivalently, of their Gaussian curvature.

One obtains this way the Gauss' Theorema Egregium which states that isometric mappings must conserve Gaussian curvatures. In turn, any change of Gaussian curvature will indicate a change of metrics, i.e., a stretch. This way be illustrated on the elementary surface $\xi = \frac{1}{2}c_1x^2 + \frac{1}{2}c_2y^2 + \text{h.o.t.}$ considered above by noticing that, for equal curvature $c_1 = c_2 = c$, and thus for nonzero $G = c^2$, the axisymmetric mapping from the plane to this surface which would preserve the perimeter of the circle of radius r would inevitably stretch its radius (Fig. 7.5b).

7.3.3 Sheet Equilibrium and Föppl–von Kármán's Equation

Consider a weakly distorted sheet from its planar state $(x,y,0)$, its distortion being described by the equation $z = \xi(x,y)$, and focus attention to an elementary part of it with normals $\mathbf{n}$ at the boundaries. Its equilibrium condition requires mechanical equilibrium on the in-plane directions and on the normal direction:

* In-plane directions: no gain of momentum is allowed from the flux of stretching stresses $\bar{\sigma} = \underline{\bar{\sigma}}\mathbf{n}$ at its boundaries: $\text{div}(\underline{\bar{\sigma}}) = 0$. This requirement for a 2D in-plane tensor $\underline{\bar{\sigma}}$ imposes that it derives from a scalar potential, the Airy potential χ, such that $\bar{\sigma}_{ij} = (-1)^{i+j}\partial^2\chi/\partial x_i\partial x_j$.
* Normal direction: for a bent sheet, the normal component of the net contribution of stretching stresses applied at its boundaries must equilibrate that provided by bending. The former is proportional to h and may be expressed with the Airy potential. This yields the first Föppl–von Kármán equation:

$$B\triangle^2\xi - h[\xi,\chi] = 0 \tag{7.1}$$

where the first term denotes bending contribution, the second the stretching contribution, and the brackets, the Poisson brackets $[U,V] = \partial^2_{ii}U\partial^2_{jj}V + \partial^2_{ii}V\partial^2_{jj}U - 2\partial^2_{ij}U\partial^2_{ij}V$.

The second Föppl–von Kármán equation corresponds to a compatibility condition for stresses to derive from strains induced by an actual displacement. This condition corresponds to the above relationship $c-(a+b)/2 = 0$ for isometric mappings and, more generally, $c-(a+b)/2 = G$ for mappings inducing a Gaussian curvature. Relating strains to stresses and then to the Airy potential yields:

$$\triangle^2\chi + E[\xi,\xi] = 0 \tag{7.2}$$

where $[\xi,\xi] = G$ for weakly distorted surfaces.

These two equations describe the relationships between stresses (χ) and geometry (ξ) for weakly distorted sheets at equilibrium.

7.4 Experiment

The experiment aims at compressing a thin sheet while keeping one of its principal curvature fixed at a nonzero value. For this, one seeks to compress a sheet in between parallel cylinders so that their curvature radius R fixes one sheet's principal curvature [14]. To this end, the sheet is clamped by two of its sides on a cylinder along a direction normal to the cylinder generatrix. This will then force the fold axes to adopt the cylinder curvature (Fig. 7.6).

7.4.1 Setup

The compression setup is shown in Fig. 7.7. It consists in a fixed upper plate and a moving lower plate in between which the sheet to compress is placed. The lower plate is pushed up by a piston placed at the middle of the system but is blocked by three stepper motors before touching the upper plate. These motors thus enable to monitor the gap Y between the compressing plates to an accuracy of a tenth of microns.

The upper plate is taken transparent so as to allow visualization from above. For the present experiment, the plates, usually flat, have been replaced by cylinders made of plexiglass for the upper plate and of polycarbonate for the lower plate. Thin rulers enable the sheet to be clamped on the curved sides of the lower cylindrical plate. Visualization of the sheet form has been achieved by illuminating it from the sides with two different colors, red and blue. This way, the image recorded by a camera fixed on top of the setup could make the difference between right and left sides of the sheet folds, thus improving the contrast and the understanding of the sheet form.

The sheet is made of polycarbonate and has the following dimensions: length $l = 155\,\text{mm}$, width $L = 190\,\text{mm}$, and thickness h ranging from $0.05\,\text{mm}$ to $0.5\,\text{mm}$. It is clamped along its length onto the bottom cylinder. As the distance between the clamping arches, $X = 180\,\text{mm}$, is smaller than the sheet width L, the sheet is thus already buckled before compression (Fig. 7.8). Finally, the cylinder curvature radius, $R = 50\,\text{cm}$, is taken large compared to the few millimeters gap Y and thus to the fold sizes, so as to make the configuration locally close to a compression between parallel plates.

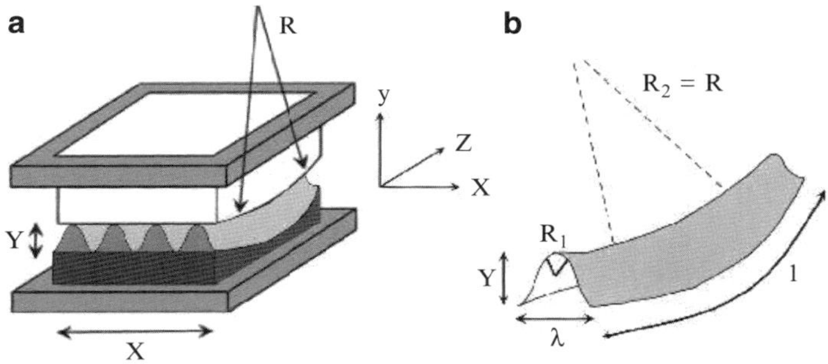

Fig. 7.6 A compressing device (**a**) involves cylindrical plates distant from a controlled gap Y with, in between, a sheet clamped on their curved sides. (**b**) By iterative buckling, gap reduction yields the generation of folds of ever smaller size λ, whose axis is bent by the cylinders. This results in two principal curvatures $c_1 \sim \lambda^{-1}$, $c_2 \sim R^{-1}$ and thus in Gaussian curvature and in-plane stretching

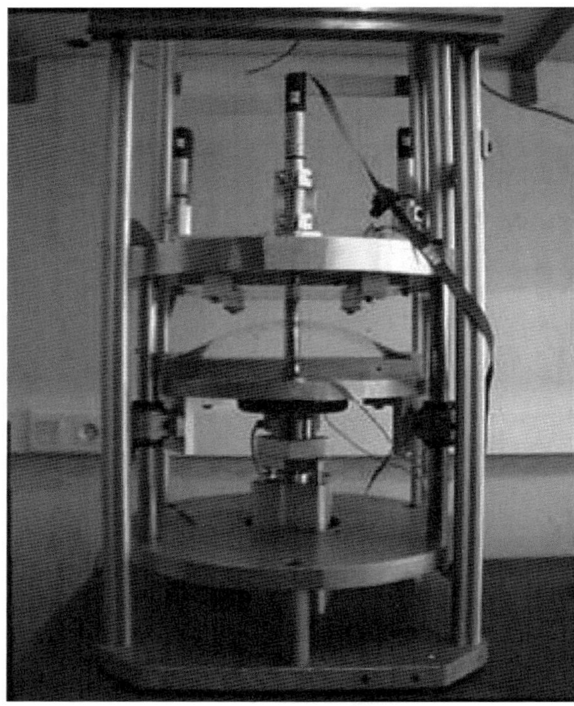

Fig. 7.7 Snapshot of the setup used to perform and study sheet compression. A bottom plate is pushed up by a piston and blocked by stepper motors, leaving a gap Y to a fixed top plate. Sheets are clamped on one of the plates and compressed as the gap Y decreases. Visualization is achieved from above thanks to a transparent top plate

7.4.2 Compression Route

The cylinder curvature radius being large, the compression route shares some analogy with the Elastica route [18–20]. In particular, a buckling cascade is observed with the fold number ever increasing as compression proceeds (Fig. 7.9). Following

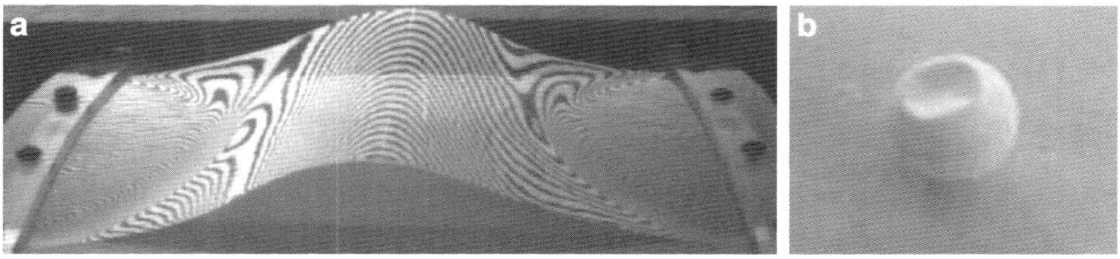

Fig. 7.8 (**a**) Clamped sheet prior compression showing two ridges at the contact lines between the sheet and the cylinder. (**b**) Buckled ping-pong ball showing ridges where large curvature and stretch are focused

the invariance of the compressing cylindrical plates on both the clamping direction (same curvature along it) and its normal (straight cylinder axis), the sheet folds, either smooth or singular, show the same shapes. In particular, their common width λ follows the gap Y in the sense that the ratio λ/Y varies in a short range that depends on the excess length L/X and which corresponds in practice here to the range $[20/3, 9]$. The largest bound corresponds to buckling folds and the lowest bound to just buckled folds. Accordingly, the number $n = X/\lambda$ of folds varies as $nY = XY/\lambda$, with nY bounded in the range $[X/9, 3X/20]$, i.e., $[20, 27]$ mm here.

Without compression, the clamped sheet shows two domains in contact with the cylinder with a fold in between (Fig. 7.8a). The frontier between them corresponds to an elastic defect, the so-called "ridge", on which curvature is focused. It is analogous to that observed on a buckled ping-pong ball (Fig. 7.8b) [11, 21] and on which stretching yields plastic deformation. However, it is actually weaker in the sense that no plastic transition is triggered here.

As compression proceeds together with the resulting successive bucklings, the sheet shows first ridges at its contact with cylinders (Fig. 7.9a, b), as on the uncompressed state of Fig. 7.8. However, on further compression, another kind of defect appears, the d-cone (Fig. 7.9c, d, e).

This kind of defect corresponds to those found when distorting a planar sheet by pressing it with a sharp tip [10, 12]. All the stress is then focused on this tip, leaving the remaining of the sheet unstretched. One may observe that the sheet is not axisymmetric with respect to the tip axis but shows a folded circumference that makes the difference with a cone. This traces back to the fact that the cone is not a developable surface, i.e., that it cannot be continuously mapped onto a plane without cutting it somewhere. This indicates that some Gaussian curvature is in order, not on the cone sides since they are curved on a single direction but at the cone tip where all the stretch is thus concentrated [9, 10]. However, in comparison, the planar sheet making a d-cone is actually developable since it was initially planar. The difference between both is the folded part of the distorted planar sheet which, if removed, could yield an actual cone.

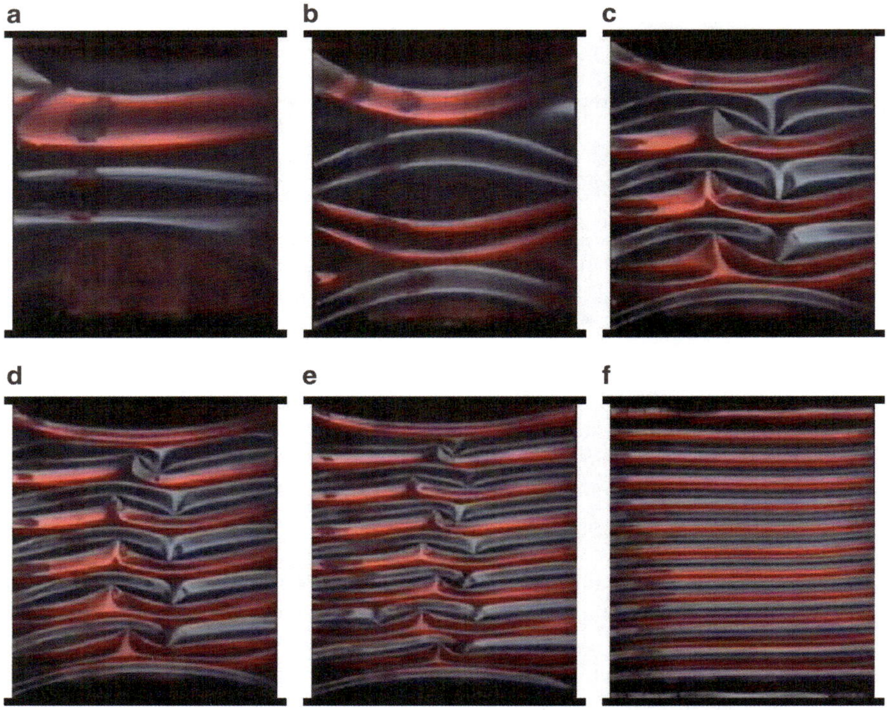

Fig. 7.9 Experimental pictures taken from above; $h = 180$ microns. Colors refer to left (*red, bright*) or right (*blue, bright*) fold sides. Clamped sides are shown by black ticks. Compression increases from (**a**) to (**f**). (**a**) Single fold with ridge. (**b**) Two folds with ridges. (**c**) (**d**) (**e**) Four, six, and eight folds with ridges and d-cones. (**f**) Regular state of twelve folds involving *no* singularity

These d-cones thus involve a large curvature at the tip of the sharp tongue they display (Fig. 7.9c, d, e) with, therefore, a large stretch there. In contrast, they enable the stretch to be removed from the remaining of the sheet, as on the canonical example of a sheet pressed with a tip [10].

The number of d-cones increases with the fold number, each fold displaying its d-cone (Fig. 7.9a, b). Accordingly, focusing stress on the tip of elastic defects seems to be the nominal mean for self-organizing the sheet so as to adapt compression. Viewed this way, there should be no reason to self-organize differently when compressing further. However, somewhat surprisingly, beyond a fold number $n = 11$, all d-cones *spontaneously disappear* from the bulk leaving a smooth, regular, state, free of elastic defect (Fig. 7.9f) [14]. Then, further increasing simply yields further buckling with no longer any defect occurrence.

It should be noticed that the regular state made of parallel defect-free folds nevertheless involves stretch and nonzero Gaussian curvature G since the fold axes are bent by the cylinders. However, the difference with the defect state is that this

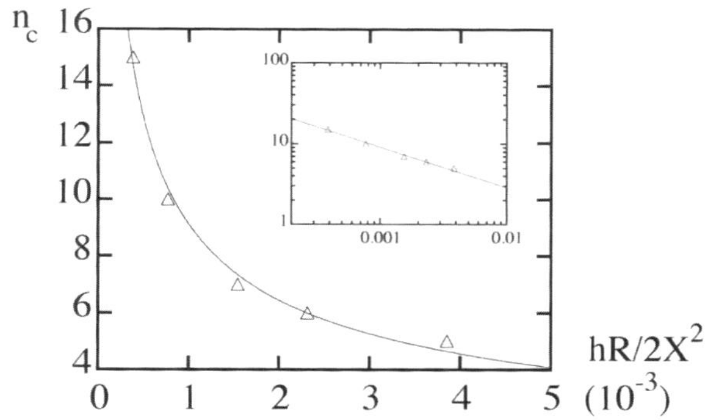

Fig. 7.10 Critical number of folds n_c at the uncrumpling transition for different sheet thicknesses h. Continuous line corresponds to $2\pi\gamma_c n_c = (hR/2X^2)^{-1/2}$ with $\gamma_c = 0.55$ as best fitting coefficient. Inset: same data in log-log scales

stretch is distributed on the whole sheet instead of being concentrated in localized areas. The morphological transition on defect appearance/disappearance therefore corresponds to a transition from a condensed stretch to a distributed stretch or, equivalently, to a stress-defocusing process. The remaining of this study is devoted to understand it.

7.4.3 *Defocusing Scaling*

At a defect core, the sheet can no longer be viewed as a 2D surface (or a collection of superimposed 2D surfaces) and must be considered as three-dimensional. This means that the sheet thickness should parametrize the defect occurrence or disappearance and thus the stress-defocusing phenomenon. Similarly, the fact that both stretch and curvature are involved in the sheet organization converges to the same conclusion since they scale differently with the thickness h. These remarks thus invite us to vary the sheet thickness and address the resulting modification of stress defocusing.

Varying h by a factor of 10, from 0.05 to 0.5 mm at otherwise same length and width, we observed similar routes exhibiting the same qualitative events and only displaying quantitative variations. In particular, the fold number n_c at which stress defocusing occurs has been found to vary with h as a power law: $n_c \propto h^{-1/2}$ (Fig. 7.10). The fact that n_c scales with h gives confidence for using this relationship to infer relevant information on stress defocusing. In particular, as the relevant characteristic scales playing on n_c are the cylinder radius R, the distances X between arches, the sheet width L, and the sheet thickness h, one may expect from dimensionality a scaling relationship of the kind $n_c \propto (R/h)^{\alpha}(X/h)^{\beta}(L/h)^{\gamma}$. The objective of the modeling will thus be to determine these exponents and recover the observed fact that their sum is $1/2$.

7.5 Energy Criterion for Stress Focusing and Scalings

In the introduction, we have argued that energy could help in clarifying the origin of stress defocusing and more generally the self-organization of elastic sheets within the prescribed boundaries. The argument consists in identifying the observed state with the less energetic state, leaving aside the issue regarding the path required to change state and, therefore, possible metastability. Here, we would like to make this argument quantitative so as to recover the location of stress defocusing on the compression route and especially the power law variation $n_c(h)$.

For this, we thus address the energy criterion for defocusing and express it by using scaling arguments [14].

7.5.1 Energy Criterion for Stress Focusing or Defocusing

Our goal is to compare the sheet elastic energy E in a stress-focused state and in a stress-defocused state. This energy expresses as the integral over the sheet surface of its energy density e: $E = \int_{\text{sheet}} e\ ds$. Here e can be decomposed in a bending contribution e_{b}, a stretching contribution e_{s}, and a defect contribution e_{d}, the two formers being referring to the sheet except the defect core and the latter to these defect cores. To facilitate the comparison, we shall denote the stress-focused state with the superscript "f" and the stress-defocused state with the superscript "d."

As the less energetic state is favored, the criterion reads:

- Stress focusing: $E^f \ll E^d$.
  There is an energetic gain to focus stress in defects.
- Stress defocusing: $E^d \ll E^f$.
  There is an energetic gain to defocus stress on the whole sheet.
- Stress focusing/defocusing transition: $E^d \sim E^f$.
  Both energies being similar, there is no clear advantage in one or the other state.

In the defocused states, the defect energy density vanishes by definition: $e_{\text{d}}^d = 0$. On the other hand, the absence of multiple characteristic scales on these states allows the scaling of the evolutions with compression to be determined from the Föppl–von Kármán equations for both the bending and the stretching energy densities, e_{s}^d, e_{b}^d. Accordingly, in defocused states, one should be able to follow the evolution of the sheet energy E^d with the fold number n, as compression proceeds.

By comparison, in the focused states, a similar determination is delicate owing to a more complex geometry of the sheet states and to the difficulty in expressing the energy density e_{d}^f on a defect without solving for its inner structure. However, denoting ρ the size of the defect core, its energy is about $B(\rho/h)^{1/3}$ [13] and, as $\rho \sim h$, of the order of B, i.e., of the sheet bending energy. Accordingly, we may omit this defect energy in the evaluation of E^f without noticeable implication on the determination of the transition. In particular, as we expect a variation of several orders of magnitude of the balance between the energies of the focused and defocused states, order one differences between the different terms are irrelevant. On the other hand, the stretching energy density e_s^f in the focused states is for sure largely reduced compared to its defocused value, so that it no longer stands as the dominant energy density. We shall then assume that it is of the same order of the bending energy density e_b^f which, on the other hand, should remain comparable to its defocused value: $e_\mathrm{s}^f \sim e_\mathrm{b}^f \sim e_\mathrm{b}^d$. Accordingly, we shall thus consider the bending energy as representative of the order of magnitude of the energy density e^f of the focused state: $e^f \sim e_\mathrm{b}^f \sim e_\mathrm{b}^d$.

The transition criterion $E^f \sim E^d$ then reads $e_\mathrm{b}^d \sim e_\mathrm{s}^d + e_\mathrm{b}^d$ or, equivalently, $e_\mathrm{s}^d \sim e_\mathrm{b}^d$. Interestingly, it is thus expressed on the defocused state only, which we know and can evaluate. Its physical meaning is that defects actually relax the stretching energy density E_s on the whole sheet, leaving the bending energy E_b plus the defect energy E_d, which are of the same order. This is obviously energetically favorable when the stretching energy is dominant on regular states, i.e., when $e_\mathrm{s}^d >> e_\mathrm{b}^d$, in which case defects should occur. On the opposite, for $e_\mathrm{s}^d \sim e_\mathrm{b}^d$, regular states should be maintained. The criterion for a transition between focused and defocused states is thus $e_\mathrm{s}^d / e_\mathrm{b}^d \sim O(1)$.

7.5.2 Scalings

The compression route shows two imbricated phenomena: buckling and defect occurrence/disappearance. The former yet occurs on compression between parallel plates, i.e., for $R = \infty$, and the latter is specific of a finite R. Let us address them successively, first within the Elastica and then within the Föppl–von Kármán equations.

Elastica involves no Gaussian curvature and no stretch. The only source of stresses is thus bending via flexural terms. As a consequence, there is no longer a scaling competition between stretching ($\propto h$) and bending ($\propto h^3$), so that the sheet thickness h only serves to gauge stresses and forces without implication on the sheet state. In particular, the sheet behavior satisfies scale invariance with respect to h.

This scale invariance enables us to relate states by zooming them in and out. Consider a n-fold solution obtained after several bucklings (e.g., a fourfold solution in Fig. 7.3-left). Each fold is physically equivalent to the onefold solution found prior to buckling. Both are connected by a zoom such that the fold of the fourfold solution whose width is X/n is mapped onto the onefold whose width is X, i.e., by a zoom factor of $X/(X/n) = n$. Accordingly, the folds corresponding to characteristic lengths $(Y/n, X/n)$ and (Y, X) are geometrically similar (and also dynamically similar as shown in [18–20]). This, in particular, states that the fold width λ and the fold height Y scale like n^{-1}.

Let us now consider curved plates, i.e., a large but finite R, and determine how scalings operate in a situation where stretching and bending compete on a defocused state. In particular, our objective is to express the evolution of the stretching and bending energy densities as compression proceeds.

Regarding the bending energy density, $e_b = B[2C^2 - (1-\nu)G]$, one may notice that, when integrated over the whole sheet, the contribution of the Gaussian curvature is constrained by the Gauss–Bonnet theorem [22] which states that the integral of G over a compact surface is related to a topological invariant, the Euler characteristics of the surface, and to the boundary integral of the geodesic curvature at the surface boundary. On the other hand, the small curvature of the cylinders negligibly changes the sheet form at its boundaries so that the boundary integral and finally the surfacic integral of G hold a value close to the one they have for a compression between planes. However, as the Gaussian curvature vanishes in this case, its net integral contribution would be zero for such a compression between plane and, by extension, for the present compression between cylinders. For this reason, we shall skip the contribution of G to the bending energy density e_b hereafter and reduce it to $e_b = 2BC^2 \sim Eh^3C^2$.

Regarding the stretching energy density e_s, its definition $\delta e_s = h\bar{\sigma}_{ij}\delta\bar{\gamma}_{ij}$ with $\bar{\gamma} \sim \bar{\sigma}/E$ shows that its scalings follow those of the combination $h\bar{\sigma}^2/E$ where $\bar{\sigma}_{ij} = (-1)^{i+j}\partial^2\chi/\partial x_i\partial x_j$, the Airy potential satisfying $\triangle^2\chi + EG = 0$. To evaluate them, let us model the regular folds by the surface $\xi(x,z) = \xi_0(x) + z^2/2R$, where $\xi_0 \approx (Y/2)\sin(2\pi x/\lambda)$ is a λ-periodic function and where the dependence on z conveys the fold curvature imposed by the cylinders. Spatial derivatives then extract the length scale $\Lambda = \lambda/(2\pi)$ so that $\Lambda^{-4}\chi \sim EG$, $\sigma \sim \Lambda^{-2}\chi$ and, finally, $e_s \sim Eh\Lambda^4G^2$.

Altogether, this yields the transition criterion $e_s/e_b = O(1)$ with $e_s/e_b \sim \gamma^4 = O(1)$ and $\gamma = \Lambda(G/hC)^{1/2}$. We stress that it applies on any compression route of thin sheets.

On the original compression route between cylinders studied here, one has $c_1 \sim Y/\Lambda^2$, $c_2 \sim 1/R$ with $c_1 \gg c_2$ and thus $C \sim c_1/2$ and $G/C \sim 2c_2$. This, with $\lambda \sim n^{-1}$, yields $\gamma = (\lambda/2\pi)(2/hR)^{1/2} \sim n^{-1}$ and thus an order parameter for the transition varying sensitively with the fold number n as $e_s/e_b \sim \gamma^4 \sim n^{-4}$. This sensitivity to the fold number then relativizes the role of prefactors in the above scaling relationships and therefore supports the analysis.

Calling γ_c the value of γ at the transition, the above expression of γ yields the critical fold number at the transition: $n_c = (hR/2X^2)^{-1/2}/2\pi\gamma_c$. One thus recovers the experimental scaling evidenced in Fig. 7.10 with a computed value $\gamma_c \sim 0.76$ of order one, as expected for the transition.

7.6 Phase Diagram and Nature of Singularities

Following the above criterion, we are now able to determine the conditions required for focusing or defocusing stress in a sheet and synthesize them in a phase diagram [14]. This will be especially useful to interpret the three canonical compression routes that are compared in Sect. 7.2. However, before turning attention to this quantitative view, it is instructive to realize that the surprising phenomenon of defocusing is a simple logical consequence of scale invariance.

7.6.1 Scale-Invariance and Defocusing

Three characteristic lengths are in order on a sheet state: (1) a fold morphological scale, the fold width λ, for instance, (2) the cylinder curvature radius R, and (3) the sheet thickness h. These are in particular the three variables which enter the transition criterion.

The fact that, in contrast to Elastica, h is a relevant scale here forbids the different states of a given compressed sheet to be geometrically similar, since any homothety would ask to change the sheet thickness. This, however, does not forbid to use scale invariance to compare, at fixed thickness h, the evolutions undergone when varying λ with respect to R or vice versa.

In particular, let us consider the change of fold width λ: $(\lambda, R, h) \to (\alpha\lambda, R, h)$. Owing to the absence of intrinsic scale in elasticity, i.e., to its scale-invariant nature, similar states may be obtained by changing the scale of *all* lengths by the *same* factor. Accordingly, the latter state $(\alpha\lambda, R, h)$ is equivalent to $(\lambda, R/\alpha, h/\alpha)$.

Moreover, in our configuration where a principal curvature c_1 is much larger than the other c_2, the Föppl–von Kármán Eqs. (7.1) and (7.2) involve a scale invariance focused on the couple of scales (R, h). This may be evidenced by noticing that, as $c_1 >> c_2$, the laplacian operator reduces to $\triangle\xi \approx \partial^2\xi/\partial x^2$, so that $\triangle^2\xi \approx \partial^4\xi/\partial x^4$. Note that, in these relationships, the equality is even actually achieved within the modeling of the sheet's surface $\xi(x,z) = \xi_0(x) + z^2/2R$. As, from relation (7.2), the Airy potential follows the modulations of the sheet's surface, the same conclusion may be drawn for it: $\triangle^2\chi \approx \partial^4\chi/\partial x^4$. In this instance, it then appears that, as $B \sim Eh^3$, the Föppl–von Kármán Eqs. (7.1) and (7.2), respectively, agree with the following scaling relationships $\chi x^2 \sim h^2 z^2$ and $\chi z^2 \sim x^2 \xi^2$ where variables are used here to denote their scale (e.g. $\partial^4\chi/\partial x^4 \sim \chi/x^4$ or $[\xi, \chi] \sim \xi\,\chi\, x^{-2}\, z^{-2}$). These relationships are equivalent to $\chi \sim x^2 \sim hz$, since by definition $\xi \sim z$. A class of scale change which satisfies these scaling constraints is the following: $(x, z, \xi, \chi, h) \to (x, \beta^{-1}z, \beta^{-1}\xi, \chi, \beta h)$. It means that an increased thickness h, $h \to \beta h$, goes together with an anisotropic zoom $(x, z) \to (x, \beta^{-1}z)$. As the curvature R^{-1} corresponds here to $\partial^2\xi/\partial z^2 \sim z^{-1}$, one gets $R \sim z$ and thus the following change for R, $R \to \beta^{-1}R$. This therefore corresponds to an actual invariance for both x (and thus λ) and the combination Rh: $x \to x$, $Rh \to (\beta^{-1}R)(\beta h) = Rh$. An echo of this property is found in the fact that, besides λ, the variables R and h enter the transition parameter γ through this combination Rh.

According to this additional symmetry, the state $(\lambda, R/\alpha, h/\alpha)$ is also physically equivalent to $(\lambda, R/\alpha^2, h)$. So are therefore the states $(\alpha\lambda, R, h)$ and $(\lambda, R/\alpha^2, h)$ which interestingly display the same thickness h. Accordingly a decrease of λ/R at *fixed h* can be equally obtained in two physically equivalent ways:

(1) By increasing R at *fixed* λ and h: $(\lambda, R, h) \to (\lambda, R/\alpha^2, h)$, $\alpha < 1$
(2) By decreasing λ at *fixed* R and h: $(\lambda, R, h) \to (\alpha\lambda, R, h)$, $\alpha < 1$

With this in mind, we now compare two compression routes, the first route referring to the sole variation of R at fixed λ and h, and the second route referring to the sole variation of λ at fixed R and h (Fig. 7.11). One recognizes in the first route the bending of a fold axis and in the second route, the compression between cylinders worked out here. Both however refer to a variation of the ratio λ/R at fixed h and should thus tell the same story.

- First route: fold bending
  Bending the axis of a fold turns out increasing its principal curvature $c_2 = 1/R$ from zero at a fixed λ, i.e., at a fixed principal curvature c_1. As one may easily figure out, this yields the generation of defects, actually d-cones, beyond some critical value of c_2.
  This route, which corresponds to decreasing the ratio $c_1/c_2 = \lambda/R$, thus shows us that stress focusing should be encountered this way.
- Second route: compression between cylinders
  Compressing a sheet between cylinders turns out decreasing the fold width λ by iterated buckling at fixed R, i.e., at fixed c_2. This corresponds to increasing the principal curvature $c_1 \sim \lambda^{-1}$ at fixed c_2 and, therefore, to increasing the ratio $c_1/c_2 \sim \lambda/R$. From this point of view, this route corresponds to an opposite direction of compression as compared to the first route. Both routes being similar as viewed with respect to the ratio c_1/c_2 at fixed h, this means that one should encounter defocusing on the second route as surely as one encounters stress focusing on the first route.

This simple reasoning naturally explains why compression between cylinders should yield defocusing despite the general increase of stress undergone by the sheet (Fig. 7.11). It emphasizes that the thing which matters regarding the occurrence or not of defects is not the global amount of stress but its balance between stretch and bending on regular states. In particular, reducing the fold width by iterated buckling renders the curvature radius of their axis apparently larger, as compared to their width. This corresponds to decreasing the effective bending of their axis, i.e., to rendering them more straight, until this bending becomes small enough for defocusing to occur.

7.6.2 Scalings and Phase Diagram for Singularities

The transition criterion $\gamma = \Lambda(G/hC)^{1/2} = \gamma_c = O(1)$ derived in Sects. 7.5.1 and 7.5.2 applies to any states of thin sheets. It thus enables us to determine the domains where focused or defocused stress are in order.

Before applying this to work out a phase diagram for stress focusing, we would like to turn from the geometrical expression of the criterion in terms of mean

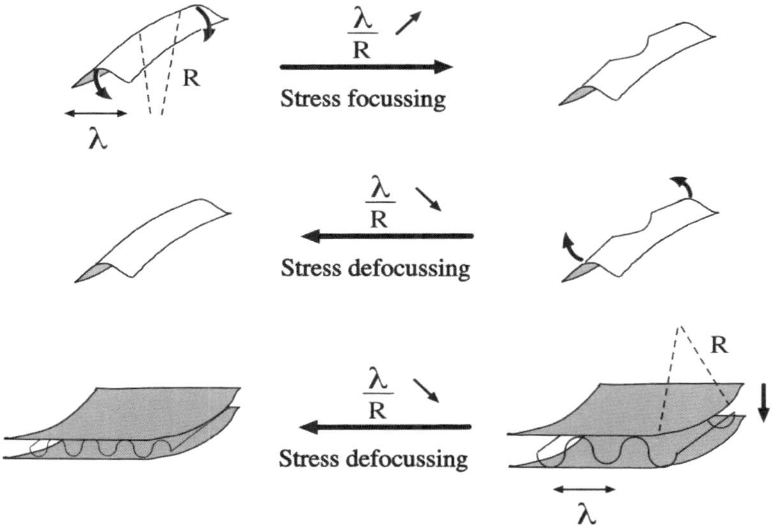

Fig. 7.11 Equivalence between fold de-bending and compression between cylinders. Fold bending increases λ/R and yields stress focusing. Fold de-bending therefore reduces λ/R by variation of R and yields stress defocusing. Similarly, compression between cylinders together with iterated buckling makes λ/R also decrease but by variation of λ. As for fold de-bending, this must therefore also yield defocusing

curvature C and Gaussian curvature G to an expression in terms of typical strains relative to curvature γ_C or stretch γ_G. The decomposition of strains in Sect. 7.3.1 shows that γ_C may be taken as hC.

The determination of stretching stresses from the Airy potential in Sect. 7.3.3 together with scaling arguments similar to those applied in Sect. 7.5.2 shows that $\gamma_G \sim \sigma/E \sim \Lambda^{-2}\chi/E$ with $\chi \sim \Lambda^4 EG$, so that $\gamma_G \sim \Lambda^2 G$. Altogether this yields the transition criterion to read $\gamma = (\gamma_G/\gamma_C)^{1/2} = \gamma_c = O(1)$.

In the variable space (γ_C, γ_G) and in logarithmic coordinates, the transition criterion thus corresponds to a straight line whose slope is $1/2$ and which is located according to the experimental value 0.55 of γ_c. Above this line, one finds focussed singular states and below it defocused regular states (Fig. 7.12).

Let us now place the compression routes in this diagram:

- Elastica
  The Elastica corresponds to $\gamma_G = 0$ and thus to a horizontal line with an ordinate repelled up to $-\infty$. Whereas we cannot reproduce it on the diagram, we may realize that it stands within the regular, defocused, domain.
- Crumpled paper
  Because of isotropy, all the length scales and especially the fold widths and the two principal curvature radii take similar values. Accordingly $\Lambda^2 \sim 1/G$ so that $\gamma_G \sim O(1)$. This compression route thus corresponds to a horizontal line located

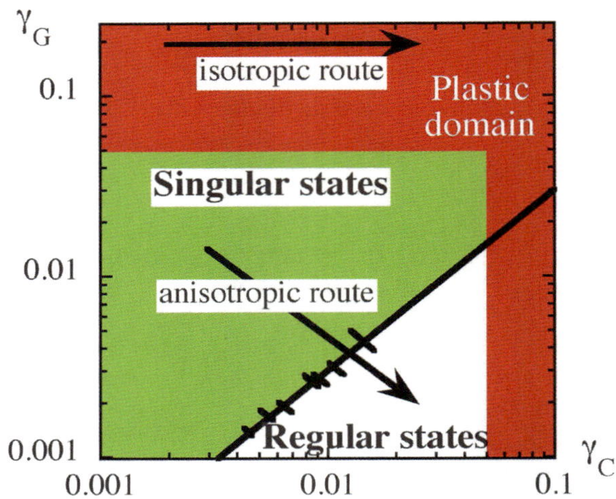

Fig. 7.12 Transition diagram in the variable space (γ_C, γ_G), where $\gamma_G = hC$ and $\gamma_G = \Lambda^2 G$ are the typical strains due to curvature and to in-plane stretching. The isotropic route always lies in the plastic domain. In contrast, anisotropic crumpling yields elastic regularization and stress defocusing before experiencing plastic deformation. The thick line corresponds to the uncrumpling transition for $\gamma_c = 0.55$ and black ticks to the observed uncrumpling transition for h varying from 50 to 500 microns, including error bars. Note that the diagram would be similar in variables $(G/C^2, hC)$

at values of ordinates about unity. It thus stands within the singular, focused, domain till the first stage of compression (Fig. 7.12). Accordingly, defects should occur at the very beginning of the compression of a paper sheet in hands, as may be directly confirmed in practice. However, they should also encounter the transition to defocusing at large compression, a fact which is not corroborated in practice, for reasons explained below.

- Compression between cylinders
  Here $c_1 \sim Y/\Lambda^2$, $c_2 \sim 1/R$ with $c_2 \ll c_1$. This yields $\gamma_C \sim hY/\Lambda^2$, $\gamma_G \sim Y/R$. In addition, following buckling, Λ and Y decrease similarly so that $\Lambda \sim Y$. This finally yields $\gamma_G \sim (h/R)\gamma_C^{-1}$, i.e., a line with slope -1 and located according to the value of h/R. As displayed in Fig. 7.12, it crosses the transition diagram from the singular, focused, domain to the regular, defocused, domain, as compression proceeds.

These compression routes, which are reported in the phase diagram of Fig. 7.12, thus well reproduce the experimental evidence, except for the crumpled paper which is found to neither defocus stress nor remove defect at large compression. This discrepancy will be explained below by plasticity.

This phase diagram provides a new vision of singularity occurrence in elasticity. Instead of being viewed as objects forced by external means, by frustration, or by boundary conditions, they simply appear here as a possible expression of a sheet state, besides an alternative one corresponding to a stress smoothly distributed on a geometrically regular state. In particular, we note that the disappearance of singularities under compression evidenced here differs from that observed when pressing a fold with a sharp tip [23] in the sense that singularities disappear here in the bulk whereas they are expelled to the sheet boundary in the latter case. This difference emphasizes the fact that stress focusing or defocusing stands here as a bulk matter, actually the spontaneous selection of a preferred phase under constraint. This therefore makes this elastic issue close to a thermodynamic issue, the different routes being simply different kinds of path followed in phase space.

Attached to this thermodynamic view is the requirement that forming singularities do not change physics irreversibly, since making a closed path in phase space should yield back to the starting phase. This is actually satisfied in this experiment of compression between cylinders since decompressing the sheet yields back to the starting planar sheet without evidence of the history. Such a reversibility is however not involved in paper crumpling since permanent scars are generated. This calls for completing the phase diagram by taking into account plasticity. Of course, as this complement will not concern the whole phase diagram but only a part of it, it does not break the thermodynamic interpretation but simply complete it with an additional phenomenon.

7.6.3 Plasticity

Plastic deformations occur at too large strains. Then the sheet escapes the linear elasticity domain within which removing the stress makes the system return to its initial strain. This therefore results in irreversible deformations that are located where the strains were too large, i.e., here, at the defect cores.

A criterion for a transition to plasticity is the occurrence of strains larger than a threshold value, typically a few percent, actually 5 % here [24]. This yields us to complete the phase diagram by restricting the linear elasticity domain to a square domain located at low strains. Interestingly, this discriminates the different routes addressed here since the isotropic compression route corresponding to crumpled paper entirely lays in the plastic domain whereas the anisotropic route provided by compression between cylinders stands in the elastic one. Their different behaviors regarding stress defocusing are then naturally explained: whereas both should display defocusing at large compression, the isotropic route will not since the plastic transition preempts the defocusing transition; however, the anisotropic route will, provided the regular states are not too much compressed before experiencing defocusing.

As for other thermodynamic transitions, this complete phase diagram opens strategic choices to achieve sheet compressions that might avoid plasticity or not, visit the singularity domain or not, keep within the regular domain or not, etc. These different strategies might correspond to interesting practical issues in mechanics, electronics, or in compaction.

7.7 Conclusion

Crushing a paper in a ball yields the generation of scars which denote a transition from a regular geometry to a singular geometry. Meanwhile the stress distribution changes from regularly distributed to condensed on singularities, a phenomenon called stress focusing. The relevant stress part in this focusing refers to the stretch induced by a nonzero Gaussian curvature $G = c_1 c_2$ where c_i denotes the principal curvatures at a point. Usually, two compression routes are investigated, the Elastica routes where $G = 0$ and the isotropic compression routes where $c_1 \sim c_2$. The former then yields no singularity whereas the latter corresponds to crumpled paper. However, as two curvatures are in order in G, intermediate routes may exist to explore the full bi-dimensionality of the issue. The objective of this study has been to manage the compression configuration so as to address one of them here. It consists in compressing a sheet between cylinders so as to increase a principal curvature c_1 only, while keeping the other one c_2 constant.

Interestingly, compressing a sheet this way generated singularities which surprisingly spontaneously all disappeared at large compression. This unintuitive stress defocusing by compression is at variance with the view that could be inherited from crumpled paper. We explained it qualitatively by showing from scale invariance of the Föppl–von Kármán equations that the reduction by buckling of the width of the folds that are bent by the cylinders is physically equivalent to decreasing the bending of the axes of folds of fixed width. As the latter route removes the singularities that could have appeared at large axis bending, it corresponds to a stress defocusing. The same phenomenon is thus in order by compressing folds between cylinders and may be qualitatively understood by the fact that cylinders seem all the more flat as the folds are small.

This interpretation has been made quantitative from scaling arguments which have been applied on a criterion for defocusing. This criterion states that, as defects relax stretch and decrease the stretching energy, it is advantageous to focus stress only if the stretching energy is large compared to the bending energy. This was synthesized on a phase diagram in which a transition line separates the domain where stress is focused from the one where it is defocused. The bidimensionality of this diagram echoes the existence of two relevant modes for stresses (stretch and bending) or for strains (Gaussian curvature and mean curvature). In particular, the different routes addressed here highlight the existence of two principal curvatures whose features vary according to the compression protocol.

To explain the formation of irreversible scars, the phase diagram has been completed with plastic domains. This shows that the plastic transition may preempt or not the defocusing transition depending on the compression route. In particular, it actually preempts it on the isotropic configuration of crumpled paper but not on the anisotropic compression routes investigated here since defocusing occurs on them prior to plasticity. Accordingly, crumpled paper appears as a misleading example of elasticity since the structures it shows, the singularities, should have disappeared if the plastic transition had not occurred. In this sense, it is a combined example of elasticity plus plasticity. On the opposite, the defocusing phenomenon exhibited here by compression between cylinders fully refers to linear elasticity from which it naturally derives thanks to scale invariance.

These results provide a renewed view of singularities according to which they are understood as the expression of an elastic singular phase in competition with a regular defocused phase. Within this thermodynamic interpretation, the bidimensionality of the phase diagram allows the elaboration of strategies regarding crushing. In particular, using cylinders instead of plates to compress sheets enabled us to remove singularities before encountering plastic transition. This could find interesting practical applications for instance for compressing more material sheets without altering them.

Acknowledgements This work, which gave rise to the publication [14], has been performed in close collaboration with Benoit Roman. I thank him for many fruitful discussions and interactions.

References

1. V. Pereira, A. Castro Neto, H. Liang, L. Mahadevan, Phys. Rev. Lett. **105**, 156603 (2010)
2. J. Genzer, J. Groenewold, Soft. Matter **2**, 310 (2006)
3. U. Seifert, Adv. Phys. **46**, 13 (1997)
4. M. Alava, K. Niskanen, Rep. Prog. Phys. **69**, 699 (2006)
5. M. Golombek, F.S. Anderson, M.T. Zuber, J. Geophys. Res. **106**, 811 (2001)
6. A. Lobkovsky, S. Gentges, H. Li, D. Morse, T. Witten, Science **270**, 1482 (1995)
7. A. Lobkovsky, Phys. Rev. E **53**, 3750 (1996)
8. A. Lobkovsky, T. Witten, Phys. Rev. E **55**, 1577 (1997)
9. M. Ben Amar, Y. Pomeau, Proc. R. Soc. Lond **A 453**, 729 (1997)
10. E. Cerda, L. Mahadevan, Phys. Rev. Lett. **80**, 2358 (1998)
11. B. Audoly, Y. Pomeau, *Elasticity and Geometry: From Hair Curls to the Non-Linear Response of Shells* (Oxford University Press, Oxford, 2010)
12. S. Chaïeb, F. Melo, J.C. Géminard, Phys. Rev. Lett. **80**, 2354 (1998)
13. T. Witten, Rev. Mod. Phys. **79**, 643 (2007)
14. B. Roman, A. Pocheau, Phys. Rev. Lett. **108**, 074301 (2012)
15. L. Landau, E. Lifshitz, *Theory of Elasticity* (Elsevier, Cambridge, England, 1986)
16. L. Euler, *Methodus Inveniendi Lineas Curvas Maximi Minimivi Propreitate Gaudentes. Additamentum I: De curvis elasticas* (Lausanne & Geneva, 1744)
17. W.A. Oldfather, C.A. Ellis, D. Brown, Isis **20**, 72 (1930)
18. B. Roman, A. Pocheau, Europhys. Lett. **46**, 602 (1999)
19. B. Roman, A. Pocheau, J. Mech. Phys. Sol. **50**, 2379 (2002)

20. A. Pocheau, B. Roman, Physica D **192**, 161 (2004)
21. A. Pogorelov, *Bendings of Surfaces and Stability of Shells* (American Mathematical Society, Providence, 1988)
22. D.J. Struik, *Lectures on Classical Differential Geometry* (Addison-Wesley, Reading, 1961)
23. A. Boudaoud, P. Patricio, Y. Couder, M. Ben Amar, Nature **407**, 718 (2000)
24. B. Du, O.C. Tsui, Q. Zhang, T. He, Langmuir **17**, 3286 (2001)

Chapter 8
Transport Properties in a Model of Quantum Fluids and Solids

Christophe Josserand

Abstract We discuss the general transport properties of the non-linear Schrödinger equation in the context of quantum fluid models. In particular, we will discuss two striking behaviors described within this model: the nucleation of quantized vortices and the non-classical rotational inertia.

8.1 Introduction: One Equation, Many Contexts

The nonlinear Schrödinger equation (NLS later on) has become since the last 50 years an emblematic equation both for mathematicians and physicists. It has been first deduced by Pitaevskii [1] and Gross [2] to model the superfluidity of helium (He^4), but, due to its generic features, the equation is relevant in many different domains as diverse as superfluid helium, Bose–Einstein condensates, water wave dynamics, or nonlinear optics for instance. On a mathematical aspect, this equation represents the "simplest" nonlinear equation that one can write with a complex function that has a conservative and Hamiltonian dynamics. In general, the NLS equation writes, in a dimensionless formulation:

$$i\frac{\partial \psi(\mathbf{x},t)}{\partial t} = -\frac{1}{2}\Delta \psi(\mathbf{x},t) + G|\psi(\mathbf{x},t)|^2 \psi(\mathbf{x},t), \tag{8.1}$$

where, in d space dimensions, we have:

C. Josserand
Institut D'Alembert, CNRS & UPMC (Univ. Paris 6), case 162,
Tour 55/65, 4, place Jussieu, 75005 Paris, France
e-mail: christophe.josserand@upmc.fr

X. Leoncini and M. Leonetti (eds.), *From Hamiltonian Chaos to Complex Systems: A Nonlinear Physics Approach*, Nonlinear Systems and Complexity 5,
DOI 10.1007/978-1-4614-6962-9_8, © Springer Science+Business Media New York 2013

$$\mathbf{x} = \sum_{i=1}^{d} x_i \mathbf{e}_i \qquad \text{and} \qquad \Delta = \sum_{i=1}^{d} \frac{\partial^2}{\partial x_i^2} ,$$

and the factor G characterizes the nonlinearity. The field $\psi(\mathbf{x},t)$ is a complex function and it describes the condensate wavefunction in superfluid helium and Bose–Einstein condensates (BEC) and also in the models of supersolids, the envelope of the electromagnetic field in nonlinear optics or the envelope of water waves in fluid mechanics for instance. In the context of quantum systems, the NLS equation is often called the Gross–Pitaevskii equation (GP later on) and we will use both of these two denominations through the review.

In this review, we will discuss some general properties and results of this equation related to transport phenomena. Indeed, since the equation corresponds to a Hamiltonian dynamics, no dissipation is present. However, apparent dissipation or also irreversible processes can be observed, exhibiting some peculiar features of the dynamics. These striking effects are related to two important properties of the equation that will be presented here in details: first, quantized vortices, which can be understood as topological defects, can naturally be described by NLS. Also, the phase space of the equation is very complex and is dominated by small scales fluctuation so that under some assumption the equation can mimic thermal dynamics. In this introduction section, different contexts where the NLS equation is relevant will be presented. Then, in Sect. 8.2, the general properties of the NLS equation will be deduced. Then, Sect. 8.3 will show how quantized vortices can be nucleated in the domain with simple dynamical configurations. Section 8.4 will discuss a particular case of the NLS equation where the interaction potential is long range so that a crystalline order can be present, important for modeling supersolids.

8.1.1 Bose–Einstein Condensates

Bose–Einstein condensation, although predicted in 1924 [3, 4], has only been observed experimentally in the mid 1990s, with, starting in 1995, the condensation of atomic gases [5–7] (and see [8, 9] for a review). The atoms are confined using an electromagnetic trap and the condensation is obtained through laser cooling and then evaporation cooling at the final stage. Through these processes, very low temperature of the order of microKelvins is reached with a sufficient high atom density so that the Bose–Einstein condensation can occur. The GP equation provides a very good model for the dynamics of the BEC. Indeed, considering that all the atoms are in the same quantum state $\psi(\mathbf{x},t)$, one obtains easily, starting with the Schrödinger equation for N particles within the dilute gas and Born approximations, that the following evolution equation for the wavefunction ψ holds:

$$i\hbar \frac{\partial \psi(\mathbf{x},t)}{\partial t} = -\frac{\hbar^2}{2m} \Delta \psi(\mathbf{x},t) + V(\mathbf{x},t)\psi(\mathbf{x},t) + Ng|\psi(\mathbf{x},t)|^2 \psi(\mathbf{x},t) . \qquad (8.2)$$

The complex field $\psi(\mathbf{x},t)$ is therefore called here the wavefunction of the condensate. $\hbar$ is the usual Planck constant and m the mass of an atom of the gas. The potential $V(\mathbf{x},t)$ represents the external trapping potential that confines the atom (it is a priori space and time dependent). N is the total number of atoms in the trap and g is the coupling parameter between the atoms, related to the scattering length a of the two atoms interaction potential, yielding

$$g = \frac{4\pi\hbar^2 a}{m}. \tag{8.3}$$

It is important to notice that a and thus g can be either positive or negative numbers depending on the gas species. As a first approximation, the magnetic external potential can be considered an harmonic potential:

$$V(\mathbf{x}) = \frac{1}{2} m \sum_{i=1}^{3} \omega_i^2 x_i^2 \ , \tag{8.4}$$

where the ω_i are the trapping frequencies that can be time dependent. Moreover, high contrast between the ω_i can be obtained so that the BEC can be almost 2D (disc shaped, where one ω_i is much bigger that the two other ones) or 1D (cigar shaped, where one ω_i is much smaller than the two other ones).
In this framework, the wavefunction ψ is normed to one, meaning:

$$\int_{\mathscr{D}} \psi(\mathbf{x},t)\psi^*(\mathbf{x},t)\mathrm{d}^d x = \int_{\mathscr{D}} |\psi(\mathbf{x},t)|^2 \mathrm{d}^d x = 1 \ , \tag{8.5}$$

where $\mathscr{D}$ is the physical domain of dimension d. The gas density $\rho(\mathbf{x},t)$ is defined by:

$$\rho(\mathbf{x},t) = N|\psi(\mathbf{x},t)|^2. \tag{8.6}$$

In this framework, the nonlinear term can be interpreted as the mean field potential describing the two-body interactions in the gas. The particle current $\mathbf{j}$,

$$\mathbf{j}(\mathbf{x},t) = \frac{iN\hbar}{2m}\left(\psi\nabla\psi^* - \psi^*\nabla\psi\right) = \rho(\mathbf{x},t)\,\mathbf{v}(\mathbf{x},t) \ , \tag{8.7}$$

allows for a consistent definition of the condensate velocity $\mathbf{v}$.

It has to be emphasized here that in this context of BEC, the GP equation offers a *quantitative* description of the dynamics, by contrast for instance with superfluid helium for which the equation has been first introduced (see below).

8.1.2 Superfluid Helium

The GP equation has been introduced independently in 1961 by Gross [2] and Pitaevskii [1] to describe the properties of superfluid helium discovered within the last 20 years [10–15]. Below the critical temperature $T_\lambda = 2.17\,\mathrm{K}$ (called the λ point because of the profile of the heat capacity of helium near that point), liquid helium (He^4) exhibits striking properties called superfluidity: in particular, its viscosity through thin capillary vanishes, leading to the decrease of the moment of inertia of a rotating vessel, the fountain effect, the existence of a second sound for instance. Because of these low temperature properties and since T_λ is very close to the Bose–Einstein condensation temperature computed for liquid helium, it has always been argued that superfluidity of helium is related to Bose–Einstein condensation, although no direct links can be drawn between these two properties. In fact, no mention of Bose–Einstein condensation is made (and needed) in the pioneering theory of Landau [13, 14] where only a "coherent state" of the atoms was needed. Moreover, Penrose and Onsager [21] have shown in 1956 that no more than 8 % of the helium atoms were condensed at $T = 0$ in helium while the superfluid fraction, defined roughly as the amount of fluid exhibiting no viscosity, was one for $T = 0$. However, assuming that the superfluid fraction could be described by a wavefunction (again called wavefunction of the superfluid), Gross and Pitaevskii obtained using general arguments (invariance, symmetries, etc.) the following equation:

$$i\hbar\frac{\partial\psi(\mathbf{x},t)}{\partial t} = -\frac{\hbar^2}{2m}\Delta\psi(\mathbf{x},t) + \int_{\mathscr{D}} U(\mathbf{x}'-x)|\psi(\mathbf{x}',t)|^2\psi(\mathbf{x},t)\mathrm{d}^3x'\,. \tag{8.8}$$

Here, m is the effective mass of the helium atom and $U(\mathbf{r})$ describes the two-body interaction potential of helium atoms. This form of the NLS equation is in fact more general than the one written above that can be retrieved by taking $U(\mathbf{r}) = G\delta(\mathbf{r})$ where δ is the delta function. In fact, although this equation can describe *qualitatively* most of the properties observed in superfluid helium, one cannot deduce it formally starting from the microscopic quantum description of the liquid atoms of helium in interaction. In particular, an important assumption to deduce NLS from the N particles Schrödinger equation is the weakness of the interactions that is not satisfied for helium. Therefore, the status of GP equation for superfluid helium has to be considered more as a phenomenological model than a microscopic description. However, because of its simplicity and generality, this model has been widely used to investigate the nucleation of vortices [16–18] or the vapor–liquid transition in quantum liquids [19, 20].

8.1.3 A Model for Supersolidity?

While liquid helium below the λ-point and BEC are superfluids (super-liquids and gases, respectively), the question of the existence of a super-solid state has been first raised in the 1950s [21], although it might first appear impossible because of the opposition between long range order (coherent state) and short range variations (the solid structure). However, as first stated by Andreev and Lifshitz [22] and then by Leggett [23] in particular, it is a priori possible to observe a nonclassical rotational inertia (NCRI later on) in quantum solids: it is the property that when put in infinitesimal rotation, the rotational inertia of the system is not equal to that obtained in usual solid rotation of the system. It is only recently that such effect has been observed in solid helium below 70 mK [24, 25], following a signal of phase transition in the heat capacity of solid helium [26]. Using a torsional oscillator, the solid helium exhibits a sudden drop in the rotational inertial of the system. Although these results have been reproduced qualitatively by six other groups [27–32], important differences have been observed in the superfluid fraction (defined as the fraction of the solid that is *not* rotating) between experiments, suggesting that the disorder in the solid is crucial [28, 33]. In addition, no evidence of a superflow has been noticed in solid He^4 [27, 34, 35] in the conditions where NCRI exists and an increase of the shear modulus for solid helium is also observed with a temperature dependence similar to the observed NCRI [36–38]. However, even if these disorder and elastic effects seem to invalidate the superfluid property of solid helium, they cannot explain alone all the experiments done where an NCRI is observed [39]. For all these reasons, the status of such a new "supersolid" state of matter is still a question of scientific debates.

However, the GP equation provides a consistent model of supersolid that presents two important features: the superfluid property through a nonzero NCRI for instance and a crystalline structure with a shear modulus typical of solids [40–42]. In order to obtain such crystal configuration with the GP equation, we have to use the nonlocal formulation (8.8) where the two-body potential $U(\mathbf{r})$ is not a dirac function. In fact, as we will discuss later on, it is sufficient that the Fourier transform of this potential is negative for some domain in the Fourier space to observe a solid structure and one can choose with no loss of physical meaning:

$$U(|\mathbf{r}-\mathbf{r}'|) = U_0\theta(a-|\mathbf{r}-\mathbf{r}'|) \tag{8.9}$$

with $\theta(\cdot)$ the usual Heaviside function ($\theta(s) = 1$ if $s > 0$ and $\theta(s) = 0$ if $s < 0$), a being the range of the potential and U_0 its strength.

Similarly, a model coupling an NLS equation with a solid structure has been also suggested [43]: there, the solid structure is "normal" and somehow decoupled from the "superfluid" part that is described by a GP equation, in the same spirit than the original Pitaevskii *à la Landau* approach. In our model, the "wavefunction of the condensate" describes quantum solids where the solid structure is intimously coupled with the superfluid features of the GP equation [44–46]. The properties of this supersolid model will be reviewed in Sect. 8.4 in more details.

8.1.4 Nonlinear Optics

When high-intensity lights propagate inside optic fibers or complex materials, for instance, nonlinear effects are observed through a dependence of the optic index with the light intensity. This can be in fact present in everyday tools such as the compact disc lasers or pointers. Considering then the slowly varying envelope of the electromagnetic field that propagates around the main frequency, one can obtain the following NLS equation [47]:

$$i\frac{\partial \psi(\mathbf{x},t)}{\partial z} = -\frac{1}{2}\left(\frac{\partial^2}{\partial t^2} + \Delta'\right)\psi(\mathbf{x},t) + f(|\psi|^2)\psi(\mathbf{x},t)\ , \tag{8.10}$$

where f is a real function that characterizes the nonlinear interaction between the light and the matter. z is the direction of propagation of the light and Δ' is the Laplacian term in the plane orthogonal to z (in the 1D case of fiber optics, only the second derivative of time is present). The diffraction terms (second derivatives) can be due to the diffraction of the light (time derivatives) but also to the dispersion (spatial derivatives normal to the direction of propagation of the light). The function f can have different forms: the usual NLS equation can be recovered in the case of the Kerr effect where the optical index is a linear function of the light intensity $I = |\psi|^2, n = n_0 + \alpha I$. Finally, in this case, the wavefunction is not normalized since the integral of $|\psi|^2$ is the total intensity.

8.1.5 Fluid Mechanics

The GP equation has been though introduced to model the dynamics of superfluids. In fact, it can also describe usual fluid mechanics problems, when the dissipation can be neglected. In particular, one can deduce an NLS equation in the case of the dynamics of surface waves: for the sake of simplicity we will focus on the 1D case where the interface is located on $z = h(x,t)$. Then similarly to the nonlinear optics case, we will develop the wave equation around its mean wavenumber k_0

$$h(x,t) = \Re(A(x,t)\mathrm{e}^{i(k_0x-\omega_0t)})\ . \tag{8.11}$$

The amplitude $A(x,t)$ is a complex function, with slow- and large-scale variations (compared to $1/\omega_0$ and $1/k_0$). The dispersion relation of the waves in *deep* water follows:

$$\omega_0 = \sqrt{gk_0}\ . \tag{8.12}$$

Neglecting the viscous effects in these large scale dynamics, the wave dynamics is described by the mass conservation coupled to the Bernoulli equation written at the interface:

$$\frac{\partial h}{\partial t} - \frac{\partial \phi}{\partial y} = -\frac{\partial h}{\partial x}\frac{\partial \phi}{\partial x}\,, \tag{8.13}$$

$$\frac{\partial \phi}{\partial t} + gh = -\frac{1}{2}\left(\left(\frac{\partial \phi}{\partial x}\right)^2 + \left(\frac{\partial \phi}{\partial y}\right)\right), \tag{8.14}$$

where ϕ is the velocity potential. Using a multi-scale approach, one can obtain the following NLS equation for the amplitude $A(x,t)$ [48]:

$$\frac{\partial A}{\partial t} = -i\frac{\omega_0}{8k_0^2}\frac{\partial^2 A}{\partial x^2} - i\frac{\omega_0 k_0^2}{2}A^2A^*\,. \tag{8.15}$$

Remark that one can also deduce this equation from the usual Korteweg–de Vries equation in a low amplitude expansion [49]. Interestingly, the NLS equation can also describe (with some important limitations though) the opposite limit of shallow water through the Bernoulli-like equation (see below).

8.2 General Properties of the NLS Equation

Prior to the study of the transport features in the systems described by the NLS equation, it is important to present the general properties of this equation. To simplify this section, we will work with no loss of validity with the dimensionless and local potential version of the NLS equation (8.1):

$$i\frac{\partial \psi(\mathbf{x},t)}{\partial t} = -\frac{1}{2}\Delta\psi(\mathbf{x},t) + G|\psi(\mathbf{x},t)|^2\psi(\mathbf{x},t)\,.$$

Two cases can be distinguished depending on the sign of G (with respect with the negative sign in front of the Laplacian term):

- $G > 0$: it is the defocusing NLS equation, where the nonlinearity tends to saturate the amplitude of the wavefunction. It is the equation that describes the dynamics of BEC in general (except for some specific gases where the scattering length can be negative, in particular, the lithium Li_7 [7]) and superfluid helium.
- $G < 0$ is for the focusing NLS equation. In that case, as it will be obvious in the Hamiltonian formulation, very high amplitude localized peaks can be formed by the dynamics, eventually leading to finite time singularities [50]. This equation is often valid in nonlinear optics where it describes the pulses propagation or for water waves in deep water where it might explain the formation of freak waves [51].

Notice that in this dimensionless version, one can have simply $G = \pm 1$.

8.2.1 Conserved Quantities and Hamiltonian Structures

Different quantities are conserved through the evolution of the NLS equation and in particular two scalar quantities, the mass or particles number

$$N = \int_{\mathscr{D}} |\psi|^2 \mathrm{d}^d x \,. \tag{8.16}$$

In the context of the BEC, it is nothing else than the normalization of the wavefunction. Such property is straightforward to demonstrate, yielding

$$\begin{aligned}
\frac{\mathrm{d}N}{\mathrm{d}t} &= \int_{\mathscr{D}} \left(\psi \frac{\partial \psi^*}{\partial t} + \frac{\partial \psi}{\partial t} \psi^* \right) \mathrm{d}^d r \,, \\
&= \int_{\mathscr{D}} i \left(\psi \left(-\frac{1}{2} \Delta \psi^* + \alpha |\psi|^2 \psi* \right) - \left(-\frac{1}{2} \Delta \psi + \alpha |\psi|^2 \psi \right) \psi^* \right) \mathrm{d}^d r \,, \\
&= \int_{\mathscr{D}} \frac{i}{2} (\psi^* \Delta \psi - \psi \Delta \psi^*) \mathrm{d}^d r = - \int_{\mathscr{D}} \mathrm{div} \left(\frac{i}{2} (\psi \nabla \psi^* - \psi^* \nabla \psi) \right) \mathrm{d}^d r, \\
&= - \int_{\partial \mathscr{D}} \mathbf{j} \cdot \mathbf{n} \,,
\end{aligned}$$

where j is similar to the current already introduced above:

$$\mathbf{j} = \frac{i}{2} (\psi \nabla \psi^* - \psi^* \nabla \psi) \,. \tag{8.17}$$

The other scalar conserved quantity, the total energy $\mathscr{H}$ is related to the Hamiltonian structure of the equation. Indeed, considering the Hamiltonian:

$$\mathscr{H} = \frac{1}{2} \int_{\mathscr{D}} (|\nabla \psi|^2 + G |\psi|^4) \, \mathrm{d}^d r, \tag{8.18}$$

one can show that the NLS equation can be written through the functional relation:

$$i \frac{\partial \psi}{\partial t} = \frac{\delta \mathscr{H}}{\delta \psi^*}, \tag{8.19}$$

so that the energy (the Hamiltonian) is conserved by the dynamics.

Similarly, but with less fundamental implications, two vector quantities are conserved by the dynamics, the total momentum **P** and the angular momentum **M**, that can be defined immediately from the expression of the current j, following

$$\mathbf{P} = \int_{\mathscr{D}} \mathbf{j} \mathrm{d}^d x, \tag{8.20}$$

and

$$\mathbf{M} = \int_{\mathscr{D}} \mathbf{x} \times \mathbf{j} \mathrm{d}^d x. \tag{8.21}$$

8.2.2 Invariances of the Equation

The solutions of the NLS equation exhibit different invariances that play an important role in the general properties of its dynamics, namely, if $\psi(\mathbf{x},t)$ is a solution of NLS, then $\psi'(\mathbf{x},t)$ is also solution, with:

- Phase invariance: $\psi'(\mathbf{x},t) = \psi(\mathbf{x},t)\mathrm{e}^{i\phi}$.
- Translational invariance: $\psi'(\mathbf{x},t) = \psi(\mathbf{x}-b,t)$ where $\mathbf{b}$ is a constant vector.
- Galilean invariance: $\psi'(\mathbf{x},t) = \psi(\mathbf{x}-V_0 t,t)\mathrm{e}^{-i\left(\frac{V_0^2 t}{2} - \mathbf{V}_0 \cdot \mathbf{x}\right)}$ where V_0 is the traveling velocity.
- Dilatation invariance: $\psi'(\mathbf{x},t) = \lambda\psi(\lambda\mathbf{x},\lambda^2 t)$. In that case the number of particles has changed.

8.2.3 Integrability, Solitons and Solitary Waves

There is a situation where there are an infinity of conserved quantities, in addition to the previous ones: indeed, in 1D, the NLS equation is integrable [52] for the cubic nonlinearity. In this case, solitons exist that propagate and interact freely between each other. For the defocusing case $G = 1$, the family of solitons in 1D can be written:

$$\psi_S(x,t) = (\nu \tanh(\nu(x-\chi t)) + i\chi)\mathrm{e}^{-it}, \tag{8.22}$$

with the condition $\nu^2 + \chi^2 = 1$. These solitons correspond to trough of depth ν^2 that propagates at constant velocity ξ, as shown in Fig. 8.1 and are called grey solitons. The particular case of $\nu = 1$ is called the black soliton and describes a kink.

Similarly, solitons exist for the focusing case ($G = -1$) and are used to describe optical pulses in nonlinear optics. They read

$$\psi_{\mathrm{op}} = \frac{\sqrt{\frac{N}{2}}}{\mathrm{ch}\left(\sqrt{\frac{N}{2}}(x-Ut)\right)} \mathrm{e}^{i\left(Ux + (\frac{N}{4} - \frac{U^2}{2})t\right)}, \tag{8.23}$$

where N is the total "mass" of the pulse and U its velocity. Pulses with different masses are shown in Fig. 8.2.

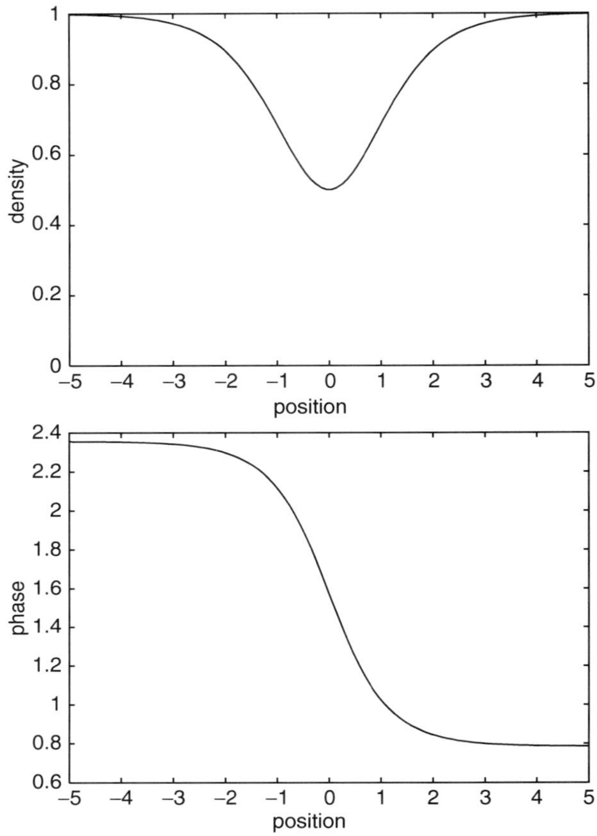

Fig. 8.1 Density (*up*) and phase (*down*) of a grey soliton

In higher spatial dimension these functions are still solution of the NLS equation, but they are no more solitons. In fact traveling solutions exist also in NLS for different nonlinearities [20,53,54] and are usually called "solitary waves" in contrast with soliton that should be used only for integrable systems.

8.2.4 Hydrodynamical Equations

The usual decomposition of the complex wavefunction ψ in the modulus–argument allows for an interesting hydrodynamical analogy. Indeed, as discussed above, the modulus square of ψ can be interpreted as a density (particle, mass density, or light intensity for instance) while the argument can be related to a velocity potential following the definition of the current $\mathbf{j}$ so that we can write

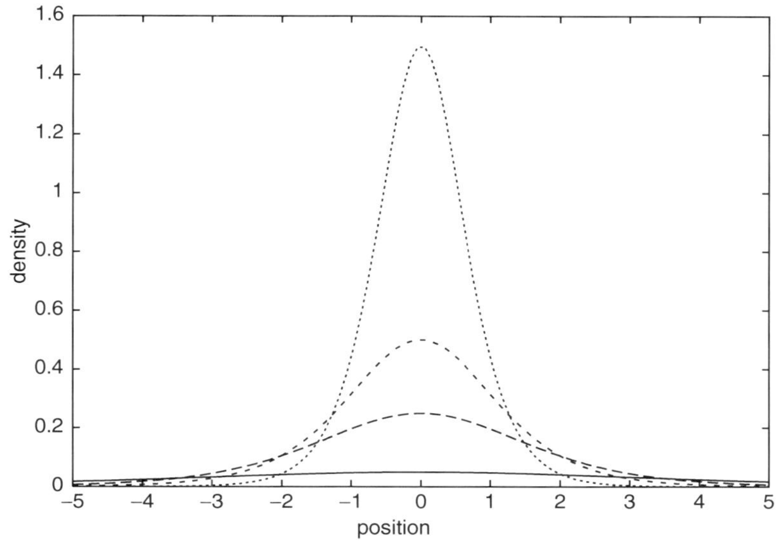

Fig. 8.2 Density ψ^2 for optical solitons Eq. (8.23) for $N = 0.1, 0.5, 1$, and 3

$$\psi(\mathbf{x},t) = \sqrt{\rho(\mathbf{x},t)}\mathrm{e}^{i\phi(\mathbf{x},t)} , \tag{8.24}$$

and we define the velocity $\mathbf{v} = \nabla\phi$. Then, the NLS equation decomposes in a set of two equations (for the real and imaginary part):

$$\frac{\partial\rho}{\partial t} + \nabla\cdot(\rho\mathbf{v}) = 0 , \tag{8.25}$$

$$\frac{\partial\varphi}{\partial t} = -\frac{1}{2}\frac{\Delta\sqrt{\rho}}{\sqrt{\rho}} + \frac{v^2}{2} + G\rho . \tag{8.26}$$

The first Eq. (8.25) is nothing else than the mass conservation equation and the second one Eq. (8.26) a Bernoulli-like equation for a compressible fluid (with the pressure $P = G\rho^2/2$). Moreover there is an additional term in the Bernoulli equation

$$-\frac{1}{2}\frac{\Delta\sqrt{\rho}}{\sqrt{\rho}},$$

by contrast with usual fluid since it involves the spatial variation of the density. This term is called the quantum pressure since it vanishes in the limit $\hbar \to 0$ in the GP equation and it can also play the role of a surface tension for small enough density variations. In fact, this term can be neglected in the hydrodynamical limit of small- and large-scale density variations since its pertinent length scales on the order of the so-called healing length: this length can be observed when seeking for a 1D solution of NLS for $x > 0$ where the density at infinity is ρ_0 while the boundary condition on $x = 0$ is $\psi = 0$. This latter boundary condition describes in particular a solid boundary in the context of BEC or superfluids. The lowest energy solution reads

$$\phi(x) = \sqrt{\rho_0}\tanh\left(\frac{x}{\sqrt{2}\xi}\right)\mathrm{e}^{-iG\rho_0 t}\ ,$$

where $\xi = 1/\sqrt{G\rho_0}$.

8.2.5 Quantized Vortices

Finally, other particular solutions of NLS of great importance are vortices: in fact, one would expect from the modulus–argument transform Eq. (8.24) that NLS describes potential flows only where vorticity cannot exist. This is in fact true except when $\rho = 0$ where the phase is undefined: these are topological defects (they can be seen as the intersection of two hypersurfaces, one where the real part of ψ is zero, the other where the imaginary part vanishes), i.e., points in 2D and lines in 3D. Since the phase can be determined within an additional $2\pi n$ factor where n is an integer, the circulation of the velocity around the defects is quantized ($2\pi n$ in dimensionless formulation, $2\pi\hbar n/m$ for the GP equation). Vortex solutions can be studied in 2D for a wavefunction, for $G = 1$ and a density at infinity $\rho_0 = 1$ with no loss of generality. They are stationary solutions in the form:

$$\psi(x,y) = f(r)\mathrm{e}^{iq\theta}\mathrm{e}^{-it}\ , \tag{8.27}$$

where r is the radius ($r^2 = x^2 + y^2$) and θ in the polar angle. Then f satisfies the following equation:

$$f'' + \frac{f'}{r} - q^2\frac{f}{r^2} + f(1 - f^2) = 0, \tag{8.28}$$

which can be solved using the shooting method. Remarkably, one can show that $f(r) \propto r^q$ for small r. Interestingly, in this case ($G = 1$ with constant density at infinity) only single charged vortices (with $|q| = 1$) are dynamically stable while a higher order vortex decomposes into single charged ones. However, when the system is confined by a strong potential (stronger than harmonic), multiple charged vortices (called giant vortices) can be observed [55–57].

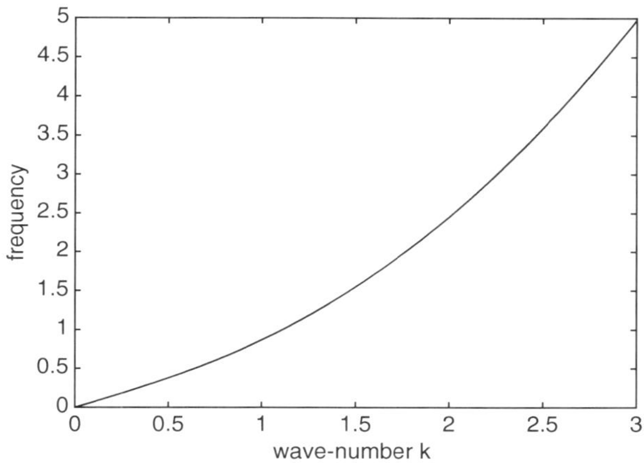

Fig. 8.3 Dispersion relation (8.29) for $G\rho_0 = 0.5$

8.2.6 Dispersion Relation, Spectrum of Excitation and Superfluidity

In the defocusing case of interest for superfluid helium and usual BEC, the NLS equation provides an emblematic model of superfluidity since it can describe a fluid flowing with no dissipation. As demonstrated by Landau [13, 14], this property is related to the excitation spectrum that is the dispersion relation of the perturbations around a constant density solution $\psi = \sqrt{\rho_0}\mathrm{e}^{-i\rho_0 t}$. Performing the linear analysis around this solution

$$\psi = (\sqrt{\rho_0} + \delta\psi \mathrm{e}^{i\omega_k - \mathbf{k}\cdot\mathbf{x}})\mathrm{e}^{-i\rho_0 t},$$

we find the well-known Bogoliubov spectrum for the frequency ω_k:

$$\omega_k^2 = G\rho_0 k^2 + \frac{1}{4}k^4. \tag{8.29}$$

This spectrum, drawn in Fig. 8.3 for $G\rho_0 = 0.5$, exhibits for large scales (small k) a phonon behavior that is at the heart of the superfluidity property [13] (with a phonon or Landau velocity $\sqrt{\rho_0}$) while the free particle spectrum ($\omega_k \sim k^2/2$) is valid for large k.

8.3 Vortex Nucleation

Quantized vortices have been observed both in superfluid and recently in Bose–Einstein condensate; for instance, vortices appear in superfluid helium in a rotating cell [58], which is in fact the unique experiment where vortices have been clearly identified; difficulties to observe vortices in superfluid ^{4}He come from the small value of the vortex core size ξ_0 which corresponds to the healing length. On the contrary, BEC offer great opportunities, since the vortex core attains a larger size and vortices have been first observed into a rotating BEC in 2000 [59]. Vortices nucleation is crucial in superfluid/BEC since it provides a "dissipative"-like flow through the motion of the vortices. As a prototype problem, we address here the vortex nucleation for flows past obstacles in the framework of the Gross–Pitaevskii dynamics. At rest, let us consider first the steady state solution with constant density at infinity that minimizes the energy: besides a layer near the obstacle due to the boundary condition, it is almost a uniform density solution. As the GP equation is Galilean invariant, it is easy to construct from this ground state another solution representing a uniform flow by boosting the rest state to a specified speed. Over the last 20 years, important progresses have been made in the case of a 2D flow around a circular disc, that is, the solution of the NLS equation with uniform flow speed and constant mass density at infinity, with a boundary condition on the surface of the disc [16, 18, 60, 61].

It has been observed in numerical simulations that beyond a certain critical speed, the flow around the disc becomes time dependent, because vortices are emitted from the disc surface as shown in Fig. 8.4. In 1D, the nucleation of vortices is replaced by the periodic release of grey solitons [62].

The release of vortices from the boundary of the disc is shown to be a consequence of a transition from a locally subsonic to supersonic flow in this specific compressible model [16]: in fact, the local Landau critical velocity (which is the same in NLS than the speed of sound) is reached and the vortex is a consequence of the resulting complex NLS dynamics. Similar vortex nucleation has been identified in BEC in the framework of a moving laser beam [63, 64]. In ordinary fluid mechanics, a transonic transition leads to the formation of a shock wave but nothing similar exists in the GP equation, because of the lack of built-in irreversibility (due to the Hamiltonian structure), something that is necessary to balance nonlinearities inside the shock wave. In the present model, the formation of shock waves is replaced by the nucleation of quantized vortices as first argued in [16]. Later on [18], the structure of the transonic transition has been studied, exhibiting a Euler–Tricomi equation for the evolution of the velocity potential. We will resume now the main results obtained in former works [16, 18, 61].

8.3.1 Around the Transonic Regime

We start with the defocusing NLS equation [(8.1) with $G = 1$]:

$$i\partial_t \psi(\mathbf{r},t) = -\frac{1}{2}\nabla^2\psi + \psi(\mathbf{r},t)|\psi(\mathbf{r},t)|^2.$$

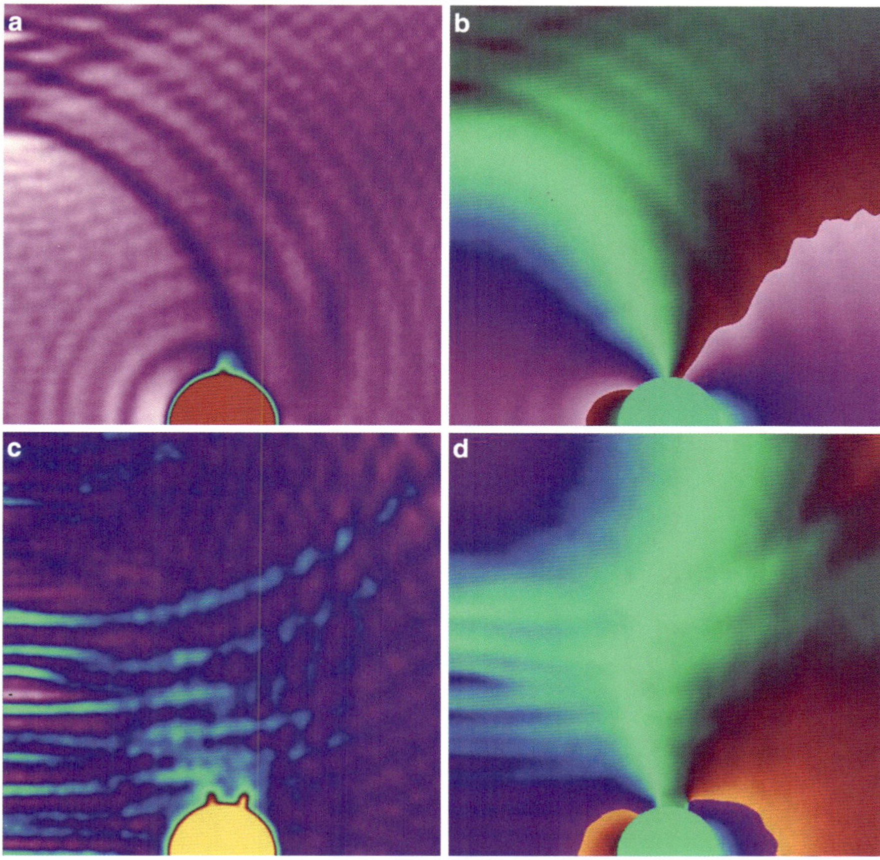

Fig. 8.4 Numerical simulation of the nonlinear Schrödinger equation for a 2D flow around half a disc: the velocity at infinity is $v_\infty = 0.442$, the mesh grid $\mathrm{d}x = 0.125$ and the radius of the disc is $R = 7.5$ ($\xi_0 = 1$). (**a**) and (**b**) respectively the modulus and the phase of the wavefunction at $t = 20$ time unit of NLS. The density and the phase go up from *bright* to *dark color*. One can see the low density around the *top* of the obstacle, due to the Bernoulli effect. (**c**) and (**d**) same functions at $t = 50.6$. A low density structure is now moving with the flow (at right of the *top* of the disc). The phase profile around this structure shows a 2π circulation: a quantized vortex. Reprinted with permission from [61], Copyright (2001) IOP

The ground state in an infinite or periodic box is the homogeneous solution: $\psi_0 = \sqrt{\rho_0}\mathrm{e}^{-i\rho_0 t}$. The long-wave and small-amplitude perturbations propagate with the sound speed $c_s = \sqrt{\rho_0}$, while the healing length is $\xi_0 \sim \frac{1}{\sqrt{\rho_0}} = \frac{1}{c_s}$.

If the flow is stationary, $\partial_t \phi$ is a constant that is defined by the conditions at infinity and mass density ρ can be computed everywhere as a function of v, as long as the quantum pressure term can be neglected, and the following equations can be deduced from Eqs. (8.25) and (8.26) for the stationary flow around a disc (of radius R, much bigger than the intrinsic length scale ξ_0) with a uniform velocity at infinity v_∞:

$$\nabla \cdot (\rho(|\nabla\phi|)\nabla\phi) = 0, \tag{8.30}$$

$$\rho(v) = \rho(|\nabla\phi|) = \rho_\infty + \frac{1}{2}(v_\infty^2 - |\nabla\phi|^2), \tag{8.31}$$

$$\hat{n}\cdot\nabla\phi = 0 \quad \text{on the disc}, \tag{8.32}$$

$$\phi = v_\infty x \quad \text{at infinity}, \tag{8.33}$$

where $\hat{n}$ is the normal vector on the disc.

Where we can assume that the density is a function of v only, Eq. (8.30) becomes

$$\partial_v(\rho(v)v)\partial_{xx}\phi + \rho(v)\partial_{yy}\phi = 0, \tag{8.34}$$

where the origin of the axis is now placed at the disc pole where the vortices are emitted, x being the local coordinate tangent to the main flow (and to the disc boundary) and y the orthogonal one. At low velocities this second-order partial differential equation is elliptic but Eq. (8.25) becomes hyperbolic beyond a critical velocity, and this happens when $\partial_v(\rho(v)v)$ vanishes, that is, when the mass current takes its largest possible value.

The condition $\partial_v(\rho(v_c)v_c) = 0$ gives the relation $v_c^2 = \frac{2}{3}\rho_0 + \frac{1}{3}v_\infty^2$ and using Eq. (8.31) the local density is $\rho_c = v_c^2$ so that the local sound speed is simply v_c and Eq. (8.25) becomes hyperbolic exactly at the transonic transition. When v_∞ increases, the property of ellipticity of Eq. (8.25) is broken first at the poles of the disc, leading to the nucleation of two vortices, one at each pole. As time goes on, these vortices are convected downstream by the flow and they induce a counterflow that reduces the velocity on the surface of the disc so that the local velocity at the pole is below the critical speed: the ellipticity of the equation for the velocity potential is then restored. But the vortex dipole is pulled farther and farther downstream, the counter streaming effect diminishes, until the velocity at the pole reaches again the critical value. Then two new vortices are emitted and so on, as shown in Fig. 8.5. This whole process corresponds thus to a more or less periodic release of vortices from the obstacle.

8.3.2 The Euler–Tricomi Equation in the Transonic Region

We will perform now a detailed analysis near the pole for a local velocity v_0 close to v_c and we can write the velocity potential in this vicinity in the form $\phi = v_0 x + \frac{v_c}{3}\chi$, where χ describes the local variation of the velocity. Equations (8.30) and (8.31) read in this framework:

$$-(\varepsilon + \partial_x\chi)\partial_{xx}\chi + \partial_{yy}\chi = 0 \tag{8.35}$$

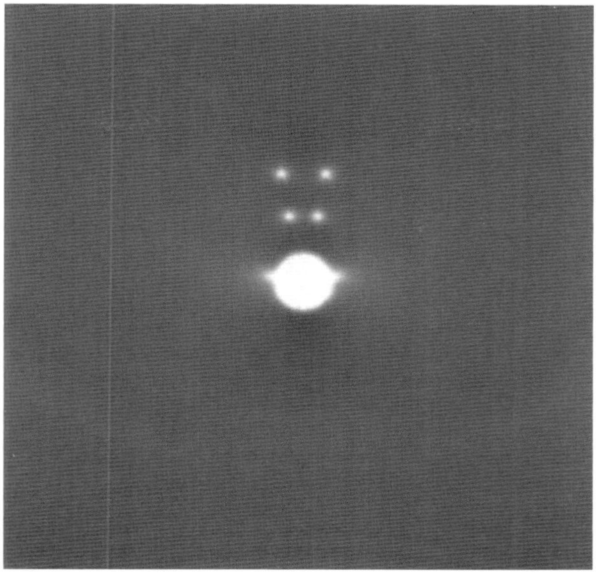

Fig. 8.5 Density snapshot for the solution of NLS with a velocity at infinity above the critical velocity ($v_\infty = 0.450$, for $\xi_0/R = 1/12$) for vortex nucleation. The flow goes from *bottom* to *top*. A pair of vortices have been nucleated long ago already. They have traveled along the flow, until a new pair could be released. One can also observe that another pair of vortices will soon be emitted from the disc. The color scale is such that the density increases from *white* ($\rho = 0$) to *black* ($\rho = 1$). Reprinted with permission from [61], copyright (2001) IOP

with $\varepsilon = 3(v_0 - v_c)/v_c$ and the expansion of the boundary conditions (8.32) near the pole reads at dominant order:

$$\partial_y \chi = -M_a \frac{x}{R} \quad \text{at} \quad y = -\frac{x^2}{2R}, \tag{8.36}$$

with the local Mach number $M_a = \frac{3v_0}{v_c}$, a constant close to 1. At this order, the following scalings for the different quantities can be obtained:

$$x \sim R\varepsilon^{3/2} \quad , \quad y \sim R\varepsilon, \quad \text{and} \quad \varphi \sim R\varepsilon^{5/2}. \tag{8.37}$$

A solution of Eq. (8.35) satisfying the boundary condition can be found:

$$\chi_0 = -M_a \frac{xy}{L} \tag{8.38}$$

so that the first-order correction of the velocity field is still time independent. To understand the transition to supersonic flow, we have to expand the velocity potential to the next order, yielding $\chi = \chi_0 + \varphi$

$$-\left(\varepsilon - M_a \frac{y}{R}\right) \partial_{xx}\varphi + \partial_{yy}\varphi = 0, \tag{8.39}$$

with the following boundary conditions:

$$\partial_y \varphi = -M_a \frac{x^3}{R^3}; \quad \text{at} \quad y = -\frac{x^2}{2R}. \tag{8.40}$$

The resulting Eq. (8.39) is nothing else than the Euler–Tricomi (ET) equation that is usually deduced for shock dynamics in compressible fluids, although in our approach it is directly in the physical space by contrast to the hodograph variables in classical fluids [65]. The E–T equation is interpreted as follows: $-(\varepsilon - M\frac{y}{R})$ represents a generic tangential velocity profile of an ideal flow near a body, since the local main speed diminishes as y increases and the Mach number is exactly one at $y = \varepsilon\frac{R}{M_a}$. The physical domain is thus now decomposed into a supersonic region near the obstacle and a subsonic one elsewhere. One can directly deduce the following scalings for the ET equation

$$x \sim R\varepsilon^{3/2} \quad , \quad y \sim R\varepsilon, \quad \text{and} \quad \varphi \sim R\varepsilon^{11/2}. \tag{8.41}$$

Notice that the spatial scalings remain the same and only the velocity correction has a different scaling.

8.3.3 From the Euler–Tricomi Equation to Vortex Nucleation?

At this stage, the Euler–Tricomi equation can be solved using first a particular polynomial solution satisfying the boundary conditions:

$$\phi_0 = -M_a \frac{x^3 y}{R^3} - \varepsilon M_a \frac{xy^3}{R^3} + M_a^2 \frac{xy^4}{2R^4}, \tag{8.42}$$

and we end up to the next order again to the ET equation:

$$-\left(\varepsilon - M_a \frac{y}{R}\right) \partial_{xx}\varphi + \partial_{yy}\varphi = 0 \tag{8.43}$$

with homogenous boundary conditions

$$\partial_y \varphi = 0; \quad \text{at} \quad y = 0.$$

The general solution of this problem, including the next orders (nonhomogenous) correction due to the quantum pressure and the nonlinear terms in particular, is very technical and can be found in great details in [18,61]. Here, we propose a qualitative analysis of the ET equation that suggests the vortex nucleation mechanisms: indeed, an important feature of the ET equation is that its solution becomes multivalued in the supersonic domain. In fact, it can be understood as the formation of a phase jump that grows with time. For small phase difference, the quantum pressure and/or the nonlinear terms should regularize the velocity potential. However, when the phase jump (of order π) is high enough, it is quite natural to expect that the nucleation of a vortex is favored. Although this qualitative scenario in reducing can be used to explain how the ET gives rise to strong phase variations, it has to be said that it has not been formally deduced from the equation for the phase written above. In particular, there are no crystal clear links between a short-scale phase variation and the nucleation of a vortex and such nucleation mechanism is still lacking. Instead, an amplitude equation for the phase correction has been found that exhibits a saddle-node bifurcation in good qualitative agreement with the initial scenario proposed in [16].

8.4 Nonclassical Rotational Inertia in a Supersolid Model

Below we focus on the NLS equation (8.8) as a model of supersolid, valid at $T = 0$, which is fully explicit. Moreover, it exhibits all the properties requested for this state of matter, but it is sufficiently simple to allow in depth calculations. This is a model in the true sense because it cannot be deduced from the equations of atomic motion pertinent for solid helium! It tries only to keep the most fundamental properties of this quantum solid to understand its behavior. This model has many advantages: the first one is that it is a model of supersolid in the sense of Leggett [23], i.e., it shows NCRI as well as an absence of superflow induced by pressure gradient; second, it may be easily implemented in numerical simulations where superfluids as well as ordinary solid behavior may be tested; finally, some of the predictions can be compared with experiments.

Since the work of Kirzhnits and Nepomnyashchii [44] and of Schneider and Enz [66] in the early 1970s the transition from liquid to solid helium has been regarded as a manifestation of an instability as the roton minimum in the energy–momentum spectrum touches the zero energy line. Although this idea has been circulating for many years, it is only in [40] that a mean field model consisting of the Gross–Pitaevskii equation has been proposed [1, 2]. It exhibits a roton minimum in the dispersion relation, already introduced in 1993 [67], that presents a first-order phase transition to a crystalline state as the roton minimum decreases. In [41, 42] a theory for the long-wave perturbations to the ground state is obtained. The mechanical equilibrium was studied under external constraints as steady rotation or external stress and the model displays an apparently paradoxical behavior: the numerical simulations [41] clearly show that in the NLS model, nonclassical rotational inertia is observed as well a regular elastic response to external stress or forces without any flow of matter, as in experiments [27, 34, 68, 69].

The existence of a nonclassical rotational inertia in the limit of small rotation speed does not require defects nor vacancies (in full agreement with Leggett's ideas) and no superflow under small (but finite) stress nor external pressure

(continued)

(continued)
gradient. The only matter flow for finite stress is due to plasticity being facilitated by the eventual presence of defects.

In addition, a new propagating "sound" mode is found besides the usual longitudinal and transverse phonons in classical crystals. The speed of this mode is smaller than the usual elastic sound wave speed, since it is related to the superfluid fraction $f^{ss} = \rho^{ss}/\rho$ (defined in more details below) following $c \sim \sqrt{f^{ss}}$ the superfluid fraction . This slow mode $f^{ss} \sim 10^{-4}$ is related to the phonon contribution in Bose–Einstein condensates.

We will discuss here the following aspects of this model:

(i) The ground state of the mean field model is a crystal, that is, a periodic pattern (a hexagonal one in 2D and a *hcp* in 3D).
(ii) The existence of NCRI.
(iii) Usual solid elastic properties.

This section is a light version extracted from lecture notes published for the Warsaw School of Statistical Physics [70].

8.4.1 Properties of the Model

Our model is based upon the original form of the Gross–Pitaevskii (G–P) equation, which is a non-linear and non-local partial differential equation (8.8). To have stability of long-wave fluctuations, the two-body interaction potential, which depends on the relative distance, should satisfy that $\int U(|\mathbf{r}|)\,\mathrm{d}\mathbf{r}$ is finite and positive. Moreover, we shall require also that the Fourier transform

$$\hat{U}_k = \int U(|\mathbf{r}|)\,\mathrm{e}^{i\mathbf{k}\cdot\mathbf{r}}\mathrm{d}\mathbf{r} \tag{8.44}$$

is bounded for all $\mathbf{k}$, and, as we will see later, $\hat{U}_k$ should become negative in some bandwidth in the k space.

The author does not know any experimental situation where this nonlocal Gross–Pitaevskii equation may be formally deduced, although it has been suggested recently that a BEC of dipolar atoms would exhibit such peculiar interactions [71, 72]: finally, whatever the context (supersolid model or long range interaction BEC), this equation is very interesting to study for our knowledge of many-body systems.

Keeping first the physical units, Eq. (8.8) follows the usual properties of the NLS equation, in particular:

- Translation invariance–phase invariance, similarly $\psi(\mathbf{r},\mathbf{t})\mathbf{e}^{\mathbf{i}\alpha}$ with α any real constant is a solution.

- Galilean invariance.
- Hamiltonian structure yielding.

$$H[\psi,\psi^*] = \frac{\hbar^2}{2m}\int |\nabla\psi(\mathbf{x})|^2 \,\mathrm{d}\mathbf{x} + \frac{1}{2}\int\int U(|\mathbf{x}-\mathbf{x}'|)|\psi(\mathbf{x})|^2|\psi(\mathbf{x}')|^2 \,\mathrm{d}\mathbf{x}\mathrm{d}\mathbf{x}', \tag{8.45}$$

  which is positive if $U(s) \geq 0$ for all s.
- The Hamiltonian H, the number of particles $N = \int |\psi(\mathbf{r})|^2 \,\mathrm{d}\mathbf{r}$, and the linear momentum $\mathbf{P} = -\frac{i\hbar}{2}\int(\psi^*\nabla\psi - \psi\nabla\psi^*)\mathrm{d}x$ are conserved by the dynamics with adequate boundary conditions.[1]
- The Hamiltonian is convex for real values of ψ, that is, if one defines $E[\psi^2] \equiv H[\psi,\psi]$ for a real wavefunction ψ, then [73]

$$E[\psi^2 = \lambda\psi_1^2 + (1-\lambda)\psi_2^2] \leq \lambda E[\psi_1^2] + (1-\lambda)E[\psi_2^2].$$

- The hydrodynamical formulation yields

$$\frac{\partial\rho}{\partial t} + \frac{\hbar}{m}\nabla\cdot(\rho\nabla\phi) = 0 \tag{8.46}$$

  and

$$\hbar\frac{\partial\phi}{\partial t} + \frac{\hbar^2}{2m}(\nabla\phi)^2 + \int U(|\mathbf{x}-\mathbf{x}'|)|\rho(\mathbf{x})|^2 \,\mathrm{d}\mathbf{x}' + \frac{\hbar^2}{4m}\left(\frac{(\nabla\rho)^2}{2\rho^2} - \frac{\nabla^2\rho}{\rho}\right) = 0. \tag{8.47}$$

Remarks:

1. The Hamiltonian takes the form in the polar variables:

$$H[\rho,\phi] = \frac{\hbar^2}{2m}\int\left(\frac{1}{4\rho(\mathbf{r})}|\nabla\rho(\mathbf{r})|^2 + \rho(\mathbf{r})|\nabla\phi(\mathbf{r})|^2\right)\mathrm{d}\mathbf{r} + \frac{1}{2}\int U(|\mathbf{r}-\mathbf{r}'|)\rho(\mathbf{r})\rho(\mathbf{r}')\,\mathrm{d}\mathbf{r}\mathrm{d}\mathbf{r}'. \tag{8.48}$$

2. According to the energy Eq. (8.48), the ground-state solution is real (up to a constant phase) and any nonuniform phase increases the ground-state energy and one can show that in general the ground state cannot vanish.

8.4.1.1 Bogoliubov Spectrum with Rotons

Given ρ the mean density defined as the number of particles per unit length, surface or volume in one, two or three space dimensions, respectively, the homogenous and steady (up to a phase frequency) function $\psi_0 = \sqrt{\rho}\mathrm{e}^{-i\frac{E_0}{\hbar}t}$, where $E_0 = \rho\hat{U}_0$,

[1]Note that a Galilean boost with a speed $\mathbf{v}$ changes the energy $H' = H + \mathbf{P}\cdot\mathbf{v} + \frac{1}{2}mNv^2$ and the momentum $\mathbf{P}' = \mathbf{P} + mN\mathbf{v}$ as usually in classical mechanics.

is a solution of the dynamics, where $\hat{U}_0 = \int U(|\mathbf{r}|)\mathrm{d}\mathbf{r}$ (more generally the Fourier transform of $U(\cdot)$ is $\hat{U}_k = \int U(|\mathbf{r}|)\mathrm{e}^{i\mathbf{r}\cdot\mathbf{k}}\mathrm{d}\mathbf{r}$).

As discussed in the next section, one can show that this solution is stable and makes the ground state for small enough n. Indeed, small perturbations around this uniform solution are dispersive waves:

$$\psi(\mathbf{r},t) = \psi_0 + (u_{\mathbf{k}}\mathrm{e}^{i(\mathbf{k}\cdot\mathbf{r}-\omega_k t)} + v_{\mathbf{k}}\mathrm{e}^{-i(\mathbf{k}\cdot\mathbf{r}-\omega_k t)})\mathrm{e}^{-i\frac{E_0}{\hbar}t},$$

where $\mathbf{k}$ and ω_k satisfy the Bogoliubov dispersion relation or spectrum [74]

$$\hbar\omega_k = \sqrt{\left(\frac{\hbar^2 k^2}{2m}\right)^2 + \frac{\hbar^2 k^2}{m}\rho\hat{U}_k}. \tag{8.49}$$

Assuming that the potential scales as $\hat{U}_0$ and contains a single length scale a, the spectrum depends then only on a single dimensionless parameter that is the product of a de Boer kind of parameter (that measures the ratio between the particle interaction and a zero point energy) with the dimensionless parameter ρa^D that characterizes the density, defining

$$\Lambda = \frac{ma^2}{\hbar^2}\rho\hat{U}_0. \tag{8.50}$$

Notice that the existence of a single dimensionless parameter is a simplification of the model since in real solids, one has at least two independent dimensionless numbers ρa^D and $\frac{ma^{2-D}}{\hbar^2}\hat{U}_0$.

Then, the Bogoliubov dispersion relation gives for the dimensionless frequency $\tilde{\omega}_\Lambda$

$$\tilde{\omega}_\Lambda(s) = \sqrt{\frac{s^4}{4} + \Lambda s^2 u_D(s)}, \tag{8.51}$$

where $u_D(s) = \hat{U}_{s/a}/\hat{U}_0$ depends only on the interparticle potential and of space dimension.

For some analytical results and for the numerics later on, we will choose the soft core interaction [67] with no loss of generality, that is Eq. (8.9):

$$U(|\mathbf{r}-\mathbf{r}'|) = U_0\theta(a - |\mathbf{r}-\mathbf{r}'|)$$

with $\theta(\cdot)$ Heaviside function: $\theta(s) = 1$ if $s > 0$ and $\theta(s) = 0$ if $s < 0$.

The Fourier transform of this special interaction potential reads

$$\hat{U}_k = \hat{U}_0 u_D(ka) \tag{8.52}$$

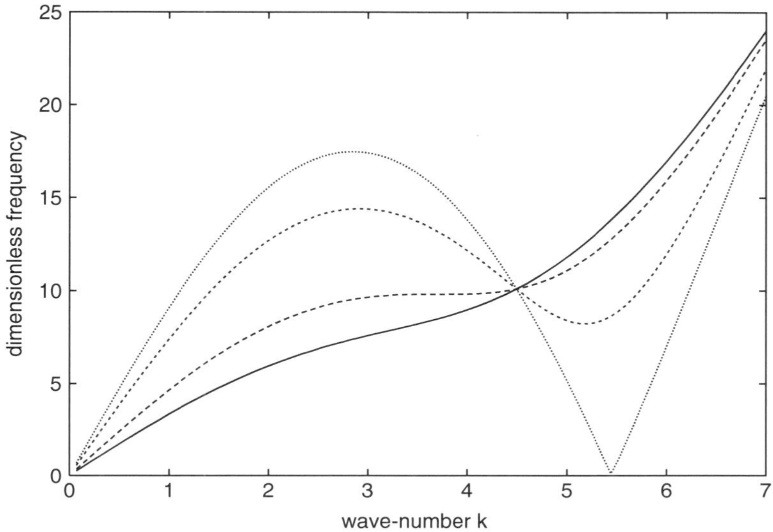

Fig. 8.6 Various dimensionless spectrum shapes for different values of $\Lambda = 12$, 23.43, 60 and 90.95 in three space dimensions

with

$$\hat{U}_0 = \begin{cases} 2aU_0 & 1-D, \\ \pi a^2 U_0 & 2-D, \\ \frac{4\pi}{3} a^3 U_0 & 3-D, \end{cases}$$

and

$$u_D(s) = \begin{cases} \frac{\sin s}{s} & 1-D, \\ 2\frac{J_1(s)}{s} & 2-D, \\ \frac{3}{s^2}\left(\frac{\sin s}{s} - \cos s\right) & 3-D, \end{cases},$$

where $J_1(x)$ is a Bessel function. Note that U_0 is the amplitude of the interaction (with units of energy) while $\hat{U}_0 = \int U(\mathbf{r})\,\mathrm{d}^D\mathbf{r}$, with units of energy$\times$volume ($\hat{U}_0 \propto a^D U_0$).

This spectrum has a phonon part for long-wave fluctuations that propagate with a speed $c_s = \frac{\hbar}{ma}\sqrt{\Lambda}$ and a so-called roton spectrum at smaller scale with the following characteristics (see Fig. 8.6):

(i) If $0 < \Lambda < \Lambda_s$ the spectrum ω_k grows monotonically with k; therefore there is no roton minimum.
(ii) For $\Lambda = \Lambda_s$ an inflexion point appears at k_s and the energy is given by $\hbar\omega_s = \frac{\hbar^2}{ma^2}\tilde{\omega}_s$.
(iii) If $\Lambda_s < \Lambda < \Lambda_c$ the spectrum exhibits a roton minimum $k_r a = s_r$ which is an implicit function of Λ:

$$\Lambda = -\frac{s_r^3}{2s_r u_D(s_r) + s_r^2 u_D'(s_r)}. \tag{8.53}$$

(iv) For $\Lambda = \Lambda_c$, the spectrum reaches the axis for k_c as at the edge of the phonon branch in solids. A reasonable value for k_c is $k_c = 5.45/a$ ($\approx 2.1\text{Å}^{-1}$ for HeII). This picture suggests that the existence of a roton minimum in HeII is, probably, a reminiscence or a ghost of the solid state as we already suggested [40].
(v) If $\Lambda > \Lambda_c$ the spectrum becomes pure imaginary in a finite bandwidth in k, implying the occurrence of a linear instability, leading to a periodic pattern of density modulation.

The parameters considered $\Lambda_s, \Lambda_c, \Delta_s, \Delta_c, k_s$ and k_c depend explicitly on the space dimensions and can be computed easily [40, 67].

8.4.2 Ground State of the Gross–Pitaevskii Model

As Λ increases, the roton characteristics are enhanced and we expect that there is a critical value $\Lambda_s < \Lambda_{c_1} < \Lambda_c$ for which the system crystallizes, that is, has a ground state with a density periodic in space. The increase of Λ may be realized, for instance, by keeping constant the range a and the magnitude U_0 and increasing the density n. Such a density increase might be achieved in a physical system by increasing the pressure and/or by cooling. Crystallization due to the roton minimum can be expected near the real solid phase since solid helium exists only at nonzero pressure. The transition occurs when the roton minimum is near the k-axis for zero frequency. If we use Landau's notation for rotons: $\hbar\omega_k = \Delta + \frac{\hbar^2}{2\mu}(|k| - k_r)^2$ for $k \approx k_r$. In our picture Δ, k_r and μ are nontrivial functions of Λ. However, Δ decreases and the roton minimum k_r increases, as Λ increases. One should also remark that besides the details of the model, the functions $\Delta(\Lambda)$, $k_r(\Lambda)$ and, $\mu_r(\Lambda)$ are known and, ultimately, must be determined experimentally.

As discussed above, the phase of the ground state is always uniform in space, even when this state shows modulations of the density. A physical consequence is that the ground state has zero momentum $\mathbf{P}$. For low Λ the ground state is a homogeneous solution, a superfluid (without positional order). The uniform ground state ($\psi = \sqrt{\rho}$), however, cannot be stable for any Λ, as argued already in [44–46, 66]. Considering a small perturbation around a uniform solution $\rho(\mathbf{r}) = \rho_0 + \tilde{\rho}(\mathbf{r})$ and $\phi(\mathbf{r}) = -\mathbf{E_0}/\hbar\mathbf{t} + \tilde{\phi}(\mathbf{r})$ the Hamiltonian yields

$$H_2 = \frac{1}{2}\int \left[\left(\frac{\hbar^2 k^2}{4mn} + \hat{U}_k \right) |\tilde{\rho}_k|^2 + \frac{\hbar^2 k^2}{4m}\rho \left|\tilde{\phi}_k\right|^2 \right] \mathrm{d}\mathbf{k};$$

one sees that if

$$\frac{\hbar^2 k^2}{4mn} + \hat{U}_k < 0,$$

then, the uniform solution is no more linearly stable. Therefore a periodic structure is expected at least as the roton minimum reaches the zero frequency axis (i.e., $\Lambda > \Lambda_c$). In [66] the possibility of a linear instability was only considered, although the transition is subcritical (first order) in two and three space dimensions [40, 44–46]. Indeed, by decreasing the roton minimum Δ, there is a critical value Δ_{c_1} such that for $\Delta < \Delta_c$, the ground state shows a periodic modulation of density in space.

From now on we shall consider the dimensionless form of the nonlocal Gross–Pitaevskii equation, Λ being the only parameter, defined by Eq. (8.50), and one can write the Hamiltonian Eq. (8.45) following $H/\mathscr{D} = \frac{\hbar^2}{ma^2}\rho\,\mathscr{E}$ with

$$\mathscr{E} = \frac{1}{\Omega}\left[\int_\Omega \frac{1}{2}|\nabla\psi(\mathbf{x})|^2\,\mathrm{d}\mathbf{x} + \frac{\Lambda}{2}\int_\Omega\int_\Omega \tilde{U}(|\mathbf{x}-\mathbf{x}'|)|\psi(\mathbf{x})|^2|\psi(\mathbf{x}')|^2\,\mathrm{d}\mathbf{x}\mathrm{d}\mathbf{x}'\right], \tag{8.54}$$

$$1 = \frac{1}{\mathscr{D}}\int_\Omega |\psi(\mathbf{x},t)|^2\mathrm{d}^D\mathbf{x}. \tag{8.55}$$

Here $\mathscr{D}$ is the total volume of the system in D-space dimension so that the energy density $\mathscr{E}$ converges because the double integral is performed on a compact support (or very localized shape) of the nonlocal interaction $\tilde{U}(|\mathbf{x}-\mathbf{x}'|)$.

In the following, we shall estimate modulation of periodic solutions in one, two, and three space dimensions based on a variational approach.

8.4.2.1 Weak Amplitude Periodic Modulation in 1D

In one space dimension the minimization of the energy leads to a supercritical (i.e., continuous second order) transition from a homogeneous (liquid phase) solution to a periodic (solid phase) solution. Following the standard perturbation analysis near threshold in the study of pattern formation of lamellar structures[75].

If $\Lambda \sim \Lambda_c$ a weak amplitude development with a single wave number selection is possible: we consider the wavefunction that is normalized in a period $\lambda = 2\pi/k_c$ according to the normalization condition (8.55)

$$\psi(x) = \frac{1}{\sqrt{1+2|A|^2}}\left(1 + A\mathrm{e}^{ik_c x} + A^*\mathrm{e}^{-ik_c x}\right). \tag{8.56}$$

Introducing this trial function into Eq. (8.54) one finds the energy per unit length:

$$\mathscr{E} = \frac{1}{2}\frac{2k_c^2|A|^2}{(1+2|A|^2)} + \frac{\Lambda}{2}\left(1 + \frac{8|A|^2\hat{U}_{k_c}}{(1+2|A|^2)^2} + \frac{2|A|^4\hat{U}_{2k_c}}{(1+2|A|^2)^2}\right).$$

The minimum of this quantity gives the value for the modulus of the complex amplitude A:

$$|A|^2 = -\frac{k_c^2 + 4\Lambda \hat{U}_{k_c}}{2(k^2 + \Lambda(\hat{U}_{2k_c} - 4\hat{U}_{k_c})}. \tag{8.57}$$

The structure displays a periodic modulation with a wave number $k_c = 4.078\ldots$. Therefore, setting $k = k_c$ into Eq. (8.57) one may calculate this amplitude as a function of Λ:

$$|A_c|^2 = \frac{-8\sin(k_c)}{k_c^3(8 - \cos(k_c))}(\Lambda - \Lambda_c) \approx 0.011(\Lambda - \Lambda_c).$$

8.4.2.2 First Order Transition in Two and Three Space Dimensions

In two and three dimensions, the phase transition is subcritical (first order) and its analysis is more complex so that we will restrict our discussion here to numerical simulations. Indeed, numerically, the ground state can be reached by considering the dissipative version of the G–P equation, called the Ginzburg–Landau (G–L) equation that can be understood as a imaginary time evolution of the G–P equation. It writes in dimensionless form

$$\frac{\partial \psi}{\partial t} = \frac{1}{2}\nabla^2 \psi - \Lambda \psi(\mathbf{x}) \int U(|\mathbf{x} - \mathbf{x}'|)|\psi(\mathbf{x}')|^2 \mathrm{d}\mathbf{x}' + \mu \psi\,, \tag{8.58}$$

where μ is the chemical potential to impose the mean density (or total mass) of the system. The NLS and G–L equations have the same stationary solutions and ground states, but the dissipative G–L dynamics converges to a local minimum of the free energy. Thus, starting with noisy initial conditions and running numerically the G–L dynamics, one can achieve a ground state. We use two types of numerical methods, a pseudo-spectral one when periodic boundary conditions can be considered. Figure 8.7 shows the ground state in 1 and 2 dimensions with periodic boundary conditions. A regular 1D crystal is observed and in 2D a hexagonal pattern is formed, minimizing the free energy of the system.

In three space dimensions the most stable configuration is the *hcp* crystalline structure as it can be seen in Fig. 8.8.

8.4.2.3 Ground-State in the Large Λ Limit

The crystal structures with a weak modulation of density have been described in 1-D for moderate values of Λ, say $\Lambda \sim \Lambda_{c_1}$. In fact, the ground state defined as the minimum of the energy functional Eq. (8.54) exists and is unique, because of the convexity, for any Λ. For large Λ, the potential energy in Eq. (8.54) requires small ψ while the mass normalization Eq. (8.55) forbids ψ to be small everywhere. Therefore the energy minimization leads to a periodic structure with zones where

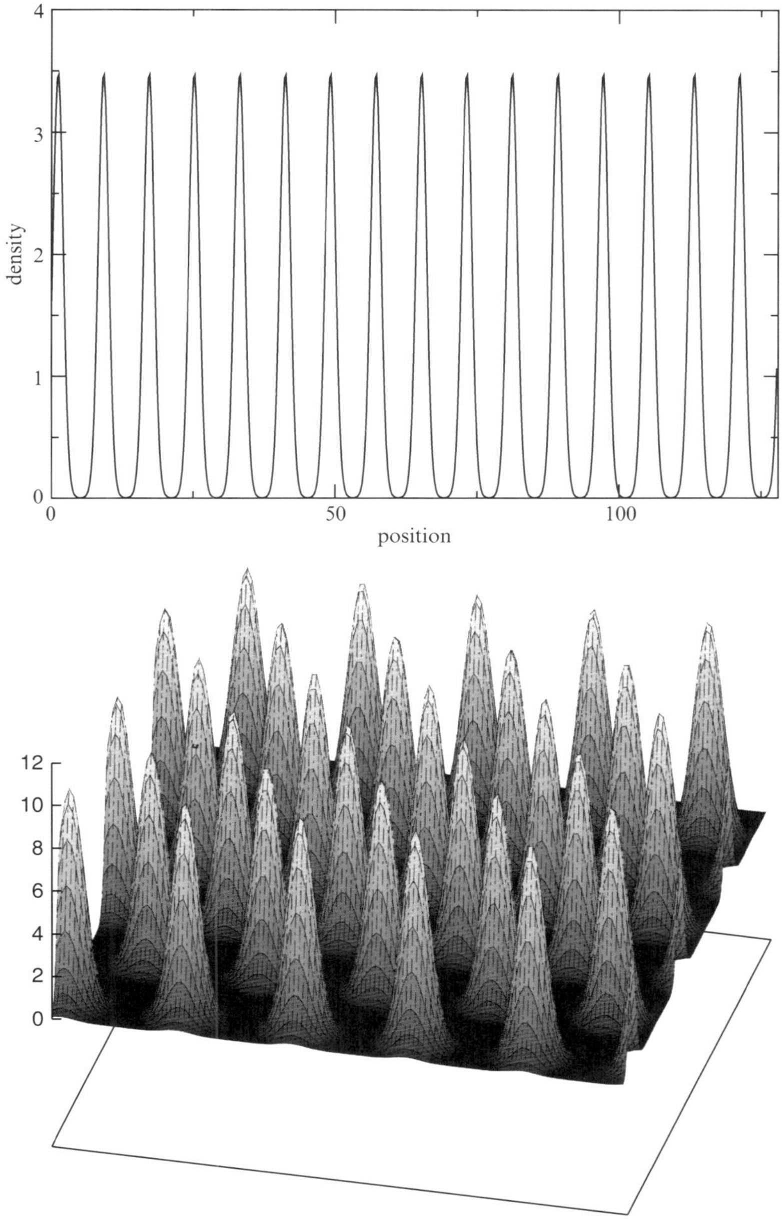

Fig. 8.7 Ground state of the G–P equation obtained numerically using the dissipative G–L equation: (*up*) in 1D, where a periodic crystal of density peaks is formed (here $\Lambda = 43.2$); (*down*) in 2D, it leads to a periodic hexagonal pattern ($\Lambda = 107$). Same figure than in [70]

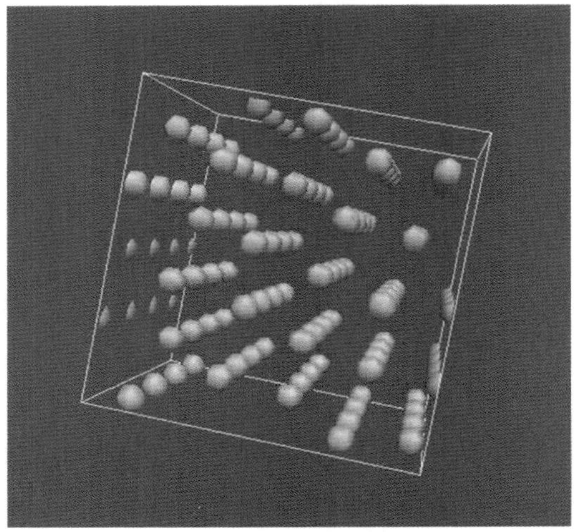

Fig. 8.8 Ground state of the G–P equation obtained numerically using the dissipative G–L equation in 3D, exhibiting a regular *hcp* crystal ($\Lambda = 429$)

$\psi \approx 0$ balancing zones where $\psi \gg 1$. Recently it has been proven in [76] that in the limit $\Lambda \to \infty$, the minimization of Eq. (8.54) with the restriction Eq. (8.55) is equivalent to a close packing arrangement of rods, discs, and spheres in one, two, and three space dimensions, respectively.

In the particular case of 1D, it is shown that in the large Λ limit $\psi \neq 0$ only in a small zone : $x \in (-\delta, \delta)$ of the whole period: $(-\lambda/2, \lambda/2)$, where the Euler–Lagrange condition deduced from Eq. (8.54) together with Eq. (8.55) leads to the Hemholtz equation in the domain $(-\delta, \delta) : -\psi''(x) = \mu\psi$. Finally the minimization of the energy gives δ and the wave number λ of the periodic structure.

Following this approach, with Sepúlveda [77], we estimated such a ground state for $\Lambda \gg 1$ (the extension to higher dimensions seems natural but the computations are harder). We sketch below the corresponding results. We consider the energy and normalization Eqs. (8.54) and (8.55) in a single period with the trial function in the unit cell

$$\psi(x) = \begin{cases} 0 & x \in [-\lambda/2, -\delta], \\ \sqrt{\frac{\lambda}{\delta}} \cos\left(\frac{\pi x}{2\delta}\right) & x \in [-\delta, \delta], \\ 0 & x \in [\delta, \lambda/2]. \end{cases} \tag{8.59}$$

The integrals may be computed more or less easily because of the interaction term. The minimization yields a relation between δ, the wavelength λ, and the dimensionless Λ:

$$1 + 2\delta - \lambda = 2\left(\frac{15}{\pi^2}\right)^{\frac{1}{5}} \frac{\delta}{(\lambda(\lambda-1)^2\Lambda)^{\frac{1}{5}}}. \tag{8.60}$$

This variational result is in complete agreement with the numerics and a more complete discussion on this problem may be found in [77].

8.4.3 A Model Combining Elastic and Superfluid Properties

As a built-in superfluid model, it is expected that the Gross–Pitaevskii equation exhibits superfluid properties in presence of a crystal lattice. This has been shown first in [40] where the superflow of such supersolid model was observed around a cylindrical obstacle with a critical velocity above which vortices were nucleated, similarly than for the superfluid model [16].

In fact, it is possible to use this model to mimic the torsional oscillator experiment where NCRI has been experimentally observed [24]. In [41], we have observed the NCRI effect using a 2D numerical simulation where a square sample is put under rotation. Measuring the rotational inertia, we have demonstrated that a part of the total mass was decoupled from the rotational motion, as shown in Fig. 8.9a. We use a relaxation algorithm (based on the Ginzburg–Landau equation which consists of an imaginary time evolution of the G–P equation as explained above) to converge towards an equilibrium state of the system (close to ground state). The numerical simulation is performed using a pseudo-spectral method and the system is put under rotation by considering the equation in a constant rotating frame. Figure 8.9b shows the evolution of the NCRIF in the limit of zero angular velocity as function of Λ (here by varying ρU_0), indicating that the superfluid fraction decreases as the pressure increases in agreement with Leggett's argument that the superfluid fraction should decrease with the minimal value of the wavefunction [23, 76, 77].

A more accurate way to compute the NCRI can be obtained by considering a small fraction of the sample only. In such case, neglecting the centrifugal acceleration term (which would induced a correction in ω^2 on the solution of the wavefunction), the rotation can be approximated by a Galilean boost in the azimuthal direction. The limit case of a Galilean boost of a 1D system can be understood in this context as the rotation of a very thin torus of the crystal. Using this method, one can obtain numerically the NCRI in 1 and 2 spatial dimensions using this method, allowing an accurate measure of the superfluid fraction in this model of supersolid [77, 81]. Thus, by measuring the linear momentum of the solution, one can define equivalently the supersolid density fraction tensor f_{ss} as

$$f_{ss}\mathbf{v} = \mathbf{v} - \frac{\mathbf{P}}{N}. \tag{8.61}$$

This tensor is in fact almost proportional to the identity tensor with the prefactor defining the supersolid fraction.

Figure 8.10a shows the NCRI as function of the Galilean boost velocity in 2D, showing a decrease of the NCRI as the velocity increases. Moreover, in this configuration, rapid variations of the NCRI appear regularly and can be. As shown in Fig. 8.10b, this sudden decrease of the superfluid fraction is related to the nucleation of quantum vortices which are known to lead to the loss of superfluidity in superfluids [78].

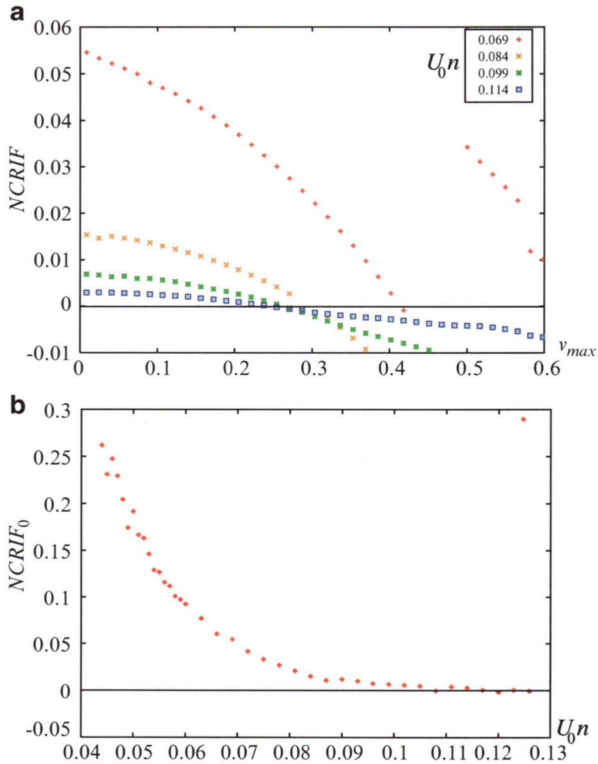

Fig. 8.9 2D numerical simulation of the dimensionless equation with 128×128 modes in a square cell of 96×96 units for different values of ρU_0; the potential range is $a = 4.3$. (**a**) The $NCRIF \equiv 1 - L_z'(\omega)/\langle I_{rb}\rangle$ *vs.* the local maximum speed $v_{\max} = \omega L/\sqrt{2}$ for $\rho U_0 = 0.069, 0.084, 0.099$ and 0.114 (res., $\Lambda = 74.1$, 90.2, 106.3, and 122.4). Here $\langle I_{rb}\rangle$ is the converging inertia moment computed numerically for large nU_0 at $\omega = 0$. Note that the jump in $NCRIF$ for $\rho U_0 = 0.069$ ($\Lambda = 74.1$) corresponds to the nucleation of a vortex in the system. (**b**) NCRIF at $\omega = 0$ as a function of ρU_0. We have verified that (**a**) and (**b**) are almost independent of the box size. This figure reprinted with permission from [41], copyright (2007) by the American Physical Society

The crystal structure that forms naturally in this model exhibits also classical elastic properties. Obviously, the elastic response of the system is due to the rapid density variation of the mean field in the crystal. Using a homogenization technique [79, 80] that separates the small-scale short time variation at the level of a density peak with the long wavelength slow macroscopic modes, a macroscopic model can be deduced from the NLS equation [41, 42]. In addition, this technique provides an explicit protocol to compute the superfluid tensor as it has been checked in details in [81]. Instead of giving the complicate derivation of the elastic macroscopic equation, we will rather present how the elastic behavior manifests in the numerics. In [41] a gravity-driven supersolid flow is investigated. As early

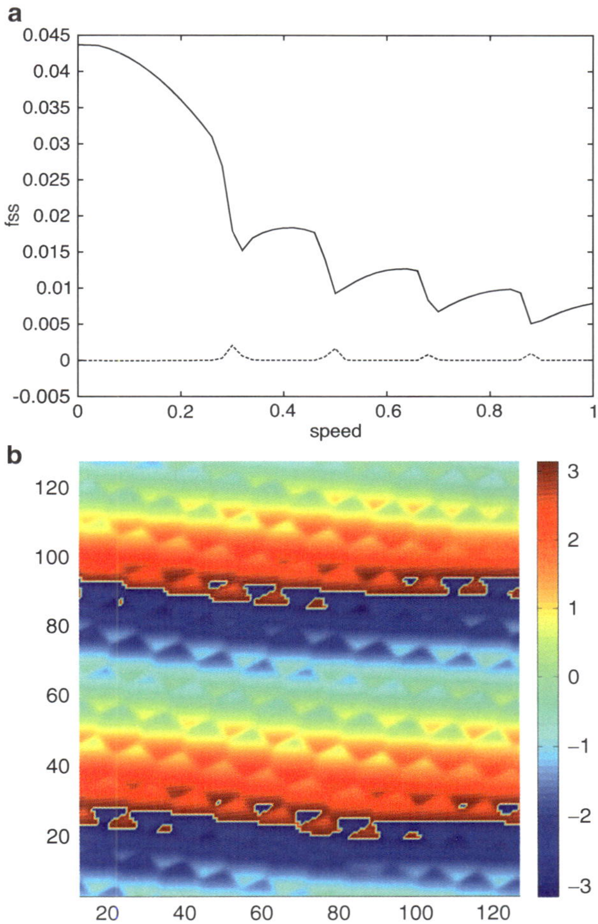

Fig. 8.10 (**a**) Supersolid fraction as a function of the boost velocity when the crystal is submitted to a Galilean boost. The *continuous line* shows the component of the supersolid tensor parallel to the velocity while the *dashed line* is for the orthogonal direction. (**b**) Phase of the wavefunction for the boost velocity $v_b = 0.4$. It shows twice a 2π phase jump (the phase is defined with a multiplier of 2π) indicating that vortices are present in the sample

suggested by Andreev et al. [82] an experiment of an obstacle pulled by gravity in solid helium could be a proof of supersolidity. Different versions of this experiment failed to show any motion [83–85]; therefore a natural question arises: how can we reconcile the NCRI experiment by Kim and Chan and the absence of pressure or gravity-driven flows?

In fact, our supersolid model (and it seems that supersolid helium too) reacts in different ways under a small external constraint such as stress, bulk force, or rotation in order to satisfy the equation of motion and the boundary conditions. For instance, if gravity (or pressure gradient) is added, then the pressure $\mathscr{E}'(\rho)$ balances the external "hydrostatic" pressure mgz while the elastic behavior of the solid balances the external uniform force per unit volume $m\rho g$. Therefore no quantum nor classical flows are needed to satisfy the mechanical equilibria. However, a flow is possible only if the stresses are large enough to display a plastic flow as it happens in ordinary solids, something that has nothing in common with superfluid property.

In conclusion, we have shown a fully explicit model of supersolid that displays either solid-like behavior or superflow depending on the external constraint and on the boundary conditions on the reservoir wall. Our numerical simulations clearly show that, within the same model, a nonclassical rotational inertia is observed as well a regular elastic response to external stress or forces without any flow of matter similarly than in the macroscopic model and in agreement with the experiments [25, 85].

8.5 Conclusion

We have shown here how complex transport phenomena can arise in the NLS equation which is eventually a rather simple Hamiltonian conservative equation valid in many contexts. In particular, we have focused on two specific aspects: the formation of topological defects (quantized vortices) that change dramatically the dynamics and the supersolid features that exhibit the model when a long range potential is taken so that nonhomogenous ground states exist. Another interesting feature of the model was not discussed here: it is the property of this Hamiltonian system to thermalize [53] when an ultraviolet cutoff is present, as naturally done in numerical simulation. In that case, the statistical equilibrium can lead even to a wave condensation [86] (it is a Bose–Einstein condensation of the "classical" waves of NLS), something apparently observed recently in nonlinear optics [87].

References

1. L.P. Pitaevskii, Sov. Phys. JETP **13**, 451 (1961)
2. E.P. Gross, J. Math. Phys. **4**, 195 (1963)
3. S.N. Bose, Z. Phys. **26**, 178 (1924)
4. A. Einstein, Sitzber. Kgl. Preuss. Akad. Wiss. 261 (1924)
5. M.H. Anderson et al., Science **269**, 198 (1995)
6. K.B. Davis et al., Phys. Rev. Lett. **75**, 3969 (1995)
7. C.C. Bradley, C.A. Sackett, J.J. Tollet, R.G. Hulet, Phys. Rev. Lett. **75**, 1687 (1995)
8. C.J. Pethick, H. Smith, *Bose-Einstein Condensation in Dilute Gases* (Cambridge University Press, Cambridge, 2002)
9. L. Pitaevskii, S. Stringari, *Bose-Einstein Condensation* (Clarendon, Oxford, 2003)
10. P.L. Kapitza, Nature **141**, 74 (1938)
11. J.F. Allen, A.D. Misener, Nature **141**, 75 (1938)

12. J.F. Allen, H. Jones, Nature **141**, 243 (1938)
13. L.D. Landau, J. Phys. U.S.S.R. **5**, 71 (1942)
14. L.D. Landau, J. Phys. U.S.S.R. **11**, 91 (1947)
15. E.L. Andronikashvili, J. Phys. USSR **10**, 201 (1946)
16. T. Frisch, Y. Pomeau, S. Rica, Phys. Rev. Lett. **69**, 1644 (1992)
17. C. Josserand, Y. Pomeau, Europhys. Lett. **30**, 43–48 (1995)
18. C. Josserand, Y. Pomeau, S. Rica, Physica D **134**, 111 (1999)
19. C. Josserand, Y. Pomeau, S. Rica, Phys. Rev. Lett. **75**, 3150–3153 (1995)
20. C. Josserand, S. Rica, Phys. Rev. Lett. **78**, 1215–1218 (1997)
21. O. Penrose, L. Onsager, Phys. Rev. **104**, 576 (1956)
22. A.F. Andreev, I.M. Lifshitz, Sov. Phys. JETP. **29**, 1107 (1969)
23. A.J. Leggett, Phys. Rev. Lett. **25**, 1543 (1970)
24. E. Kim, M.H.W. Chan, Nature (London) **427**, 225 (2004)
25. E. Kim, M.H.W. Chan, Science **305**, 1941 (2004)
26. J.M. Goodkind, Phys. Rev. Lett. **89**, 095301 (2002)
27. J. Day, T. Herman, J. Beamish, Phys. Rev. Lett. **95**, 035301 (2005)
28. A.S. Rittner, J.D. Reppy, Phys. Rev. Lett. **97**, 165301 (2006)
29. Y. Aoki, J.C. Graves, H. Kojima, Phys. Rev. Lett. **99**, 015301 (2007)
30. A. Penzev, Y. Yasuta, M. Kubota, J. Low Temp. Phys. **148**, 677 (2007)
31. M. Kondo, S. Takada, Y. Shibayama, K. Shirahama, J. Low Temp. Phys. **148**, 695 (2007)
32. B. Hunt, E. Pratt, V. Gadagkar, M. Yamashita, A.V. Balatsky, J.C. Davis, Science **324**, 632 (2009)
33. A.S. Rittner, J.D. Reppy, Phys. Rev. Lett. **98**, 175302 (2007)
34. J. Day, J. Beamish, Phys. Rev. Lett. **96**, 105304 (2006)
35. A.S.C. Rittner, W. Choi, E.J. Mueller, J.D. Reppy, Phys. Rev. **B 80**, 224516 (2009)
36. J. Day, J. Beamish, Nature **450**, 853 (2007)
37. J. Day, O. Syshchenko, J. Beamish, Phys. Rev. B **79**, 214524 (2009)
38. X. Rojas, A. Haziot, V. Bapst, S. Balibar, H.J. Maris, Phys. Rev. Lett. **105**, 145302 (2010)
39. A. Haziot, X. Rojas, A.D. Fefferman, J.R. Beamish and S. Balibar, Phys. Rev. Lett. **110**, 035301 (2013)
40. Y. Pomeau, S. Rica, Phys. Rev. Lett. **72**, 2426 (1994)
41. C. Josserand, Y. Pomeau, S. Rica, Phys. Rev. Lett. **98**, 195301 (2007)
42. C. Josserand, Y. Pomeau, S. Rica, Euro. Phys. J. S.T. **146**, 47–62 (2007)
43. P.W. Anderson, Science **324**, 631 (2009)
44. D.A. Kirzhnits, Y.A. Nepomnyashchii, ZETF **59**, 2203 (1970) [Sov. Phys. -JETP 32, 1191 (1971)]
45. Y.A. Nepomnyashchii, Theor. Math. Phys. **8**, 928 (1971)
46. Y.A. Nepomnyashchii, A.A. Nepomnyashchii, Theor. Math. Phys. **9**, 1033 (1971)
47. S. Dyachenko, A.C. Newell, A. Pushkarev, V.E. Zakharov, Physica D **57**, 96–160 (1992)
48. V. Hakim, in *Hydrodynamics and Nonlinear Instabilities*, eds. by C. Godrèche, P. Manneville (Cambridge University Press, Cambridge, 1998), pp. 295–386
49. E.A. Kusnetsov, S.K. Turitsyn, Zh. Eksp. Teor. Fiz. **94**, 119–129 (1988) [Sov. Phys. JETP **67**, 1583 (1988)]
50. N.E. Kosmatov, V.F. Shvets, V.E. Zakharov, Physica D **52**, 16–35 (1991)
51. A. Chabchoub, N.P. Hoffman, N. Akhmediev, Phys. Rev. Lett. **106**, 204502 (2011)
52. V.E. Zakharov, A.B. Shabat, Soviet Phys. JETP **34**, 62 (1972)
53. R. Jordan, C. Josserand, Phys. Rev. E **61**, 1527 (2000)
54. D.H. Peregrine, J. Austral. Math. Soc. Ser. B **25**, 16–43 (1983)
55. E. Lundh, Phys. Rev. A **65**, 043604 (2002)
56. C. Josserand, Chaos **14**, 875 (2004)
57. A. Aftalion, I. Danaila, Phys. Rev. A **69**, 033608 (2004)
58. E.J. Yarmchuk, M.J.V. Gordon, R.E. Packard, Phys. Rev. Lett. **43**, 214 (1979)
59. K.W. Madison, F. Chevy, W. Wohlleben, J. Dalibard, Phys. Rev. Lett. **84**, 806 (2000)
60. Y. Pomeau, S. Rica, Comptes Rendus Acad. Sc. (Paris), **316** Série II, 1523 (1993)

61. C. Josserand, Y. Pomeau, Nonlinearity **14**, R25-R62 (2001)
62. V. Hakim, Phys. Rev. E **55**, 2835, (1997)
63. C. Raman, M. Köhl, R. Onofrio, D.S. Durfee, C.E. Kuklewicz, Z. Hadzibabic, W. Ketterle, Phys. Rev. Lett. **83**, 2502 (1999)
64. R. Onofrio, C. Raman, J.M. Vogels, J.R. Abo-Shaeer, A.P. Chikkatur, W. Ketterle, Phys. Rev. Lett. **85**, 2228 (2000)
65. L.D. Landau, E.M. Lifshitz, *Fluid Mechanics* (Pergamon Press, Oxford, 1987)
66. T. Schneider, C.P. Enz, Phys. Rev. Lett. **27**, 1186 (1971)
67. Y. Pomeau, S. Rica, Phys. Rev. Lett. **71**, 247 (1993)
68. E. Kim, M.H.W. Chan, Phys. Rev. Lett. **97**, 115302 (2006)
69. A.C. Clark, J.T. West, M.H.W. Chan, Phys. Rev. Lett. **99**, 135302 (2007)
70. G. Düring, C. Josserand, Y. Pomeau and S. Rica, Theory of real supersolids. In: *Proceedings of the 4th Warsaw School of Statistical Physics*, Warsaw University Press, 2012
71. N. Henkel, R. Nath, T. Pohl, Phys. Rev. Lett. **104**, 195302 (2010)
72. F. Cinti, P. Jain, M. Boninsegni, A. Micheli, P. Zoller, G. Pupillo, Phys. Rev. Lett. **105**, 135301 (2010)
73. R. Benguria, H. Brezis, E.H. Lieb, Commun. Math. Phys. **79**, 167 (1981)
74. N.N. Bogoliubov, J. Phys. U.S.S.R. **11**, 23 (1947)
75. L. Pismen, *Patterns and Interfaces in Dissipative Dynamics'* (Springer, Berlin Heildelberg, 2006)
76. A. Aftalion, X. Blanc, R.L. Jerrard, Phys. Rev. Lett. **99**, 135301 (2007)
77. N. Sepúlveda, C. Josserand, S. Rica, Phys. Rev. **B 77**, 054513 (2008)
78. P. Mason, C. Josserand, S. Rica, Phys. Rev. Lett. **109**, 045301 (2012)
79. A. Bensoussan, J.L. Lions, G. Papanicolaou, *Asymptotic Analysis in Periodic Structures* (North-Holland Amsterdam, 1978)
80. E. Sánchez-Palencia, *Non-Homogeneous Media and Vibration Theory*. Lecture Notes in Physics, vol. 127 (Springer, Berlin, 1980)
81. N. Sepúlveda, C. Josserand, S. Rica, Eur. Phys. J. B **78**, 439–447 (2010)
82. A.F. Andreev, K. Keshishev, L. Mezov-Deglin, A. Shalinikov, Zh. Eksp. Teor. Fiz. Pis'ma Red. **9**, 507 (1969) [JETP Lett. **9**, 306 (1969)]
83. D.S. Greywall, Phys. Rev. **B 16**, 1291 (1977)
84. G. Bonfait, H. Godfrin, B. Castaing, J. Phys. (Paris) **50**, 1997 (1989)
85. S. Sasaki, R. Ishiguro, F. Caupin, H.J. Maris, S. Balibar, Science **313**, 1098 (2006)
86. C. Connaughton, C. Josserand, A. Picozzi, Y. Pomeau, S. Rica, Phys. Rev. Lett. **95**, 263901 (2005)
87. C. Sun, S. Jia, C. Barsi, S. Rica, A. Picozzi, J.W. Fleischer, Nature Phys. **8**, 471–475 (2012)

Part IV
Beyond Physics: Examples of Complex Systems

Chapter 9
Spatial and Temporal Order Beyond the Deterministic Limit: The Role of Stochastic Fluctuations in Population Dynamics

Duccio Fanelli

Abstract Modeling the self-consistent dynamics of an ensemble made of microscopic constituents can be tackled via deterministic or alternatively stochastic viewpoints. The latter enables one to respect the discrete nature of the scrutinized medium, a possibility which is conversely prevented when dealing with the former idealized approximation. As we shall here discuss, stochastic finite-size fluctuations can drive the emergence of regular spatiotemporal cycles that persist for moderate and even large sizes of the population and which are not captured within the mean-field descriptive scenario. The van Kampen system-size expansion is an elegant mathematical approach that allows one to investigate the key role played by the inherent stochasticity. We here provide a pedagogical introduction to such a method and discuss its application to a model of autocatalytic reactions.

9.1 Introduction

Investigating the dynamical evolution of microscopic entities in mutual interaction constitutes a rich and fascinating problem of paramount importance and cross-disciplinary interest [1]. Molecules, with their chemical properties and distinct diffusive abilities, can be ideally grouped into homogeneous families, whose concentrations vary continuously with position and time, as follows the governing dynamics [2, 3]. Similarly, families of organisms (animals, plants) can be identified in any ecological system, competition, and cooperation driving their interlaced evolution [1, 4]. Analogous concepts translate to the realm of social science

D. Fanelli (✉)
Dipartimento di Energetica, University of Florence, via S. Marta 3 50139, Florence, Italy

Dipartimento di Fisica e Astronomia, University of Florence,
via Sansone 1 50019 Sesto Fiorentino, Florence, Italy
e-mail: duccio.fanelli@unifi.it

X. Leoncini and M. Leonetti (eds.), *From Hamiltonian Chaos to Complex Systems: A Nonlinear Physics Approach*, Nonlinear Systems and Complexity 5,
DOI 10.1007/978-1-4614-6962-9_9,

applications and human communities models. In general terms, and irrespectively of the specific context of investigation, it is customary to refer to a population as to a macroscopic, extended group composition of a large sea of homologous microscopic actors. From biology to biomedicine, passing through physics and chemistry, the study of population dynamics is often tackled via a simplistic approach: the scrutinized families are assumed to be composed of an infinite collection of constitutive elements. Correlations are then neglected so to favor a mean-field description, which in many cases enables for straightforward analytical progress. The system is hence treated in the continuum limit and the interactions link the families as a whole. However, single individual effects, stemming from the intimate discreteness of the analyzed medium, can prove crucial by modifying significantly the mean-field predictions and so opening up the perspective for alternative explanations of a wide gallery of experimental observations [5, 6]. It has been shown in fact that the stochastic component of the microscopic dynamics, resulting from the aforementioned discreteness and thus associated to finite-size corrections, can induce regular macroscopic patterns, both in time and space [7–13]. The effect of the graininess materializes in an endogenous source of disturbance, also termed demographic noise, opposed to other perturbations that can be imagined to persist in the continuum limit. The fact that the demographic noise, intrinsic to the system, can spontaneously drive the emergence of regular structures, reflecting a degree of temporal and spatial macroscopic order, is in some respect counterintuitive and intriguing per se. In this chapter we shall provide an introductory description to population dynamics, highlighting the different approaches, as outlined above. We will in particular discuss a simple birth/death process, making explicit the distinct philosophies that inspire the deterministic and stochastic paradigms, and introduce the relevant mathematical concepts. Then we will move forward by reviewing a specific case study, for which both temporal and spatial order manifests, as mediated by the microscopic stochastic component of the dynamics.

9.2 On the Deterministic and Stochastic Viewpoints

The study of the dynamical evolution of interacting species of homologous quantities defines the field of population dynamics [1, 14], which, as previously emphasized, finds particularly relevant applications in life science [2, 3]. Population is indeed a technical wording which encompasses distinct fields of investigations ranging from, e.g., the level of expression of a protein in a cell to the number of animals in a finite ecosystem [1, 4]. The classical approach to population dynamics relies on characterizing quantitatively the densities of species through a system of ordinary differential equations which incorporates for the specific interactions being at play. Pure competition, predator-prey interactions, or even cooperative effects could be translated into dedicated interaction terms [1, 4] via a straightforward application of the law of mass action. Specific delays might be required to account for the processing time which is often necessary to react to an external stimulus or signal, a paradigmatic problem of many biological pathways. More than one independent variable is often to be assumed, which in turn implies dealing with systems of partial differential equations. As an example, when tracing the dispersion of a diffusing chemical compound, space and time are to be explicitly represented into the mathematical description. All these phenomena can be tackled via the population viewpoint by focusing on the mutual evolution of the families in which the elementary constituents are ideally grouped. It is customary to refer to this level of description as to the *deterministic theory*. Noise and other disturbances can be eventually hypothesized to alter the ideal deterministic, hence reproducible, dynamics but always act as a macroscopic bias.

As opposed to this formulation, a different level of modeling can be invoked focusing instead on the *individual-based description* [5, 6] which is intrinsically *stochastic*. This amounts to characterizing the microscopic dynamics via transition probabilities governing the interactions among individuals and with the surrounding environment. This approach has been recently adopted in various contexts such as predator-prey interactions, metabolic reactions, and epidemic models. The stochasticity of the systems stems from the microscopic finiteness/discreteness of the dynamical variables involved.

Deterministic and stochastic pictures, conceptually alternative, yield to different descriptions of a scrutinized phenomenon. It is therefore of interest to highlight similarities, and/or discrepancies, in the associated predictions. A viable method that enables one to bridge the gap between the deterministic and stochastic scenarios is the celebrated van Kampen's system-size expansion [6]. The idea goes as follows. Start from a stochastic, individual-based model, which formally corresponds to dealing with a master equation for the probability of photographing the system in a given configuration at a specific time. Then, perform a perturbative expansion with respect to a small parameter which encodes for the amplitude of fluctuations, or in other terms, the finite size of the system (e.g., total number of molecules or organisms). At the leading order of the perturbative calculation one recovers the mean-field equations, namely, the deterministic description alluded above. Including the next-to-leading order corrections, one obtains a description of the fluctuations, as a set of linear stochastic differential equations. Such a system can be analyzed exactly, so allowing one to quantify the differences between the stochastic formulation and its deterministic analogue. Let us emphasize again that fluctuations do not arise from an externally imposed noise source. It is the intimate discreteness of the system which results in an unavoidable intrinsic noise, a key contribution to the dynamics that has to be considered in any sensible model of natural phenomena, where a finite, though large, number of actors are simultaneously at play. These are important aspects, often omitted in the literature, and, due to their common origin, bear intriguing traits of universality across various disciplinary fields. Importantly, the inner stochastic component, also termed demographic noise, can yield to regular spatiotemporal patterns [7–13], signaling a degree of cooperativity and collective organization which instead lacks in the corresponding mean-field description.

The fact that fluctuations can be enhanced by a resonant effect was conjectured by Bartlett [15] in the context of the modeling of measles epidemics and later elaborated upon by Nisbet and Gurney [16], who called these stochastically induced oscillations, quasi-cycles. However, it is only in the last few years that these effects have been explained in rigorous terms, yielding to quantitative understanding of the phenomenon [7]. In the following, we shall elaborate on these important, and rather general, facts by selecting one specific case study. This is a scheme of autocatalytic reaction thoroughly studied in [11, 12]. The analysis presented in [11, 12] will be reviewed all along this chapter. Before that, next section is devoted to introducing the concept of master equation and to briefly discussing the van Kampen expansion technique.

9.3 The Van Kampen Expansion Applied to a Simple Birth/Death Stochastic Model

Consider a microscopic element X, which belongs to a given population, hereafter referred to as a species. Such an element can eventually die, leaving behind an empty space, called E. This simple event is exemplified by the following chemical equation:

$$X \xrightarrow{d} E \tag{9.1}$$

where d is the reaction rate for a death to occur. Similarly, the spontaneous production of an individual element of type X is ruled by the chemical reaction

$$E \xrightarrow{b} X \tag{9.2}$$

where b stands for the birth reaction rate. Further, let us assume that the number of microscopic entities, including the empties, totals in N, at time $t = 0$. N is clearly a conserved quantity of the dynamics if the system is forced to obey to the above chemical rules: every time one element of type X (resp. E) disappears, it gets immediately replaced by one element E (resp. X), so keeping the global population, sum of all individuals X and E, unchanged. Let us call n the number elements of type X and n_E the number of vacancies. Hence, $n_E = N - n$ and the system is fully specified once the integer n is being assigned.

The process that obeys to chemical equations (9.1) and (9.2) is stochastic. Mathematically, it can be described in terms of a master equation that governs the evolution of the probability $P(n,t)$ of seeing the system in a given configuration n at time t. To write such an equation one needs to quantify the transition rates $T(n'|n)$ from an initial state n to a final one, labeled with n' and compatible with the former.

The transition rate associated to, e.g. the chemical equation (9.1) can be readily evaluated as the product of (i) the probability P_1 of selecting one element of type X with (ii) the reaction constant d, which ultimately quantifies the probability that the selected individual eventually dies. Assume the individual entities, both the empties and the material elements X, to be uniformly distributed inside the volume that hosts the system. Then, $P_1 = n/N$ and this immediately yields to

$$T(n-1|n) = d\frac{n}{N} \tag{9.3}$$

Similarly, the transition rate associated to reaction (9.2) can be evaluated as

$$T(n+1|n) = b\frac{n_E}{N} = b\left(1 - \frac{n}{N}\right)$$

Under the above assumptions the system is Markov and the master equation for the probability $P(n,t)$ reads

$$\begin{aligned} \frac{\mathrm{d}P(n,t)}{\mathrm{d}t} &= -T(n-1|n)P(n,t) + T(n|n+1)P(n+1,t) \\ &\quad -T(n+1|n)P(n,t) + T(n|n-1)P(n-1,t) \end{aligned} \tag{9.4}$$

This equation provides a self-consistent and fully rigorous representation of the stochastic model defined by chemical Eqs. (9.1) and (9.2).

Starting from this setting, one can extract information on the average behavior of the system by neglecting the finite-size fluctuations and focusing on the time evolution of the mean-field concentration $\langle n \rangle$ defined as

$$\langle n \rangle = \sum_n nP(n,t)$$

To this end, multiply by n both sides of Eq. (9.4) and sum over all possible states. The left-hand side takes the form

$$\sum_n n\frac{\mathrm{d}P(n,t)}{\mathrm{d}t} = \frac{d}{d(t/N)}\sum_n \frac{n}{N}P(n,t) = \frac{\mathrm{d}\langle n \rangle}{\mathrm{d}\tau} \tag{9.5}$$

where $\tau = t/N$. Focus now on the right hand side of Eq. (9.4), modified by the multiplicative factor n. Consider the first two terms:

$$\sum_n n\ [T(n|n+1)P(n+1,t) - T(n-1|n)P(n,t)] \tag{9.6}$$

$$= \sum_{n'}(n'-1)T(n'-1|n')P(n',t) - \sum_n nT(n-1|n)P(n,t) \tag{9.7}$$

$$= -\sum_{n'} T(n'-1|n')P(n',t)$$

Changing n' into n and recalling the definition of $T(n-1|n)$ one eventually obtains

$$-\sum_n d\frac{n}{N}P(n,t) = -d\frac{\langle n \rangle}{N} \tag{9.8}$$

Proceeding in a similar way, the last two terms in the right-hand side of the modified master equation yield to

$$b\left(1-\frac{\langle n\rangle}{N}\right) \tag{9.9}$$

Introduce now $\phi = \langle n\rangle/N$. Then collecting together the above contributions one eventually gets

$$\frac{d\phi}{d\tau} = b - (b+d)\phi. \tag{9.10}$$

The above ordinary differential equation governs the evolution of the continuum concentration ϕ. The fluctuations have been in fact dropped out by performing the ensemble average $\langle\cdot\rangle$. Equation (9.10) can be solved analytically to yield

$$\phi(\tau) = \frac{b}{d+b}\left[1-\left(1-\phi_0\frac{b+d}{b}\right)\exp\left[-(b+d)\tau\right]\right] \tag{9.11}$$

Asymptotically the system converges to a stable fixed point, $\phi^* = \frac{b}{d+b}$. The above solution constitutes an ideal representation of the exact dynamics (9.4). Finite-size corrections materialize in fact in stochastic fluctuations that can sensibly affect the observed dynamics. To bring into evidence such an important aspect one can perform stochastic simulations of the chemical scheme (9.1) and (9.2) by means of the celebrated Gillespie algorithm [17, 18]. Such an algorithm produces realizations of the stochastic dynamics which are equivalent to those obtained from the governing master Eq. (9.4). In Fig. 9.1 the result of the stochastic simulations (wiggling line, black online) is compared to the deterministic solution (9.10) (smooth line, red online). The stochastic, hence exact, dynamics follows closely the idealized profile predicted by the mean-field theory. Fluctuations are however present and reflect the probabilistic nature of the problem in its original formulation. The statistics of the disturbances is investigated in Fig. 9.2 where the histograms of the quantities $n/N-\phi$, as recorded in direct Gillespie-based simulations, are plotted for different choices of the total population amount N. The profiles are Gaussian, as revealed by visual inspection. Importantly, the width of the distributions shrinks as $1/\sqrt{N}$. Hence, the fluctuations virtually disappear in the limit of infinite system size $N\to\infty$ and consequently $\phi \equiv \lim_{N\to\infty} n/N$.

Fluctuations prove however crucial for any physical system made of a finite, though large, number of constitutive elements. Equations such as (9.4) are nonetheless difficult to analyze, and one has to rely on approximate techniques to elaborate on the role of stochasticity. The famous system-size expansion, pioneered by van Kampen [6] in the sixties, provides an elegant way of capturing the essential aspects of the discrete model, thus enabling one to appreciate the contribution of demographic, finite N, fluctuations. In the following and with reference to the simple birth/death process here considered, we will discuss the main assumptions of the method as well as its formal application. As a final result, we will be able to predict the distribution of the fluctuations, as seen in numerical simulations.

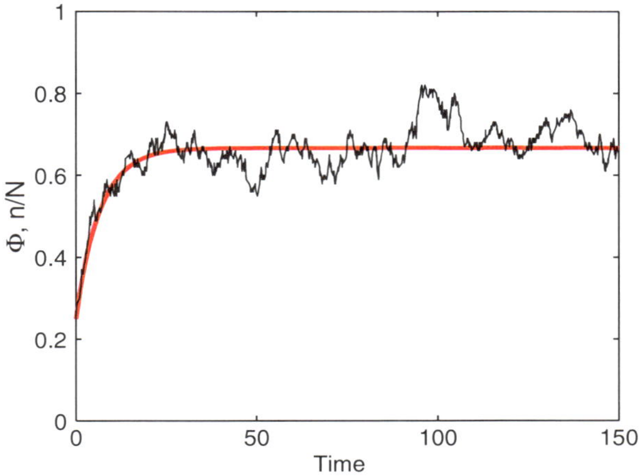

Fig. 9.1 Temporal evolution of the species concentrations. The wiggling line refers to the stochastic simulations, n/N vs. rescaled time τ. The smooth profile is the deterministic solution, $\phi(\tau)$. (9.10). Here, $b = 0.1$, $d = 0.05$, and $N = 100$

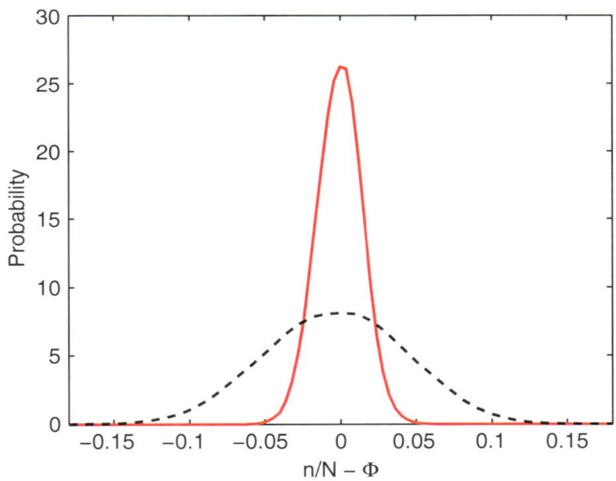

Fig. 9.2 Normalized histograms of stochastic fluctuations $n/N - \phi$ recorded from the Gillespie-based simulations. The dashed line refers to $N = 100$ and the solid line to $N = 1000$. The parameters are $b = 0.1$, $d = 0.05$

The van Kampen ansatz consists in splitting the finite-size concentration n/N into two contributions. To the continuous (mean-field) concentration ϕ, it is superposed a stochastic term which is supposed to scale as $1/\sqrt{N}$. In formulae

$$\frac{n}{N} = \phi + \frac{\xi}{\sqrt{N}} \tag{9.12}$$

where ξ is a stochastic variable. The quantity $1/\sqrt{N}$ is small for moderate or large system sizes and hence plays the role of a perturbative parameter in the van Kampen expansion.

Let us start by rewriting Equation (9.4) in the following compact form:

$$\frac{\mathrm{d}P}{\mathrm{d}t} = \left(\mathscr{E}^{+1} - 1\right) T(n-1|n)P(n,t) + \left(\mathscr{E}^{-1} - 1\right) T(n+1|n)P(n,t) \tag{9.13}$$

where the operators $\mathscr{E}^{\pm 1}$ are defined as

$$\mathscr{E}^{\pm 1} f(n) = f(n \pm 1) \tag{9.14}$$

$f(\cdot)$ being an arbitrary function of the discrete variable n. The above operators admit a straightforward expansion with respect to $1/\sqrt{N}$. A simple manipulation yields in fact to

$$\mathscr{E}^{\pm 1} = 1 \pm \frac{1}{\sqrt{N}} \frac{\partial}{\partial \xi} + \frac{1}{2N} \frac{\partial^2}{\partial \xi^2} + \cdots , \tag{9.15}$$

Hence, the first term in the right-hand side of the master equation (9.13) reads

$$\left(\mathscr{E}^{+1} - 1\right) T(n-1|n)P(n,t) = \left(\frac{1}{\sqrt{N}} \frac{\partial}{\partial \xi} + \frac{1}{2N} \frac{\partial^2}{\partial \xi^2}\right) d\left(\phi + \frac{\xi}{\sqrt{N}}\right) \Pi(\xi,t) \tag{9.16}$$

where explicit use has been made of the van Kampen ansatz (9.12). By organizing the various terms in the above expressions, one gets:

$$\frac{1}{\sqrt{N}} \left[\mathrm{d}\phi \frac{\partial}{\partial \xi} \Pi\right] + \frac{1}{N} d \left[\frac{\partial}{\partial \xi} (\xi \Pi) + \frac{1}{2} \phi \frac{\partial^2}{\partial \xi^2} \Pi\right] + \cdots \tag{9.17}$$

up to $1/N$ contributions. Similarly, for the other contribution

$$-\frac{1}{\sqrt{N}} \left[b(1-\phi) \frac{\partial}{\partial \xi} \Pi\right] + \frac{1}{N} b \left[\frac{\partial}{\partial \xi} (\xi \Pi) + \frac{1}{2} (1-\phi) \frac{\partial^2}{\partial \xi^2} \Pi\right] + \cdots \tag{9.18}$$

Using the van Kampen ansatz (9.12) in the left-hand side of the master Eq. (9.13) results in

$$\frac{\mathrm{d}P(n,t)}{\mathrm{d}t} = \frac{1}{N} \frac{\partial \Pi(\xi,\tau)}{\partial \tau} - \frac{1}{\sqrt{N}} \frac{\partial \Pi(\xi,\tau)}{\partial \xi} \frac{\mathrm{d}\phi}{\mathrm{d}\tau} . \tag{9.19}$$

where we have set $P(n,t)$ equal to $\Pi(\xi,\tau)$ and where $\tau = t/N$. Then, one can plug the contributions (9.17–9.19) into the master equation (9.13) and collect together the various terms depending on their respective order in $1/\sqrt{N}$. At the leading order, and as expected, we recover the mean-field Eq. (9.10). At the next to leading order instead we obtain the following Fokker–Planck equation [19] for the distribution of fluctuations $\Pi(\xi,t)$:

$$\frac{\partial \Pi}{\partial \tau} = (d+b)\frac{\partial}{\partial \xi}(\xi \Pi) + \frac{1}{2}(\mathrm{d}\phi + b(1-\phi))\frac{\partial^2}{\partial \xi^2}\Pi \tag{9.20}$$

The solution of the above one-dimensional Fokker–Planck equation is a Gaussian, whose first and second moments can be readily characterized.

Multiply both sides of the Fokker Planck equation by ξ and integrate over the real axis in $\mathrm{d}\xi$. A simple calculation yields to

$$\frac{\mathrm{d}\langle \xi \rangle}{\partial \tau} = -(d+b)\langle \xi \rangle \tag{9.21}$$

where $\langle \xi \rangle \equiv \int \xi \Pi \mathrm{d}\xi$. The solution of (9.21) is $\langle \xi \rangle = \langle \xi \rangle_0 \exp[-(d+b)t]$. Asymptotically, when the system settles down to its deputed equilibrium, $\langle \xi \rangle_{\mathrm{stat}} = \lim_{t\to\infty}\langle \xi \rangle = 0$.

A similar reasoning applies to the second moment. The latter is defined as $\langle \xi^2 \rangle \equiv \int \xi^2 \Pi \mathrm{d}\xi$ and obeys to the differential equation

$$\frac{\mathrm{d}\langle \xi^2 \rangle}{\partial \tau} = -2(d+b)\langle \xi^2 \rangle + [(d-b)\phi + b] \tag{9.22}$$

Assume we are interested in the statistics fluctuations around the stationary point when $\phi \to \phi^*$. Clearly, because of stationary, $\mathrm{d}\langle \xi^2 \rangle / \partial \tau = 0$ in Eq. (9.22) which implies

$$\langle \xi^2 \rangle_{\mathrm{stat}} = \frac{\mathrm{d}b}{(d+b)^2} \tag{9.23}$$

where use has been made of the expression $\phi^* = \frac{b}{d+b}$. The knowledge of the first two moments makes it possible to calculate the stationary (Gaussian) distribution $\Pi(\xi)$ and draw a direct comparison with the results of the stochastic simulations. This is done in Fig, 9.3: An excellent agreement is found. The van Kampen expansion constitutes therefore a viable strategy to quantify the impact of finite-size corrections that stem from the discrete nature of the simulated medium and beyond the customarily adopted mean-field approximation. Notice that non-Gaussian fluctuations can develop when the system is made to evolve close to an absorbing barrier. Including higher order corrections in the van Kampen expansion, beyond the next-to-leading approximation, allows one to capture the non-Gaussian traits of the distribution [20–22].

In the simple applications that we have here discussed, the stochastic fluctuations materialize in erratic disturbances of the mean-field trajectory. More complex scenarios are however possible. Surprisingly enough, in fact, the microscopic noise that is seeded by finite-size corrections can also yield to macroscopically organized patterns, both in time and space. The van Kampen technique, illustrated above with reference to a simple problem, provides us with an excellent tool to eventually explain such a peculiar behavior. In the following, building on the general ideas presented above, and with reference to a model of biological interest, we shall discuss these intriguing dynamical features. The model that we will discuss has been investigated in [11, 12] and can be seen as a minimal model of a (proto)cell.

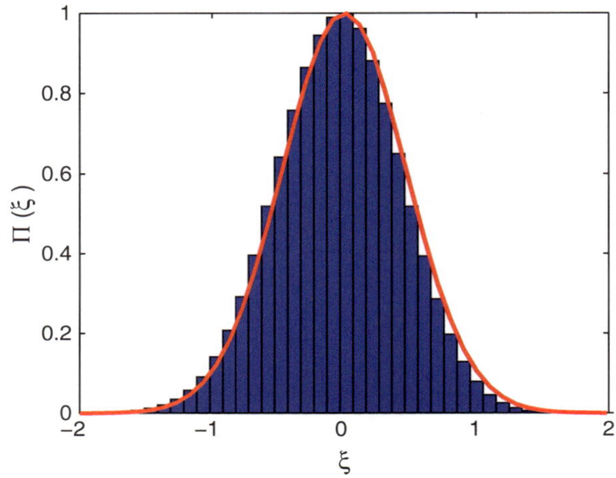

Fig. 9.3 The histogram of rescaled stochastic fluctuations ξ recorded from the Gillespie-based simulations is compared to the analytical solution of the Fokker-Planck equation (*solid line*). The agreements are excellent and point to adequacy of the van Kampen technique. Here, $N = 100$, $b = 0.1$, and $d = 0.05$

Birth and death reactions, identical to the ones hypothesized above, are still assumed to hold. The model deals however with an arbitrary large number of independent populations, which are organized in a close autocatalytic cycle. These two additional ingredients, dimensions and mutual interactions, will make the dynamics less trivial, in particular as concerns the impact of the endogeneous fluctuations.

9.4 A Model of Autocatalytic Reactions

In this section, we will review the application of the system-size expansion to a model of autocatalytic reactions, first introduced in the literature by Togashi and Kaneko [23, 24]. The results that we are going to discuss have been presented in [11, 12]. In the following, we shall start by providing a concise description of the analysis carried out for the aspatial version of the model, which proves less cumbersome from the mathematical viewpoint [11]. Then, we will turn to illustrating the extension to the spatial case, as developed in [12].

In the original scheme devised by Togashi and Kaneko, the reactions are cyclic and involve k constituents $X_1, \ldots, X_k$. The latter react according to $X_i + X_{i+1} \rightarrow 2X_{i+1}$ with $X_{k+1} \equiv X_1$, $i = 1, \ldots, k$. The chemicals are assumed to be in a container which is well stirred, but with the possibility of diffusing across the surface of the container into a particle reservoir. In [11] the above model has been slightly revisited via explicit inclusion of the null constituents E.

More specifically, the autocatalytic reaction scheme investigated in [11] reads

$$\begin{aligned} &X_i + X_{i+1} \overset{\eta_{i+1}}{\rightarrow} 2X_{i+1}, \; X_{k+1} \equiv X_1 \\ &E \overset{\beta_i}{\rightarrow} X_i; \qquad X_i \overset{\gamma_i}{\rightarrow} E\,, \; i = 1, \ldots, k \end{aligned} \tag{9.24}$$

where r_i, γ_i and β_i (with $r_{k+1} \equiv r_1$) are the rates at which the reactions take place. As explained in the preceding section, the pseudo-chemical elements E accounts for a finite carrying capacity of the scrutinized system. By denoting the size of the system with N and labeling n_i the number of elements of type X_i, then $\sum_{i=1}^{k} n_i + n_E = N$, where n_E is the number of empties E. Clearly, as an obvious consequence of the latter conservation law, n_E is always replaced by $N - \sum_{i=1}^{k} n_i$. The rate constants γ_i and β_i in Eq. (9.24) control the interactions of the system with the particle reservoir outside the container. In effect γ_i and β_i are the rates at which molecules enter and exit the system in stringent analogy with birth and death rates.

Besides their interest per se, it is speculated that autocatalytic cycles might have been fundamental, back at the origin of life, in sustaining the development of elementary cell-like entities, the so-called protocells. The shared view is that protocell's volume might have been occupied by interacting families of replicators, organized in nested autocatalytic reactions. The latter have been invoked in fact as a possible solution of the famous Eigen's paradox, a simple logic argument that implies limiting the size of self-replicating molecules to perhaps a few hundred base pairs. At odd, almost all life on Earth requires much longer molecules to encode for their genetic information. This problem is handled in living cells by the presence of enzymes which repair mutations, allowing the encoding molecules to reach sizes on the order of millions of base pairs. In primordial organisms, autocatalytic cycles might have provided the necessary degree of microscopic cooperation to prevent the Eigen's drive to self-destruction to eventually take place. In this respect, model (9.24) can constitute a sort of null model of a primordial cell. Hence, the volume where the chemicals are confined can be imagined to be delimited by the cell wall, the membrane.

In the following, we shall report about the study in [11], where the (aspatial) model introduced above has been investigated via the van Kampen perturbative technique. We will in particular show that the discreteness of the constituents that take part to the autocatalytic cycle gives rise to large sustained oscillations, even when the number of elementary units is quite large, and as opposed to mean-field predictions.

9.5 The Aspatial Model: Deterministic and Stochastic Dynamics

Let us consider the aspatial version of the autocatalytic cycle, as described by Eq. (9.24). Molecules are supposed to be uniformly stirred inside a given volume. A scalar quantity for each of the k species is therefore sufficient to unambiguously photograph the state of the system. In other terms, the state of the system is labeled by the k dimensional vector $\mathbf{n} \equiv (n_1, \ldots, n_k)$. Under the assumption that the transitions from this state to any other compatible with the former only depend on these integers, the system is Markov and can be described in terms of a master equation. As illustrated in the preceding sections, the master equation is specified if the transition rates $T(\mathbf{n}'|\mathbf{n})$ from the state $\mathbf{n}$ to the to the state $\mathbf{n}'$ are given. The assumption of a uniform distribution inside the volume implies that the probability of a reaction taking place is proportional to its rate and the number of reactant molecules. For our case, see Eq. (9.24) the transition rates take the form

$$
\begin{aligned}
T(n_1,\dots,n_i-1,n_{i+1}+1,\dots,n_k|\mathbf{n}) &= \eta_{i+1}\frac{n_i}{N}\frac{n_{i+1}}{N} \\
T(n_1,\dots,n_i+1,\dots,n_k|\mathbf{n}) &= \beta_i\left(1-\frac{\sum_{j=1}^{k} n_j}{N}\right) \\
T(n_1,\dots,n_i-1,\dots,n_k|\mathbf{n}) &= \gamma_i\frac{n_i}{N}
\end{aligned}
\tag{9.25}
$$

The master equation for the probability that the system is in state $\mathbf{n}$ at time t, $P(\mathbf{n},t)$, can be hence written as

$$
\begin{aligned}
\frac{\mathrm{d}P(\mathbf{n},t)}{\mathrm{d}t} &= \sum_{i=1}^{k}\left(\mathscr{E}_i\mathscr{E}_{i+1}^{-1}-1\right) \\
&\times\left[T(n_1,\dots,n_i-1,n_{i+1}+1,\dots,n_k|\mathbf{n})P(\mathbf{n},t)\right] \\
&+\sum_{i=1}^{k}\left(\mathscr{E}_i^{-1}-1\right)\left[T(n_1,\dots,n_i+1,\dots,n_k|\mathbf{n})P(\mathbf{n},t)\right] \\
&+\sum_{i=1}^{k}\left(\mathscr{E}_i-1\right)\left[T(n_1,\dots,n_i-1,\dots,n_k|\mathbf{n})P(\mathbf{n},t)\right]
\end{aligned}
\tag{9.26}
$$

where $\mathscr{E}_i^{\pm 1}$ are a generalization of the step operators previously introduced:

$$
\mathscr{E}_i^{\pm 1} f(\mathbf{n}) = f(n_1,\dots,n_i\pm 1,\dots,n_k) \tag{9.27}
$$

To progress in the analysis we put forward the aforementioned van Kampen ansatz that, in this case, reads

$$
\frac{n_i}{N} = \phi_i(t) + \frac{\xi_i(t)}{\sqrt{N}} \tag{9.28}
$$

$\phi_i(t)$ refers to the deterministic contribution. It labels the fraction of the molecules which are of type X_i at time t in the mean-field ($N \to \infty$) limit. The fluctuations $\xi_i(t)$, i.e., the stochastic component of the dynamics, are multiplied by the scaling factor $1/\sqrt{N}$. Inserting equation (9.28) into Eq. (9.26) allows one to expand the master equation as a power series of $1/\sqrt{N}$. By expanding the step operators (9.27) one obtains the usual expressions:

$$\mathscr{E}_i^{\pm 1} = 1 \pm \frac{1}{\sqrt{N}} \frac{\partial}{\partial \xi_i} + \frac{1}{2N} \frac{\partial^2}{\partial \xi_i^2} + \cdots \tag{9.29}$$

If we set $P(\mathbf{n},t)$ equal to $\Pi(\xi,\tau)$, one can expand the left-hand side of the master equation in analogy with what previously done, namely,

$$\frac{\mathrm{d}P(\mathbf{n},t)}{\mathrm{d}t} = \frac{1}{N}\frac{\partial \Pi(\xi,\tau)}{\partial \tau} - \frac{1}{\sqrt{N}} \sum_{i=1}^{k} \frac{\partial \Pi(\xi,\tau)}{\partial \xi_i} \frac{\mathrm{d}\phi_i}{\mathrm{d}\tau} \tag{9.30}$$

where $\tau = t/N$. Substituting Eq. (9.28) into the right-hand side of the master Eq. (9.26) and using the explicit form of the transition rates as reported in (9.25), one may group together the terms of same order in $1/\sqrt{N}$.

To leading order, the expanded master equation gives (see [11] for additional information on the algebraic, intermediate steps involved)

$$\frac{\mathrm{d}\phi_i}{\mathrm{d}\tau} = (\eta_i \phi_{i-1} - \eta_{i+1}\phi_{i+1})\,\phi_i + \beta_i \left(1 - \sum_{j=1}^{k} \phi_j\right) - \gamma_i \phi_i \tag{9.31}$$

The above equations represent a deterministic approximation to the stochastic model (9.24). Assume η_i, γ_i and β_i to be the same for each species, and so drop the index i. The continuous, time-dependent, concentration ϕ_i of species i evolves starting from the assigned initial condition and asymptotically converges to a (stable) solution ϕ^*, which can be readily obtained by setting $\mathrm{d}\phi_i/\mathrm{d}\tau = 0$. One immediately gets

$$\phi^* = \frac{\beta}{\gamma + k\beta} \tag{9.32}$$

How accurate is the deterministic approximation for the stochastic model here considered? To answer this question one can perform numerical simulations of the chemical reaction system (9.24) by use of the exact Gillespie algorithm [17, 18]. In Fig. 9.4 the outcome of the stochastic simulations (solid line) is compared to the solution of the deterministic equation (9.31) (dashed line). Once the initial transient has died out the latter tends to relax to the deputed equilibrium ϕ^*. At variance, the stochastic time series keeps on oscillating around the reference value ϕ^*. Such regular oscillations, termed quasi-cycles, manifest because of the finite-size corrections to the idealized mean-field dynamics. As we will make clear in the following, the emergence of the quasi-cycles can be successfully explained by retaining the higher order terms in the above perturbative analysis.

At next order of the perturbative development, one finds in fact the following Fokker–Planck equation:

$$\frac{\partial \Pi}{\partial \tau} = -\sum_i \frac{\partial}{\partial \xi_i}\left[A_i(\xi)\,\Pi\right] + \frac{1}{2}\sum_{i,j} B_{ij} \frac{\partial^2 \Pi}{\partial \xi_i \partial \xi_j} \tag{9.33}$$

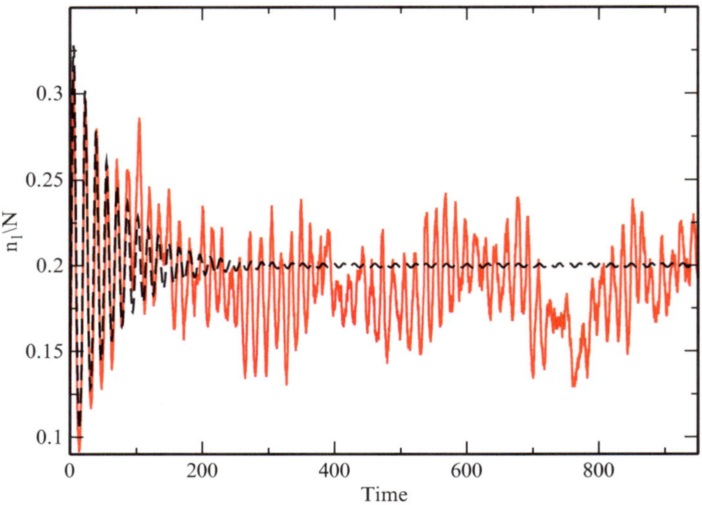

Fig. 9.4 Temporal evolution of one of the species concentrations for a system composed by $k = 4$ species and parameters set as $N = 8190$, $r_i = 10$, and $\alpha_i = \beta_i = 1/64\ \forall i$. The noisy line represents one stochastic realization obtained via the Gillespie algorithm [17, 18]. The dashed line shows the numerical solution of the deterministic system given by Eq. (9.31)

which governs the dynamics of the distribution function of fluctuations $\Pi(\xi,t)$. Here,

$$A_i(\xi) = (\eta_i\phi_{i-1} - \eta_{i+1}\phi_{i+1})\,\xi_i + \eta_i\phi_i\xi_{i-1} - \eta_{i+1}\phi_i\xi_{i+1} - \beta_i\sum_{j=1}^{k}\xi_j - \gamma_i\xi_i \tag{9.34}$$

and

$$B_{ij} = \begin{cases} -\eta_i\phi_{i-1}\phi_i, & \text{if } j = i-1 \\ \eta_{i+1}\phi_i\phi_{i+1} + \eta_i\phi_i\phi_{i-1} & \\ +\beta_i\left(1 - \sum_{j=1}^{k}\phi_j\right) + \gamma_i\phi_i, & \text{if } j = i \\ -\eta_{i+1}\phi_i\phi_{i+1}. & \text{if } j = i+1 \end{cases} \tag{9.35}$$

In Eqs. (9.34) and (9.35), $\phi_{k+1} \equiv \phi_1$ and $\xi_{k+1} \equiv \xi_1$, which follows from the cyclic nature of the model.

Since the $A_i(\xi)$ are linear functions of the ξ_j we may write them as

$$A_i(\xi) = \sum_{j=1}^{k} M_{ij}\xi_j \tag{9.36}$$

The probability distribution $\Pi(\xi,\tau)$ is therefore entirely determined by the two $k \times k$ matrices M and B, whose elements are solely functions of the mean-field concentration ϕ_j. In principle the matrices M and B are time dependent, since ϕ_j is. However, in practice we are interested in fluctuations about the stationary state, and so replace ϕ_j with its asymptotic constant analogue ϕ^*.

The Fokker–Planck Eq. (9.33) yields to the equivalent Langevin formulation:

$$\frac{d\xi_i}{d\tau} = \sum_{j=1}^{k} M_{ij}\xi_j(\tau) + \eta_i(\tau) \tag{9.37}$$

where M follows from (9.36) and η_i is a Gaussian white noise with zero mean and correlator

$$\langle \eta_i(\tau)\eta_j(\tau')\rangle = B_{ij}\delta\left(\tau - \tau'\right) \tag{9.38}$$

To bring into evidence the oscillatory nature of the fluctuations, we take the Fourier transform of Eq. (9.37):

$$\sum_{j=1}^{k} \left(-i\omega\delta_{ij} - M_{ij}\right)\tilde{\xi}_j(\omega) = \tilde{\eta}_i(\omega) \tag{9.39}$$

where the $\tilde{f}$ stands for the Fourier transform of the function f. Introducing $\Phi_{ij}(\omega) = -i\omega\delta_{ij} - M_{ij}$ the solution to Eq. (9.39) is

$$\tilde{\xi}_i(\omega) = \sum_{j=1}^{k} \Phi_{ij}^{-1}(\omega)\tilde{\eta}_j(\omega) \tag{9.40}$$

To identify the dominant frequency of the oscillating time series, one can compute the power spectrum $P_i(\omega)$ for the ith species, from Eq. (9.40). In formulae, one gets

$$P_i(\omega) \equiv \left\langle |\tilde{\xi}(\omega)|^2 \right\rangle = \sum_{j=1}^{k}\sum_{l=1}^{k} \Phi_{ij}^{-1}(\omega)B_{jl}\left(\Phi^{\dagger}\right)_{li}^{-1}(\omega) \tag{9.41}$$

In Figs. 9.5 and 9.6, the theoretical power spectra are compared to the homologous quantities calculated from averaging over many realization of the Gillespie-based simulations. The figures refer respectively to $k = 4$ and $k = 8$. One or two peaks are displayed in the power spectra, pointing to the existence of regular oscillatory behaviors in the recorded signals. Ordered temporal oscillations can therefore spontaneously emerge, driven by the stochastic component of the dynamics and as opposed to what predicted within the idealized mean-field scenario.

In the next section we will briefly turn to discussing the generalized spatial model. This setting has been studied in [12] via the van Kampen system-size expansion. We shall hereafter provide a rather compact description of the analysis, without insisting on the technical details of the calculation that can be found in [12].

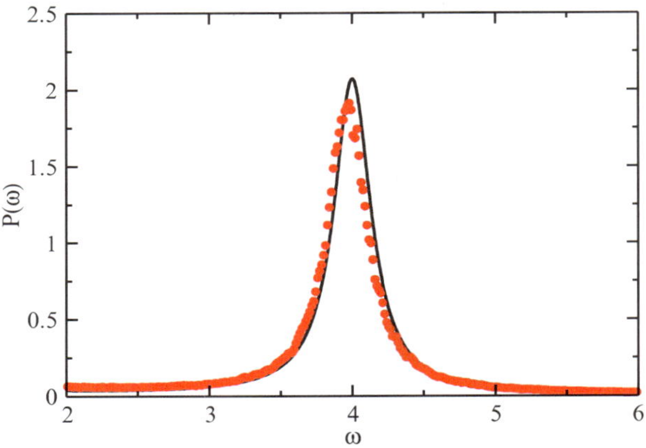

Fig. 9.5 Power spectrum of species $i = 2$ when $k = 4$. The analytical curve is shown as a solid line and the simulation (average over 500 independent realizations) as symbols. Here $r = 10$, $\gamma = \beta = 5/32$, and $N = 5,000$. Reprinted with permission from [11] Copyright (2009) by the American Physical Society

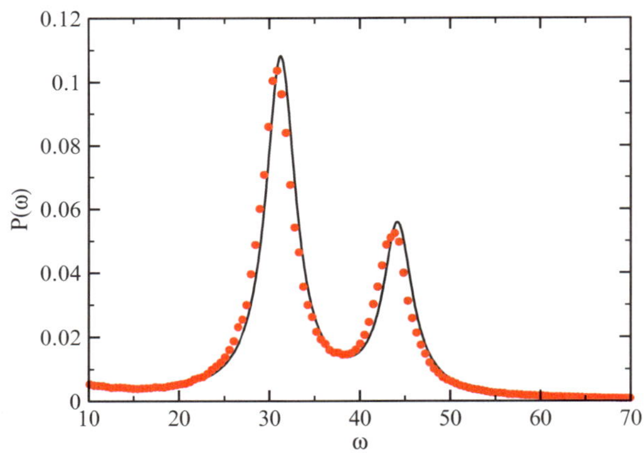

Fig. 9.6 Power spectrum of the time series for species $i = 2$ when $k = 8$. The analytical result (*solid line*) is superimposed onto the simulations (*symbols*), averaged over 500 independent realizations. Here $r = 200$, $\beta = 1.9$, $\gamma = 2$, and $N = 7,000$. Reprinted with permission from [11], Copyright (2009) by the American Physical Society

9.6 Spatial Model: Ordered Patterns Revealed by the van Kampen System Size Expansion

Model (9.24) can be also straightforwardly extended so to explicitly account for the notion of space, as done in [12]. The idea is to coarse-grain the volume where molecules are confined, by partitioning it in Ω small micro-cells, within which autocatalytic reactions do occur. Following [12], the k species are labeled X_s^j. The index s identifies the species, while $j = 1, .., \Omega$ refers to the micro-cell to which the element is bound. In analogy with the preceding discussion the reactions can be cast in the form

$$X_s^j + X_{s+1}^j \xrightarrow{\eta_{s+1}} 2X_{s+1}^j, \tag{9.42}$$

where $X_{k+1}^j = X_1^j$.

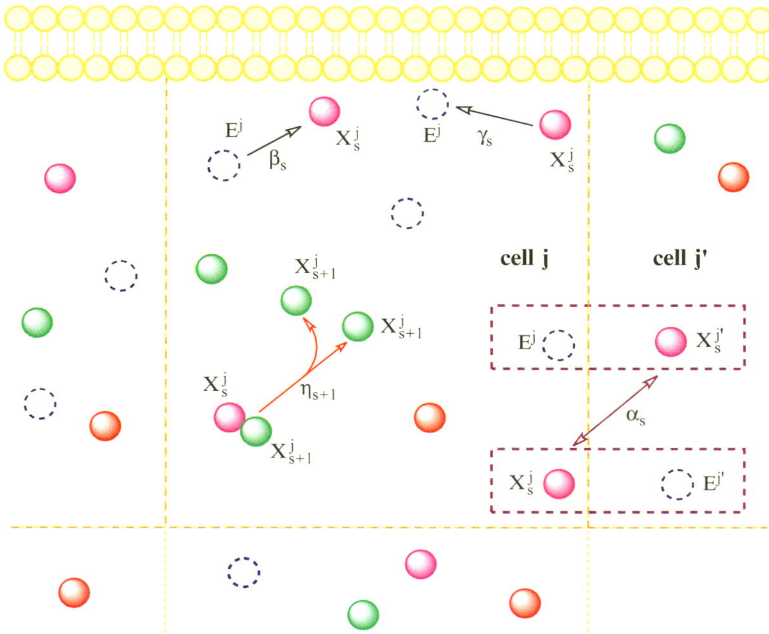

Fig. 9.7 The volume of the cell is imagined to be partitioned into Ω micro-cells. Within micro-cell j the molecular species interact according to the autocatalytic reactions specified by Eq. (9.24). In addition, the molecules can migrate from micro-cell j to its nearest neighbors, e.g., micro-cell j', as depicted in the cartoon. A molecule of type X_k^j (*full circle*) takes over a vacancy (*dashed empty circle*) of micro-cell $E^{j'}$ and so transforms into $X_k^{j'}$, leaving behind a vacancy E^j. Finally, the chemical can also diffuse in from the environment, a reaction that in turn implies changing E^j into X_k^j. The opposite holds for molecules that diffuse out into the environment. Reprinted with permission from [12], Copyright (2010) by the American Physical Society

Indeed, only the region that is adjacent to the outer boundary, the cell membrane as emphasized above, is given a detailed spatial structure. The remaining inner volume acts instead as a particle reservoir. A cartoon of the setting here imagined is depicted in Fig. 9.7. Molecules sitting in cell j can migrate towards the neighbors micro-cell j'. This is a microscopic process that obeys to the following chemical equations:

$$X_s^j + E^{j'} \xrightarrow{\alpha_i} X_s^{j'} + E^j \tag{9.43}$$

$$E^j + X_s^{j'} \xrightarrow{\alpha_i} E^{j'} + X_s^j \tag{9.44}$$

where E^j (resp. $E^{j'}$) represents vacancies in cell j (resp. j'). The capacity of each micro-cell is N: the sum of the number of molecules of each species plus the number of vacancies equals N, for every micro-cell.

Finally, a molecule X_s^j can migrate from cell j towards the (outer) environment or the inner region leaving behind an empty case E^j. Alternatively, cell j can gain a molecule X_s^j from the environment or inner region. These processes are described as

$$X_s^j \xrightarrow{\gamma_s} E^j; \qquad E^j \xrightarrow{\beta_s} X_s^j \tag{9.45}$$

When operating in this generalized setting, the mathematical analysis becomes more complex, as compared to the aspatial model. One additional index has to be forcefully introduced in the definition of the variables involved so to specify the micro-cell to which the molecules belong. In other words, the discrete concentration n_s^j is not just function of time but also sensitive to the specific spatial location. This additional degree of freedom will make it possible to eventually appreciate the emergence of spatially organized patterns. The state of the system can be characterized by the vector $\mathbf{n} = (\mathbf{n}^1, \mathbf{n}^2, \ldots, \mathbf{n}^\Omega)$ where $\mathbf{n}^j = (n_1^j, n_2^j, \ldots, n_k^j)$.

The model can be formulated in terms of a chemical master equation. Then, by applying the van Kampen perturbative scheme, one can recover the mean-field solution and determine as well the stochastic, finite N, corrections to it. The transition rates associated to the migration from one cell to the neighbor one read

$$\begin{aligned} T(n_s^j - 1, n_s^{j'} + 1 | n_s^j, n_k^{j'}) &= \frac{\alpha_s}{z\Omega} \frac{n_s^j}{N} \left(1 - \sum_{m=1}^{k} \frac{n_m^{j'}}{N}\right) \\ T(n_s^j + 1, n_s^{j'} - 1 | n_s^j, n_s^{j'}) &= \frac{\alpha_s}{z\Omega} \frac{n_s^{j'}}{N} \left(1 - \sum_{m=1}^{k} \frac{n_m^{j}}{N}\right) \end{aligned} \tag{9.46}$$

where z is the number of nearest neighbors that each micro-cell has. The reaction rates associated to the autocatalytic cycles and to the diffusion from/to the inner/outer bulk can be written as a trivial extension of the equivalent quantities obtained for the aspatial model. For this reason, these are not given here explicitly.

The master equation for the probability $P(\mathbf{n},t)$ can be cast in the form

$$\begin{aligned} \frac{\mathrm{d}P(\mathbf{n},t)}{\mathrm{d}t} &= \sum_{j=1}^{\Omega} \mathscr{T}_{\text{loc}}^{j} P(\mathbf{n},t) + \sum_{j=1}^{\Omega} \sum_{j' \in j} \mathscr{T}_{\text{mig}}^{jj'} P(\mathbf{n},t) \\ &\quad + \sum_{j=1}^{\Omega} \mathscr{T}_{\text{env}}^{j} P(\mathbf{n},t), \end{aligned} \tag{9.47}$$

where the three terms on the right-hand side refer respectively to the local chemical reactions, the migration of species between micro-cells, and the interaction with the outer/inner environment. The notation $j' \in j$ indicates that cell j' is a nearest neighbor of the cell j.

The van Kampen analysis requires introducing the ansatz

$$\frac{n_s^j}{N} = \phi_s^j + \frac{1}{\sqrt{N}}\xi_s^j \tag{9.48}$$

into the master equation and carrying out the perturbative analysis, by adopting $1/\sqrt{N}$ as a small parameter. The details of the calculations are given in [12] and we shall here solely summarize the results for what concerns the leading and next-to-leading approximations.

At the leading order, one finds the following equation for the concentration ϕ_s^j of species s in cell j:

$$\begin{aligned}\frac{d\phi_s^j}{d\tau} &= \eta_s\phi_{s-1}^j\phi_s^j - \eta_{s+1}\phi_s^j\phi_{s+1}^j \\ &+\alpha_s\left(\Delta\phi_s^j\left(1-\sum_{m=1}^{k}\phi_m^j\right)+\phi_s^j\sum_{m=1}^{k}\Delta\phi_m^j\right) \\ &+\beta_s\left(1-\sum_{m=1}^{k}\phi_m^j\right)-\gamma_s\phi_s^j\end{aligned} \tag{9.49}$$

where Δ stands for the discrete Laplacian operator $\Delta f_s^j = (2/z)\sum_{j'\in j}(f_s^{j'} - f_s^j)$. In the limit where the size of the micro-cells tends to zero, the above equations become partial differential equations, Δ converging to the more familiar Laplacian operator. Notice that Eq. (9.49) constitutes the natural generalization of Eq. (9.31) to the case where space is accounted for. Notice the cross-diffusion terms that reflect the assumption of a finite carrying capacity in each micro-cell. The importance of such additional contributions, which follows from a rigorous description of the microscopic diffusion, has been elaborated on in [25]. Assuming η_s, β_s and γ_s to be identical for all species, we can drop the index s and obtain an explicit expression for the homogeneous (uniform in space) fixed point of the dynamics, namely $\phi^* = \beta/(\gamma+k\beta)$. As expected, the latter coincides with the fixed point obtained for the aspatial model.

The next-to-leading corrections yield as usual to a Fokker–Planck equation for the distribution of fluctuations (see Eq. B1 in [12]). The Fokker–Planck is formally equivalent to a Langevin equation, which upon spatial Fourier reads

$$\frac{d\xi_s^{\mathbf{k}}}{d\tau} = \sum_r M_{\mathrm{sr}}^{\mathbf{k}}\xi_r^{\mathbf{k}} + \lambda_s^{\mathbf{k}}(\tau) \tag{9.50}$$

where

$$\langle\lambda_s^{\mathbf{k}}(\tau)\lambda_r^{\mathbf{k}'}(\tau')\rangle = \mathscr{B}_{\mathrm{sr}}^{\mathbf{k}}\Omega a^d\delta_{\mathbf{k}+\mathbf{k}',0}\delta(\tau-\tau') \tag{9.51}$$

and where $\mathbf{k}$ is the wavevector. To derive the above result it was assumed in [12] that the micro-cells form a hypercubic lattice in d-dimensions with linear spacing a. The matrix $M^{\mathbf{k}}$ is

$$M_{\mathrm{sr}}^{\mathbf{k}} = M_{\mathrm{sr}}^{(\mathrm{NS})} + M_{\mathrm{sr}}^{(\mathrm{SP})} \Delta_{\mathbf{k}} \tag{9.52}$$

where $\Delta_{\mathbf{k}}$ is the Fourier transform of the discrete Laplacian

$$\Delta_{\mathbf{k}} = \frac{2}{d} \sum_{\gamma=1}^{d} \left[\cos(\mathrm{k}_{\gamma} a) - 1\right] \tag{9.53}$$

and k_{γ} is one of γth component of the vector $\mathbf{k}$. The two matrices $M^{(\mathrm{NS})}$ and $M^{(\mathrm{SP})}$ are

$$M_{\mathrm{ss}}^{(\mathrm{NS})} = -\beta - \gamma \tag{9.54}$$

$$M_{\mathrm{sr}}^{(\mathrm{NS})} = \begin{cases} -\eta\phi^* - \beta, & \text{if } r = s+1 \\ \eta\phi^* - \beta, & \text{if } r = s-1 \\ -\beta, & \text{if } |s-r| > 1 \end{cases} \tag{9.55}$$

and

$$M_{\mathrm{ss}}^{(\mathrm{SP})} = \alpha_s \left[1 + (1-k)\phi^*\right] \tag{9.56}$$

$$M_{\mathrm{sr}}^{(\mathrm{SP})} = \alpha_s \phi^* \ \text{ if } s \neq r \tag{9.57}$$

NS stands for “non spatial,” while SP is the compact label for “spatial.” The matrix $\mathscr{B}^{\mathbf{k}}$ in Eq. (9.51) is given by

$$\mathscr{B}_{\mathrm{sr}}^{\mathbf{k}} = \mathscr{B}_{\mathrm{sr}}^{(\mathrm{NS})} + \mathscr{B}_{\mathrm{sr}}^{(\mathrm{SP})} \Delta_{\mathbf{k}} \tag{9.58}$$

where the two $k \times k$ matrices $\mathscr{B}^{(\mathrm{NS})}$ and $\mathscr{B}^{(\mathrm{SP})}$ are given by

$$\mathscr{B}_{\mathrm{ss}}^{(\mathrm{NS})} = a^d \left[\beta(1-k\phi^*) + \gamma\phi^* + 2\eta\,(\phi^*)^2\right] \tag{9.59}$$

$$\mathscr{B}_{\mathrm{sr}}^{(\mathrm{NS})} = \begin{cases} -a^d \eta\,(\phi^*)^2, & \text{if } r = s+1 \\ -a^d \eta\,(\phi^*)^2, & \text{if } r = s-1 \\ 0, & \text{if } |s-r| > 1 \end{cases} \tag{9.60}$$

and

$$\mathscr{B}_{\mathrm{ss}}^{(\mathrm{SP})} = -2a^d \alpha_s \phi^* \,(1-k\phi^*) \tag{9.61}$$

$$\mathscr{B}_{\mathrm{sr}}^{(\mathrm{SP})} = 0 \ \text{ if } s \neq r \tag{9.62}$$

As discussed above, fluctuations about the stationary state need to be taken into account, as they can be relevant even if N is relatively large. Since the model extends in space, the power spectrum of fluctuations should depend on both the spatial wavenumber $\mathbf{k}$ and the frequency ω. Defining $\Phi^{\mathbf{k}}_{\rm sr}(\omega) = (-i\omega\delta_{\rm sr} - M^{\mathbf{k}}_{\rm sr})$, one eventually obtains [12] the following compact expression for the power spectrum $P_s(\mathbf{k},\omega)$ of the fluctuations of species s:

$$\begin{aligned} P_s(\mathbf{k},\omega) &\equiv \left\langle |\xi^{\mathbf{k}}_s(\omega)|^2 \right\rangle \\ &= \Omega a^d \sum_{r=1}^{k} \sum_{u=1}^{k} \left[\Phi^{\mathbf{k}}(\omega)\right]^{-1}_{\rm sr} \mathscr{B}^{\mathbf{k}}_{\rm ru} \left[\Phi^{\mathbf{k}\dagger}(\omega)\right]^{-1}_{\rm us} \end{aligned} \tag{9.63}$$

The analysis sketched above is a straightforward, though complex, generalization of the study of [11] reviewed in the preceding section. We shall be here just concerned with presenting the main conclusion of the analysis, comparing in particular the theoretical power spectra to the homologous quantities obtained via numerical simulations. The analysis is limited to the choice $d = 1$, i.e., a one-dimensional frontier (membrane) of a two-dimensional compact domain (cell).

As reported in Fig. 9.8, a localized peak is predicted by the theory. This evidence suggests that organized spatiotemporal patterns can spontaneously emerge, as mediated by the endogenous stochasticity of the system. The plots in Fig. 9.8 refer to two distinct species and are obtained by operating in the setting with $k = 4$. The other two species display a similar degree of spatiotemporal self-organization.

The theory prediction, and thus the accuracy of the approximations involved, can be tested via direct simulations. By averaging over many independent realizations, one can calculate the power spectra of the recorded stochastic time series after Fourier transformation. The numerical power spectra are depicted in Fig. 9.9 for the same choice of parameters as in Fig. 9.8. The correspondence between the profiles is remarkably good. Spatial, as well as temporal, order can spontaneously develop as a collective amplification of the microscopic finite-size fluctuations.

9.7 Conclusion

Modeling the dynamical evolution of a large sea of mutually interacting entities is a task of great importance and cross-disciplinary interest. The customarily adopted scenario assumes dealing with continuous populations, whose concentrations change in space and time according to the governing partial or ordinary differential equations. In doing so, one neglects the intimate discreteness of the investigated medium to favor a mean-field deterministic approach. In many cases, however, the finite-size fluctuations stemming from the microscopic graininess, and therefore endogenous to the system under scrutiny, prove crucial. They can in fact amplify as follows a complex resonance mechanism and yield to organized spatiotemporal patterns. More specifically, the measured concentrations which re-

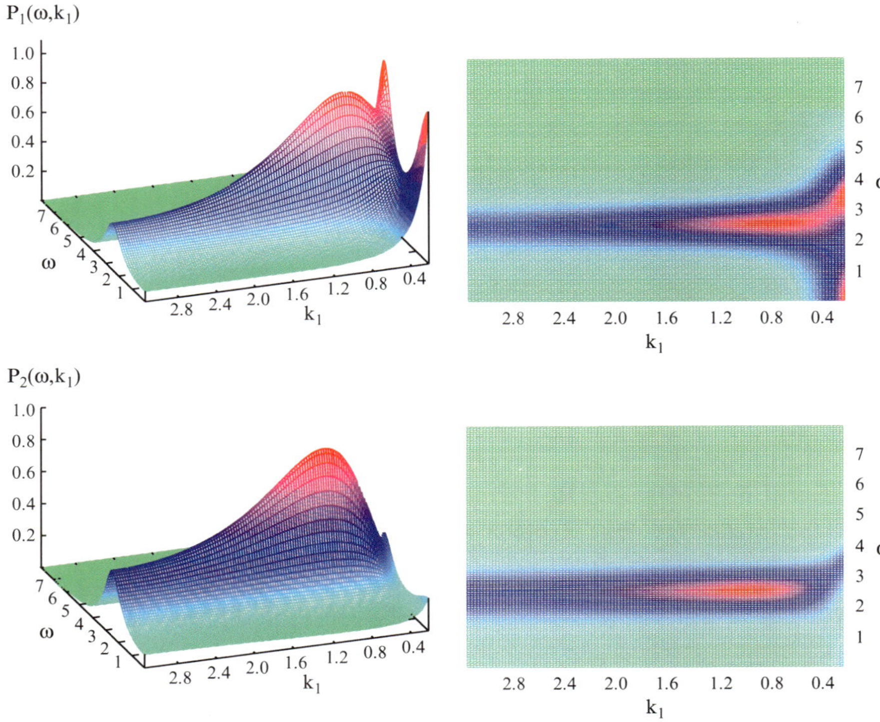

Fig. 9.8 Analytical power spectra calculated via the van Kampen system-size expansion; see [12] for details. The theoretical profiles refer to the case with $k = 4$ species and to a two-dimensional volume (one-dimensional periodic array of Ω micro-cells). Each three-dimensional plot (and its corresponding two-dimensional projection) refers to a different chemical species. A localized peak is shown, which implies the existence of regular spatiotemporal patterns. Here $\Omega = 256$, $\eta = 10$, $\beta = 5/32$, $\gamma = 5/32$, and $\alpha = [100, 0.001, 1, 500]$. Reprinted from [12]

flect the distribution of the interacting entities (e.g., chemical species, biomolecules) can oscillate regularly in time and/or display spatially patched profiles, collective phenomena which testify on a surprising degree of macroscopic order, as mediated by the stochastic component of the dynamics.

These intriguing phenomena have been recently addressed and successfully explained via rigorous analytical means. Among other techniques, the van Kampen system-size expansion can be employed to bridge the gap between the deterministic and stochastic viewpoints. In this chapter, we have provided a pedagogical introduction to such method, by considering a simple birth and death stochastic process, which accounts for the finite carrying capacity of the embedding volume. The theoretical calculations enabled us to quantify the probability distribution function of fluctuations around the stationary fixed point. The adequacy of the prediction was confirmed by direct comparison with the outcome of stochastic

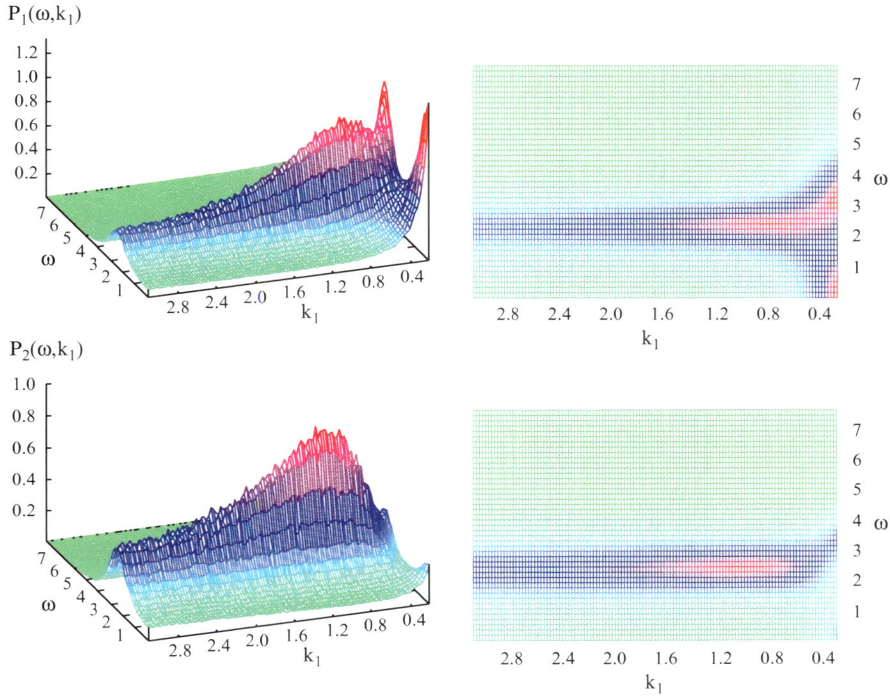

Fig. 9.9 Numerically calculated power spectra are obtained from averaging 800 realizations; see [12] for further information. Stochastic simulations are performed via the Gillespie algorithm. Parameters are set to the same values assigned when drawing the theoretical plots of Fig. 9.8. Here $N = 5,000$. Reprinted from [12]

simulations. In this case the fluctuations result in random, Gaussian-distributed disturbances around the stable fixed point of the dynamics.

More interestingly, it is the application of the van Kampen system-size expansion to a stochastic model of autocatalytic reactions. The model, introduced by Togashi and Kaneko [23] and later on revisited by Di Patti and collaborators [11], is presumably relevant for studies on the origin of life and exists into two versions, respectively: the aspatial [11] and spatial one [12]. By operating in these contexts and making use of the system-size expansion, one can show that the chemical constituents can organize in regular spatiotemporal cycles, coherent macroscopic structures that emerge from the microscopic disorder. In both cases, the perturbative scheme pioneered by van Kampen turns out to be accurate and versatile. It thus represents a powerful and reliable tool to inspect the role played by demographic fluctuations in a finite-size population, beyond the idealized mean-field approximation.

Acknowledgements A large portion of this paper is devoted to reviewing the results of [11, 12] which I coauthored with my colleagues T. Dauxois, P. De Anna, F. Di Patti, and A. McKane. I wish to thank them all.

References

1. J.D. Murray, *Mathematical Biology*, 2nd edn. (Springer, Heidelberg, Germany, 1993)
2. B. Alberts et al., *Molecular Biology of the Cell*, 5th edn. (Garland Science, New York, 2007)
3. H. Lodish et al., *Molecular Cell Biology*, 6th edn. (W.H. Freeman and Co., New York, 2008)
4. J. Maynard Smith, *Models in Ecology* (Cambridge University Press, Cambridge, 1974)
5. C.W. Gardiner, *Handbook of Stochastic Methods*, 3rd edn. (Springer-Verlag, Berlin, 2004)
6. N.G. van Kampen, *Stochastic Processes in Physics and Chemistry*, 3rd edn. (Elsevier, Amsterdam, 2007)
7. A.J. McKane, T.J. Newman, Phys. Rev. Lett. **94**, 218102 (2005)
8. A.J. McKane, J.D. Nagy, T.J. Newman, M.O. Stefanini, J. Stat. Phys. **128**, 165 (2007)
9. D. Alonso, A.J. McKane, M. Pascual, J. R. Soc. Interface **4**, 575 (2007)
10. A.J. McKane T.J. Newman, Phys. Rev. E **70**, 041902 (2004)
11. T. Dauxois, F. Di Patti, D. Fanelli, A.J. McKane, Phys. Rev. E **79**, 036112 (2009)
12. P. De Anna, F. Di Patti, D. Fanelli, A.J. McKane, T. Dauxois, Phys. Rev. E **81** 81 056110 (2010)
13. T. Biancalani, D. Fanelli, F. Di Patti, Phys. Rev. E **81**, ISSN: 1539–3755 (2010)
14. S. Strogatz, *Non Linear Dynamics and Chaos: With Applications to Physics, Biology, Chemistry and Engineering* (Perseus Book Group, Cambridge, MA, 2001)
15. M.S. Bartlett, J. R. Stat. Soc. A **120**, 48 (1957)
16. R.M. Nisbet, W.S.C. Gurney, *Modelling Fluctuating Populations* (Wiley, New York, 1982)
17. D.T. Gillespie, J. Comput. Phys. **22**, 403 (1976)
18. D.T. Gillespie, J. Phys. Chem. **81**, 2340 (1977)
19. H. Risken, *The Fokker–Planck Equation*, 2nd edn. (Springer-Verlag, Berlin, 1989)
20. C. Cianci, F. Di Patti, D. Fanelli, Europ. Phys. Lett. **96**, 50011 (2011)
21. C. Cianci, F. Di Patti, D. Fanelli, L. Barletti, preprint arXiv:1104.5668 (2011)
22. Grima R., Phys. Rev. Lett. **102**, 218103 (2009)
23. Y. Togashi, K. Kaneko, Phys. Rev. Lett. **86**, 2459 (2001)
24. Y. Togashi, K. Kaneko, J. Phys. Soc. Jpn. **72**, 62 (2003)
25. D. Fanelli, A. McKane, Phys. Rev. E **82**, 021113 (2010)

Chapter 10
An Ising Model for Road Traffic Inference

Cyril Furtlehner

Abstract We review some properties of the "belief propagation" algorithm, a distributed iterative map used to perform Bayesian inference and present some recent work where this algorithm serves as a starting point to encode observation data into a probabilistic model and to process large-scale information in real time. A natural approach is based on the linear response theory and various recent instantiations are presented. We will focus on the particular situation where the data have many different statistical components, representing a variety of independent patterns. As an application, the problem of reconstructing and predicting traffic states based on floating car data is then discussed.

10.1 Introduction

The "belief propagation" algorithm BP originated in the artificial intelligence community for inference problems on Bayesian networks [25]. It is a nonlinear iterative map which propagates information on a dependency graph of variables in the form of messages between variables. It has been recognized to be a generic procedure, instantiated in various domains like error-correcting codes, signal processing, or constraint satisfaction problems with various names depending on the context [18]: the forward-backward algorithm for hidden Markov model selection; the Viterbi algorithm; Gallager's sum-product algorithm in information theory. It has also a nice statistical physics interpretation in the context of mean-field theories, as a minimizer of a Bethe free energy [34] and a solver of the cavity equations [21] and its relation to the TAP equations in the spin-glass context [16]. A noticeable development in the recent years, related to the connection with

C. Furtlehner (✉)
INRIA Saclay – LRI, Bat. 490, Université Paris-Sud –Orsay 91405 France
e-mail: cyril.furtlehner@inria.fr

X. Leoncini and M. Leonetti (eds.), *From Hamiltonian Chaos to Complex Systems: A Nonlinear Physics Approach*, Nonlinear Systems and Complexity 5,
DOI 10.1007/978-1-4614-6962-9_10,

statistical physics, is the emergence of a new generation of algorithms for solving difficult combinatorial problems, like the survey propagation algorithm [22] for constraint satisfaction problems or the affinity propagation for clustering [6].

The subject with which this present review is dealing with is at first a statistical modelling problem. Assuming a set of high-dimensional data, in the form of sparse observations covering a finite fraction of segments in a traffic network, we wish to encode the dependencies between the variables in a probabilistic model, which turns out to be a Markov random field (MRF). We proceed in such a way as to insure that inference on this MRF with BP is optimal in some way so that it can be fast and precise at the same time, offering the possibility to address large-scale problems like inferring congestion on a macroscopic traffic network. In Sect. 10.2 we introduce the BP algorithm and review some of its properties. Section 10.3 is devoted to the general problem of encoding observation data, by addressing the inverse Ising problem. In Sect. 10.4 a traffic application is described along with the construction of an inference model. Section 10.5 is concerned with the problem of multiplicity of BP fixed points and how to turn this into an advantage when the underlying empirical distribution based on observational data is multimodal. Finally in Sect. 10.6 we present some preliminary tests of the method.

10.2 The Belief Propagation Algorithm

We consider a set of discrete random variables $\mathbf{x} = \{x_i, i \in \mathscr{V}\} \in \{1,\dots,q\}^{|\mathscr{V}|}$ obeying a joint probability distribution of the form

$$\mathscr{P}(\mathbf{x}) = \prod_{a\in\mathscr{F}} \psi_a(x_a) \prod_{i\in\mathscr{V}} \phi_i(x_i), \tag{10.1}$$

where ϕ_i and ψ_a are factors associated respectively to a single variable x_i and to a subset $a \in \mathscr{F}$ of variables, $\mathscr{F}$ representing a set of cliques and $x_a \stackrel{\text{def}}{=} \{x_i, i \in a\}$. The ψ_a are called the "factors" while the ϕ_i are there by convenience and could be reabsorbed in the definition of the factors. This distribution can be conveniently represented with a bi-bipartite graph called the factor graph [18]; $\mathscr{F}$ together with $\mathscr{V}$ define the factor graph $\mathscr{G}$, which will be assumed to be connected. The set $\mathscr{E}$ of edges contains all the couples $(a,i) \in \mathscr{F} \times \mathscr{V}$ such that $i \in a$. We denote d_a (resp. d_i) the degree of the factor node a (resp. to the variable node i). The factor graph in Fig. 10.1a corresponds, for example, to the following measure:

$$p(x_1,\dots,x_6) = \frac{1}{Z}\psi_a(x_1,x_2,x_3)\psi_b(x_4)\psi_c(x_3,x_4,x_5,x_6)$$

with the following factor nodes $a = \{1,2,3\}$, $b = \{4\}$, and $c = \{3,5,6\}$. Assuming that the factor graph is a tree, computing the set of marginal distributions, called the belief $b(x_i = x)$ associated to each variable i, can be done efficiently.

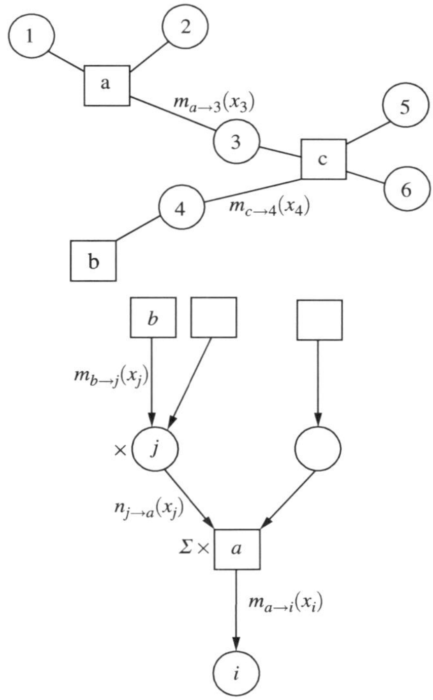

Fig. 10.1 Example of factor graph (**a**) and message propagation rules (**b**)

The BP algorithm does this effectively for all variables in one single procedure, by remarking that the computation of each of these marginals involves intermediates quantities called the messages $m_{a\to i}(x_i)$ [resp. $n_{i\to a}(x_i)$] "sent" by factor node a to variable node i [resp. variable node i to factor node a] and which are necessary to compute other marginals. The idea of BP is to compute at once all these messages, using the relation among them as a fixed point equation. Iterating the following message update rules sketched in Fig. 10.1b:

$$\begin{cases} m_{a\to i}(\mathbf{x}_i) \leftarrow \displaystyle\sum_{x_a\backslash x_i} \prod_{j\in a\backslash i} n_{j\to a}(x_j)\psi_a(x_a), \\ n_{i\to a}(x_i) \leftarrow \phi_i(x_i) \displaystyle\prod_{b\ni i} m_{b\to i}(x_i) \end{cases}$$

yields, when a fixed point is reached, the following result for the beliefs:

$$\begin{aligned} b(x_i) &= \tfrac{1}{Z_i}\phi_i(x_i)\textstyle\prod_{a\ni i} m_{a\to i}(x_i), \\ b(x_a) &= \tfrac{1}{Z_a}\psi_a(X_a)\textstyle\prod_{i\in a} n_{i\to a}(x_i). \end{aligned}$$

This turns out to be exact if the factor graph is a tree but only approximate on multiply connected factor graphs. As mentioned before, this set of beliefs corresponds to a stationary point of a variational problem [34]. Indeed, consider the Kullback-Leibler divergence between a test joint distribution $b(\mathbf{x})$ and the reference $p(\mathbf{x})$. The Bethe approximation yields the following functional of the beliefs, including the joint beliefs $b_a(x_a)$ corresponding to each factor:

$$\begin{aligned} D_{KL}(b\|p) &= \sum_{\{x\}} b(\{x\}) \log \frac{b(\{x\})}{p(\{x\})} \\ &\approx \sum_{a,x_a} b_a(x_a) \log \frac{b_a(x_a)}{\psi(x_a)\prod_{i\in a} b_i(x_i)} + \sum_{i,x_i} \log \frac{b_i(x_i)}{\phi_i(x_i)} \\ &\stackrel{\text{def}}{=} F_{Bethe} = E - S_{Bethe}. \end{aligned}$$

This is equivalent to say that we look for a minimizer of $D_{KL}(b\|p)$ in the following class of joint probabilities:

$$b(x) = \prod_a \frac{b_a(x_a)}{\prod_{i\in a} b_i(x_i)} \prod_i b_i(x_i), \qquad (10.2)$$

under the constraint that

$$\sum_{x_a \setminus x_i} b_a(x_a) = b_i(x_i) \qquad \forall a \in \mathscr{F}, \forall i \in a$$

and the approximation that

$$\sum_{x \setminus x_a} b(x) \approx b_a(x_a), \qquad \forall a \in \mathscr{F}. \qquad (10.3)$$

For a multi-connected factor graph, the beliefs b_i and b_a are then interpreted as pseudo-marginal distribution. It is only when $\mathscr{G}$ is simply connected that these are genuine marginal probabilities of the reference distribution p.

There are a few properties of BP that are worth mentioning at this point. Firstly, BP is a fast converging algorithm:

- Two sweeps over all edges are needed if the factor graph is a tree.
- The complexity scales heuristically like $KN\log(N)$ on a sparse factor graph with connectivity $K \ll N$.
- It is N^2 for a complete graph.

However, when the graph is multiply connected, there is little guarantee on the convergence [24]; even so in practice it works well for sufficiently sparse graphs. Another limit in this case is that the fixed point may not correspond to a true measure, simply because (10.2) is not normalized and (10.3) is approximate. In this sense, the obtained beliefs, albeit compatible with each other, are considered only as pseudo-marginals. Finally, for such graphs, the uniqueness of fixed points is not guaranteed, but it has been shown that:

- Stable BP fixed points are local minima of the Bethe free energy [13].
- The converse is not necessarily true [30].

There are two important special cases, where the BP equations simplify: (i) For binary variables, $x_i \in \{0,1\}$. Upon normalization, the messages are parametrized as

$$m_{a\to i}(x_i) = m_{a\to i}x_i + (1-m_{a\to i})(1-x_i),$$

which is stable w.r.t. the message update rules. Then the propagation of information reduces to the scalar quantity $m_{a\to i}$.

(ii) For Gaussian variables, the factors are necessarily pairwise of the form

$$\psi_{ij}(x_i,x_j) = \exp\big(-A_{ij}x_ix_j\big),$$
$$\phi_i(x_i) = \exp\big(-\frac{1}{2}A_{ii}x_i^2 + h_ix_i\big).$$

Since factors are pairwise, messages can be seen as sent directly from one variable node i to another j with a Gaussian form:

$$m_{i\to j}(x_j) = \exp\Big(-\frac{(x_j-\mu_{i\to j})^2}{2\sigma_{ij}}\Big).$$

This expression is also stable w.r.t. the message update rules. Information is then propagated via the 2-component real vector (x_{ij},σ_{ij}) with the following update rules:

$$\mu_{i\to j} \longleftarrow \tfrac{1}{A_{ij}}\Big(h_i + \textstyle\sum_{k\in\partial i\backslash j}\frac{\mu_{k\to i}}{\sigma_{k\to i}}\Big),$$
$$\sigma_{i\to j} \longleftarrow -\tfrac{1}{A_{ij}^2}\big[A_{ii} + \textstyle\sum_{k\in\partial i\backslash j}\sigma_{k\to i}^{-1}\big].$$

At convergence the belief takes the form

$$b_i(x) = \sqrt{\frac{\sigma_i}{2\pi}}\exp\big(-\frac{(x-\mu_i)^2}{2\sigma_i}\big)$$

with

$$\mu_i = \sigma_i\big(h_i + \sum_{j\in\partial i}\frac{\mu_{j\to i}}{\sigma_{j\to i}}\big)$$
$$\sigma_i^{-1} = A_{ii} + \sum_{j\in\partial i}\sigma_{j\to i}^{-1}$$

and the estimated covariance between x_i and x_j reads

$$\sigma_{ij} = \frac{1}{A_{ij}(1 - A_{ij}^2 \sigma_{i\to j}\sigma_{j\to i})}.$$

In this case, there is only one fixed point even on a loopy graph, not necessarily stable, but if convergence occurs, the single variables beliefs provide the exact marginals [31]. In fact, for continuous variables, the Gaussian distribution is the only one compatible with the BP rules. Expectation propagation [23] is a way to address more general distributions in an approximate manner.

10.3 The Inverse Ising Problem

Once the underlying joint probability measure is given, this algorithm can be very efficient for inferring hidden variables, but in real applications it is often the case that we have first to build the model from historical data. From now on we assume that we have binary variables. Let $\{\hat{x}_i^j, i \in \mathscr{V}_j^\star, j = 1\dots M\}$ be a set of observations where M represents the number of distinct, but possibly sparse, observations of the system as a whole and $\mathscr{V}_j^\star$ is the set of nodes observed for the jth observation. We can define an empirical measure based on these historical data as

$$\hat{\mathscr{P}}(\mathbf{x}) = \frac{1}{M}\sum_{j=1}^{M} \frac{1}{2^{M-|\mathscr{V}_j^\star|}} \prod_{i\in\mathscr{V}_j^\star} \mathbb{1}_{\{x_i=\hat{x}_i\}}.$$

As such this measure is of no use for inference and we have to make some hypothesis to find a suitable inference model. There are of course various possibilities, but a simple one is to consider that the mean and the covariance are given for respectively each variable i and each pair of variable (i,j):

$$\hat{m}_i \stackrel{\text{def}}{=} \frac{1}{\sum_j \mathbb{1}_{\{i\in\mathscr{V}_j^\star\}}} \sum_{j,\mathscr{V}_j^\star \ni i} \hat{x}_i^j,$$

$$\hat{\chi}_{ij} \stackrel{\text{def}}{=} \frac{1}{\sum_k \mathbb{1}_{\{(i,j)\subset\mathscr{V}_k^\star\}}} \sum_{k,\mathscr{V}_k^\star \supset (i,j)} \hat{x}_i^k \hat{x}_j^k - \hat{m}_i\hat{m}_j.$$

Let us introduce also the notation for the joint expectation of pairs of spins:

$$\hat{m}_{ij} \stackrel{\text{def}}{=} \hat{\mathbb{E}}(s_i s_j) = \hat{\chi}_{ij} + \hat{m}_i\hat{m}_j.$$

In this case from Jayne's maximum entropy principle [15], imposing these moments to the joint distribution leads to a model pertaining to the exponential family, that is, an Ising model for binary variables ($s_i \stackrel{\text{def}}{=} 2x_i - 1$):

$$\mathscr{P}(\mathbf{s}) = \frac{1}{Z[\mathbf{J},\mathbf{h}]} \exp\Big(\sum_i h_i s_i + \sum_{i,j} J_{ij} s_i s_j\Big),$$

where the local fields $\mathbf{h} = \{h_i\}$ and the coupling constants $\mathbf{J} = \{J_{ij}\}$ are the Lagrange multipliers associated respectively to mean and covariance constraints. They are obtained as minimizers of the dual optimization problem, namely,

$$(\mathbf{h}^\star, \mathbf{J}^\star) = argmin_{(\mathbf{h},\mathbf{J})} \log Z[\mathbf{h}, \mathbf{J}] - \sum_i h_i \hat{m}_i - \sum_{ij} J_{ij} \hat{m}_{ij}, \tag{10.4}$$

which correspond to invert the linear response equations:

$$\frac{\partial \log Z}{\partial h_i}[\mathbf{h}, \mathbf{J}] = \hat{m}_i, \tag{10.5}$$

$$\frac{\partial \log Z}{\partial J_{ij}}[\mathbf{h}, \mathbf{J}] = \hat{m}_{ij}, \tag{10.6}$$

since $\hat{m}_i$ and $\hat{m}_{ij}$ are given as input to the model. As noted, e.g., in [3], the solution is minimizing the cross entropy, a Kullback-Leibler distance between the empirical distribution based on observation and the Ising model:

$$D_{KL}[\hat{\mathscr{P}} \| \mathscr{P}] = \log Z[\mathbf{h}, \mathbf{J}] - \sum_i h_i \hat{m}_i - \sum_{i<j} J_{ij} \hat{m}_{ij} - S(\hat{\mathscr{P}}).$$

The set of Eq. (10.5, 10.6) cannot be solved exactly in general because the computational cost of Z is exponential. Approximation resorting to various mean-field methods can be used to evaluate $Z[\mathbf{h}, \mathbf{J}]$.

- A common approach is based on the Plefka expansion [26] of the Gibbs free energy by making the assumption that the J_{ij} are small. The picture is then of a weakly correlated unimodal probability measure. For example, the recent approach proposed in [3] is based on this assumption.
- A second possibility is to assume that relevant coupling J_{ij} have locally a treelike structure. The Bethe approximation mentioned in the previous section is then used with possibly loop corrections. Again this corresponds to having a weakly correlated unimodal probability measure and these kinds of approaches are referred as pseudo-moment matching methods in the literature for the reason explained in the previous section. For example the approach proposed in [17, 20, 32, 33] is based on these assumptions.
- In the case where a multimodal distribution is expected, then a model with many attraction basin is to be found and a Hopfield-like model [4, 14] is likely more relevant in this case.

10.3.1 Gibbs Free Energy

To simplify the problem it is customary to make use of the Gibbs free energy, i.e., the Legendre transform of the free energy, to impose the individual expectations $\mathbf{m} = \{\hat{m}_i\}$ for each variable:

$$G[\mathbf{m},\mathbf{J}] = \mathbf{h}^T(\mathbf{m})\mathbf{m} - F[\mathbf{h}(\mathbf{m}),\mathbf{J}]$$

(with $F[\mathbf{h},\mathbf{J}] \stackrel{\text{def}}{=} -\log Z[\mathbf{h},\mathbf{J}]$, $\mathbf{h}^T\mathbf{m}$ is the ordinary scalar product) where $\mathbf{h}(\mathbf{m})$ depends implicitly on $\mathbf{m}$ through the set of constraints

$$\frac{\partial F}{\partial h_i} = -m_i. \tag{10.7}$$

Note that by duality we have

$$\frac{\partial G}{\partial m_i} = h_i(\mathbf{m}), \tag{10.8}$$

and

$$\left[\frac{\partial^2 G}{\partial m_i \partial m_j}\right] = \left[\frac{\partial^2 F}{\partial h_i \partial h_j}\right]^{-1} = [\chi]^{-1}_{ij}, \tag{10.9}$$

i.e., the inverse susceptibility matrix. Finding a set of J_{ij} satisfying this last relation along with (10.8) yields a solution to the inverse Ising problem since the m's and χ's are given. Still a way to connect the couplings directly with the covariance matrix is given by the relation

$$\frac{\partial G}{\partial J_{ij}} = m_{ij}. \tag{10.10}$$

10.3.2 Plefka's Expansion

The Plefka expansion is used to expand the Gibbs free energy in power of the couplings J_{ij} assumed to be small. Multiplying all couplings J_{ij} by α yields the following cluster expansion:

$$G[\mathbf{m},\alpha\mathbf{J}] = \mathbf{h}^T(\mathbf{m},\alpha)\mathbf{m} - F[\mathbf{h}(\mathbf{m},\alpha),\alpha\mathbf{J}] \tag{10.11}$$

$$= G_0[\mathbf{m}] + \sum_{n=0}^{\infty} \frac{\alpha^n}{n!} G_n[\mathbf{m},\mathbf{J}], \tag{10.12}$$

where each term G_n corresponds to cluster contributions of size n in the number of links J_{ij} involved, and $\mathbf{h}(\mathbf{m},\alpha)$ depends implicitly on α in order to is always fulfilling (10.7). This precisely is the Plefka's expansion, and each term of the expansion (10.1 and 10.12) can be obtained by successive derivation of (10.11). We have

$$G_0[\mathbf{m}] = \sum_i \frac{1+m_i}{2} \log \frac{1+m_i}{2} + \frac{1-m_i}{2} \log \frac{1-m_i}{2}.$$

Letting

$$H_J = \sum_{i<j} J_{ij} s_i s_j,$$

using (10.7), the first derivative of (10.11) w.r.t α gives

$$\frac{\mathrm{d}G[\mathbf{m}, \alpha\mathbf{J}]}{\mathrm{d}\alpha} = -\mathbb{E}_\alpha\big(H_J\big),$$

while the second reads

$$\frac{\mathrm{d}^2 G[\mathbf{m}, \alpha\mathbf{J}]}{\mathrm{d}\alpha^2} = -\mathbb{E}^c_\alpha\big(H_J^2\big) - \sum_i \frac{\mathrm{d}h_i(\mathbf{m}, \alpha)}{\mathrm{d}\alpha} \mathbb{E}^c_\alpha\big(H_J s_i\big).$$

In these expressions, it is the connected part of the expectation, noted

$$\mathbb{E}^c[XY] \stackrel{\text{def}}{=} \mathbb{E}[XY] - \mathbb{E}[X]\mathbb{E}[Y],$$

which appears when deriving on the free energy. To get successive derivative of $\mathbf{h}(\mathbf{m}, \alpha)$ one can use (10.8). Another possibility is to express the fact that $\mathbf{m}$ is fixed,

$$\begin{aligned} \frac{\mathrm{d}m_i}{\mathrm{d}\alpha} = 0 &= -\frac{\mathrm{d}}{\mathrm{d}\alpha} \frac{\partial F\left[\mathbf{h}(\alpha), \alpha\mathbf{J}\right]}{\partial h_i} \\ &= \sum_{i,j} h'_j(\alpha) \mathbb{E}^c_\alpha(s_i s_j) + \mathbb{E}^c_\alpha(H_J s_i), \end{aligned}$$

giving

$$h'_i(\alpha) = -\sum_j [\chi^{-1}]_{ij} \mathbb{E}^c_\alpha(H_J s_j).$$

To get the first two terms in the Plefka's expansion we need to compute these quantities at $\alpha = 0$:

$$\begin{aligned} \mathbb{E}^c\big(H_J^2\big) &= \sum_{i<k,j} J_{ij} J_{jk} m_i m_k (1 - m_j^2) + \sum_{i<j} J_{ij}^2 (1 - m_i^2 m_j^2), \\ \mathbb{E}^c\big(H_J s_i\big) &= \sum_j J_{ij} m_j (1 - m_i^2), \\ h'_i(0) &= -\sum_j J_{ij} m_j \end{aligned}$$

(by convention $J_{ii} = 0$ in these sums). The first and second orders then finally reads

$$G_1[\mathbf{m},\mathbf{J}] = -\sum_{i<j} J_{ij} m_i m_j, \qquad G_2[\mathbf{m},\mathbf{J}] = -\sum_{i<j} J_{ij}^2 (1-m_i^2)(1-m_j^2),$$

and correspond respectively to the mean field and to the TAP approximation. Higher order terms have been computed in [10].

10.3.3 Linear Response Approximate Solution

At this point we are in position to find an approximate solution to the inverse Ising problem, either by inverting Equation (10.9) or (10.10). To get a solution at a given order n in the coupling, solving (10.10) requires G at order $n+1$, while it is needed at order n in (10.9).

Taking the expression of G up to second order gives

$$\frac{\partial G}{\partial J_{ij}} = -m_i m_j - J_{ij}(1-m_i^2)(1-m_j^2)$$

and (10.10) leads directly for the basic mean-field solution to

$$J_{ij}^{MF} = \frac{\hat{\chi}_{ij}}{(1-\hat{m}_i^2)(1-\hat{m}_j^2)}.$$

At this level of approximation for G, using (10.8) we also have

$$h_i = \frac{1}{2}\log\frac{1+m_i}{1-m_i} - \sum_j J_{ij} m_j + \sum_j J_{ij}^2 m_i (1-m_j^2),$$

which correspond precisely to the TAP equations. Using now (10.9) gives

$$\frac{\partial h_i}{\partial m_j} = [\chi^{-1}]_{ij} = \delta_{ij}\Big(\frac{1}{1-m_i^2} + \sum_k J_{ik}^2 (1-m_k^2)\Big) - J_{ij} - 2J_{ij}^2 m_i m_j.$$

Ignoring the diagonal terms, the TAP solution is conveniently expressed in terms of the inverse empirical susceptibility,

$$J_{ij}^{TAP} = \frac{\sqrt{1-8\hat{m}_i\hat{m}_j[\hat{\chi}^{-1}]_{ij}} - 1}{4\hat{m}_i\hat{m}_j}, \tag{10.13}$$

where the branch corresponding to a vanishing coupling in the limit of small correlation, i.e., small $\hat{\chi}_{ij}$ and $[\hat{\chi}^{-1}]_{ij}$ for $i \neq j$, has been chosen.

10.3.4 Bethe Approximation

In this case we remark first that when the graph formed by the observed correlations $\hat{\chi}_{ij}$ is a tree, then the form (10.2) of the joint probability corresponding to the Bethe approximation yields actually an exact solution to the inverse problem (10.4):

$$\mathscr{P}(\mathbf{x}) = \prod_{i<j} \frac{\hat{p}_{ij}(x_i,x_j)}{\hat{p}(x_i)\hat{p}(x_j)} \prod_i \hat{p}_i(x_i),$$

where the $\hat{p}$ are the single and pair variables empirical marginal given by the observations. Rewriting this expression as an Ising model yields actually the following parameters:

$$h_i = \frac{1-d_i}{2} \log \frac{\hat{p}_i^1}{\hat{p}_i^0} + \frac{1}{4} \sum_{j \in i} \log\Big(\frac{\hat{p}_{ij}^{11}\,\hat{p}_{ij}^{10}}{\hat{p}_{ij}^{01}\,\hat{p}_{ij}^{00}}\Big), \tag{10.14}$$

$$J_{ij} = \frac{1}{4} \log\Big(\frac{\hat{p}_{ij}^{11}\,\hat{p}_{ij}^{00}}{\hat{p}_{ij}^{01}\,\hat{p}_{ij}^{10}}\Big), \tag{10.15}$$

while the partition function simply reads

$$Z_{Bethe}[\hat{p}] = \exp\Big(-\frac{1}{4} \sum_{ij} \log\big(\hat{p}_{ij}^{00}\,\hat{p}_{ij}^{01}\,\hat{p}_{ij}^{10}\,\hat{p}_{ij}^{11}\big) - \sum_i \frac{1-d_i}{2} \log(\hat{p}_i^0\,\hat{p}_i^1)\Big), \tag{10.16}$$

and where the $\hat{p}$'s are parametrized as

$$\hat{p}_i^{\tau} \stackrel{\text{def}}{=} \hat{p}(x_i = \tau) = \frac{1}{2}\big(1 + m_i(2\tau - 1)\big), \tag{10.17}$$

$$\begin{aligned} \hat{p}_{ij}^{\tau_i \tau_j} &\stackrel{\text{def}}{=} \hat{p}(x_i = \tau_i, x_j = \tau_j), \\ &= \frac{1}{4}\big(1 + m_i(2\tau_i - 1) + m_j(2\tau_j - 1) + m_{ij}(2\tau_i - 1)(2\tau_j - 1)\big) \end{aligned} \tag{10.18}$$

are the empirical frequency statistics given by the observations for $m \equiv \hat{m}$. The corresponding Gibbs free energy can then be written explicitly using (10.14, 10.15, 10.16). Concerning the linear response, we get from (10.14)

$$\begin{aligned} \frac{\partial h_i}{\partial m_j} = \Big[&\frac{1-d_i}{1-m_i^2} \\ &+ \frac{1}{16} \sum_{k \in \partial i} \Big(\big(\frac{1}{\hat{p}_{ik}^{11}} + \frac{1}{\hat{p}_{ik}^{01}}\big)\big(1 + \frac{\partial m_{ik}}{\partial m_i}\big) + \big(\frac{1}{\hat{p}_{ik}^{00}} + \frac{1}{\hat{p}_{ik}^{10}}\big)\big(1 - \frac{\partial m_{ik}}{\partial m_i}\big)\Big)\Big] \delta_{ij} \\ &+ \frac{1}{16} \Big(\big(\frac{1}{\hat{p}_{ij}^{11}} + \frac{1}{\hat{p}_{ij}^{10}}\big)\big(1 + \frac{\partial m_{ij}}{\partial m_i}\big) + \big(\frac{1}{\hat{p}_{ij}^{00}} + \frac{1}{\hat{p}_{ij}^{01}}\big)\big(1 - \frac{\partial m_{ij}}{\partial m_i}\big)\Big)\Big] \delta_{j \in \partial i}. \end{aligned}$$

Using (10.15), we can also express

$$\frac{\partial m_{ij}}{\partial m_i} = -\frac{\frac{1}{\hat{p}_{ij}^{11}} + \frac{1}{\hat{p}_{ij}^{01}} - \frac{1}{\hat{p}_{ij}^{10}} - \frac{1}{\hat{p}_{ij}^{00}}}{\frac{1}{\hat{p}_{ij}^{11}} + \frac{1}{\hat{p}_{ij}^{01}} + \frac{1}{\hat{p}_{ij}^{10}} + \frac{1}{\hat{p}_{ij}^{00}}},$$

so that with little assistance of maple, we may finally reach the expression [2]

$$[\hat{\chi}^{-1}]_{ij} = \Big[\frac{1-d_i}{1-m_i^2} + \Sigma_{k\in\partial i}\frac{1-m_k^2}{(1-m_i^2)(1-m_k^2)-\chi_{ik}^2}\Big]\delta_{ij} - \frac{\chi_{ij}}{(1-m_i^2)(1-m_k^2)-\chi_{ij}^2}\delta_{j\in\partial i} \qquad (10.19)$$

equivalent to the original one derived in [32] albeit written in a different form, more suitable to discuss the inverse Ising problem. This expression is quite paradoxical since the inverse of the $[\chi]_{ij}$ matrix, which coefficients appear on the right-hand side of this equation should coincide with the left-hand side, given as input of the inverse Ising problem. The existence of an exact solution can therefore be checked directly as a self-consistency property of the input data $\hat{\chi}_{ij}$ for a given pair (i,j) either:

- $[\hat{\chi}^{-1}]_{ij} \neq 0$, then this self-consistency relation has to hold and J_{ij} is given by (10.15) using $\chi_{ij} = \hat{\chi}_{ij}$.
- $[\hat{\chi}^{-1}]_{ij} = 0$ then $J_{ij} = 0$ while $\hat{\chi}_{ij}$ can be nonzero, because (10.15) does not hold in that case.

Finally complete consistency of the solution is checked on the diagonal elements in (10.19). If full consistency is not verified, this equation can nevertheless be used to find approximate solutions. Remark that if we restrict the set of Equation (10.19), e.g., by some thresholding procedure, in such a way that the corresponding graph is a spanning tree, then, by construction, $\chi_{ij} \equiv \hat{\chi}_{ij}$ will be solution on this restricted set of edges, simply because the BP equations are exact on a tree. The various methods proposed, for example, in [20, 33] actually correspond to different heuristics for finding approximate solutions to this set of constraints. As noted in [2], a direct way to proceed is to eliminate χ_{ij} in the equations obtained from (10.15) and (10.19):

$$\chi_{ij}^2 + 2\chi_{ij}(m_i m_j - \coth(2J_{ij})) + (1-m_i^2)(1-m_j^2) = 0,$$

$$\chi_{ij}^2 - \frac{\chi_{ij}}{[\chi^{-1}]_{ij}} - (1-m_i^2)(1-m_j^2) = 0.$$

This leads directly to

$$J_{ij}^{Bethe} = -\frac{1}{2}atanh\Big(\frac{2[\hat{\chi}^{-1}]_{ij}}{\sqrt{1+4(1-\hat{m}_i^2)(1-\hat{m}_j^2)[\hat{\chi}^{-1}]_{ij}^2} - 2\hat{m}_i\hat{m}_j[\hat{\chi}^{-1}]_{ij}}\Big). \qquad (10.20)$$

Note that J_{ij}^{Bethe} and J_{ij}^{TAP} coincide at second order in $[\hat{\chi}^{-1}]_{ij}$.

10.4 Application Context

10.4.1 Road Traffic Inference

Once the underlying joint probability measure is given, the BP algorithm can be very efficient for inferring hidden variables, but in real applications it is often the case that we have first to build the model. This is precisely the case for the application that we are considering concerning the reconstruction and prediction of road traffic conditions, typically on the secondary network from sparse observations. Existing solutions for traffic information are classically based on data coming from static sensors (magnetic loops) on main arterial roads. These devices are far too expensive to be installed everywhere on the traffic network and other sources of data have to be found. One recent solution comes from the increasing number of vehicles equipped with GPS and able to exchange data through cellular phone connections, for example, in the form of so-called floating car data (FCD). Our objective in this context is to build an inference schema adapted to these FCD, able to run in real time and adapted to large-scale road networks, of size ranging from 10^3 to 10^5 road segments. In this respect, the BP algorithm seems well suited, but the difficulty is to construct a model based on these FCD. To set an inference schema, we assume that a large amount of FCD sent by probe vehicles concerning some area of interest are continuously collected over a reasonable period of time (one year or more) such as to allow a finite fraction (a few percents) of road segments to be covered in real time. Schematically the inference method works as follows:

- Historical FCD are used to compute empirical dependencies between contiguous segments of the road network.
- These dependencies are encoded into a graphical model, which vertices are (segment,timestamps) pairs attached with a congestion state, i.e., typically CONGESTED/NOT CONGESTED.
- Congestion probabilities of segments that are unvisited or sit in the short-term future are computed with BP, conditionally to real-time data.

On the factor graph, the information is propagated both temporally and spatially. In this perspective, reconstruction and prediction are on the same footing, even though prediction is expected to be less precise than reconstruction.

10.4.2 An Ising Model for Traffic

10.4.2.1 Binary Latent State and Traffic Index

When looking at standard traffic information system, the representation of the congestion network suggests two main traffic states: uncongested (green) or congested (red) as shown in Fig. 10.2. If we take seriously this seemingly empirical represen-

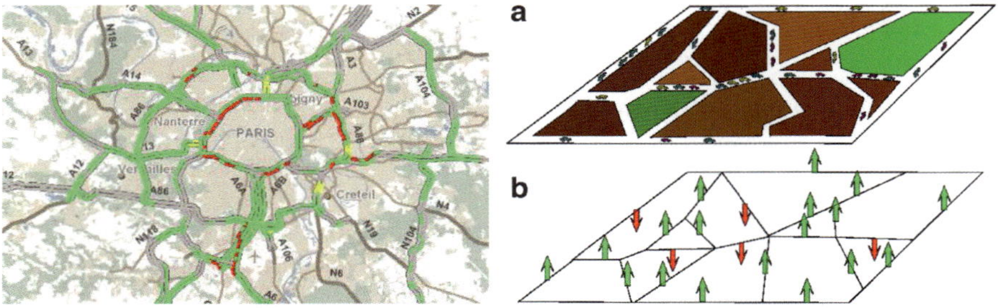

Fig. 10.2 Underlying Ising modelling of traffic configurations

tation, we are asking the question: Is it possible to encode traffic data on the basis of a binary latent state $s_{i,t} \in \{-1,1\}$ (Ising) corresponding to congested/non-congested state? As a corollary, what is the proper criteria to define the congested/uncongested state and for which purpose? In some recent work we have proposed an answer to this question [9, 19]. As said before, static sensors and probe vehicles deliver real-valued information, i.e., respectively, speed and density, and speed and travel time. For each segments, we may potentially collect a distribution $\hat{f}$ of travel time and it is not clear how to decide from this distribution, whether a link is congested or not given a newly observed travel time. A straightforward possibility is to consider the mean travel time or, even more robust, the median travel time as a separator of the two states. The way we actually see this encompasses this possibility but is not limited to it. The idea is to define the latent binary state τ $(= \frac{1+s}{2})$ associated to some travel time x in an abstract way through the mapping:

$$\Lambda(x) \stackrel{\text{def}}{=} P(\tau = 1|x).$$

This means that an observation x is translated into a conditional probability for the considered segment to be congested. This number $\Lambda(x) \in [0,1]$ represents our practical definition for the *traffic index*. Using Bayes rules and the Boolean notation $\bar{\tau} \stackrel{\text{def}}{=} 1 - \tau$, we obtain

$$P(x|\tau) = \Big(\frac{\Lambda(x)}{p_\Lambda}\tau + \frac{1-\Lambda(x)}{1-p_\Lambda}\bar{\tau}\Big)\hat{f}(x), \tag{10.21}$$

where $p_\Lambda \stackrel{\text{def}}{=} P(\tau = 1)$. The normalization constraint imposes

$$p_\Lambda = \int \Lambda(x)\hat{f}(x)\mathrm{d}x. \tag{10.22}$$

A certain amount of information can be stored in this mapping. A special case mentioned before corresponds to having for Λ a step function, i.e.,

$$\Lambda(x) = \mathbb{1}_{\{x>x^*\}}, \tag{10.23}$$

with an adjustable parameter corresponding to the threshold x^*. Another parameter free possibility is to use the empirical cumulative distribution:

$$\Lambda(x) = \hat{F}(x) \stackrel{\text{def}}{=} P(\hat{x} < x). \tag{10.24}$$

Now, given a map Λ, an obvious way to convert back a probability $u = P(\tau = 1)$ into a travel time consists then simply in using, when it exits, the inverse map:

$$\hat{x} = \Lambda^{-1}(u). \tag{10.25}$$

Actually another legitimate way to proceed is based on the conditional probability (10.21) to yield the following estimator:

$$\hat{x} = argmin_y\, \mathbb{E}\big(\|x-y\|_r\big),$$

where the expectation is taken from the probability distribution

$$P(x) = P(x|\tau = 1)u + P(x|\tau = 0)(1-u)$$

and where $\|x-y\|_r$ represents the loss function, measuring the error between the prediction x and the actual value y. In this last case, the natural requirement that we seek for Λ is that the mutual information between x and τ be maximal. This reads

$$\begin{aligned}
I(x,\tau) &\stackrel{\text{def}}{=} \sum_{\tau\in\{0,1\}} \int \mathrm{d}x P(x,\tau)\log\frac{P(x,\tau)}{P(x)P(\tau)},\\
&= \int \mathrm{d}x\big(\Lambda(x)\log\Lambda(x) + (1-\Lambda(x))\log(1-\Lambda(x))\big)\hat{f}(x) - h(p_\Lambda),\\
&= \int \mathrm{d}u\, h\big[\Lambda\big(\hat{F}^{-1}(u)\big)\big] - h(p_\Lambda),
\end{aligned}$$

after introducing the binary information function $h(x) \stackrel{\text{def}}{=} x\log x + (1-x)\log(1-x)$. In this form, h being convex, reaching its maximum at $x = 0$ and $x = 1$, its minimum at $x = 1/2$, it is then straightforward to obtain that the step function (10.23) with $x^* = \hat{F}^{-1}(1/2)$ corresponding to the median observation is the limit function which maximizes $I(x,\tau)$. If instead we use the inverse map Λ^{-1}, the mutual information between x and τ is not relevant. Without any specific hypothesis on the distribution of beliefs that BP should generate, a simple requirement is then to impose a minimum information i.e., a maximum entropy contained in the variable $u = \Lambda(x)$, in which probability density is given by

$$\begin{aligned}
\mathrm{d}F(u) &\stackrel{\text{def}}{=} \int \delta\big(u - \Lambda(x)\big)\mathrm{d}\hat{F}(x),\\
&= \frac{\mathrm{d}\hat{F}}{\mathrm{d}\Lambda}\big(\Lambda^{-1}(u)\big).
\end{aligned}$$

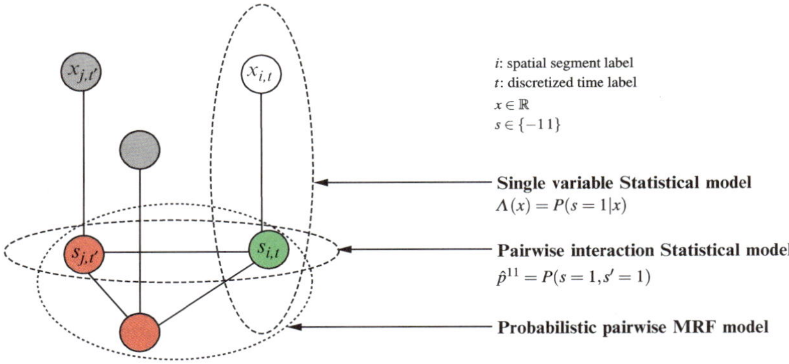

Fig. 10.3 Sketch of the Ising based inference schema

Using this and the change of variable $x = \Lambda^{-1}(u)$ yields the entropy

$$S[u] = -\int d\hat{F}(x)\frac{d\hat{F}}{d\Lambda}(x) = -D_{KL}(\hat{F}\|\Lambda)$$

expressed as the opposite of the relative entropy between F and Λ. Without any further constraint, this leads to the fact that $\Lambda = F$ is the optimal mapping. In both cases, additional constraints come from the fact that we want a predictor $\hat{x}$ minimizing a loss function $\|\hat{x} - x\|_r$ which depends on the choice of the Euclidean norm $\mathbb{L}_r$ (see [19] for details).

10.4.2.2 Global Inference Model

In fact the mapping between real-valued observations and the binary latent states is only one element of the model. The general schema of our Ising-based inference model is sketched in Fig. 10.3. It can be decomposed into four distinct pieces:

- A single variable statistical model translating real-valued observations into binary latent states
- An pairwise statistical model of the dependency between latent states
- A MRF model to encode the network of dependencies
- The belief propagation algorithm to decode a partially observed network

It is based on a statistical description of traffic data which is obtained by spatial and temporal discretization in terms of road segments i and discrete time slots t corresponding to time windows of typically a few minutes, leading to consider a set of vertices $\mathscr{V} = \{\alpha = (i,t)\}$. To each vertex is attached a microscopic degree of freedom $x_\alpha \in E$, as a descriptor of the corresponding segment state (e.g., $E = \{0,1\}$, 0 for congested and 1 for fluid). The model itself is based on historical data in form of empirical marginal distributions $\hat{p}(x_\alpha)$, $\hat{p}(x_\alpha, x_\beta)$, giving reference states and statistical interactions between degrees of freedom. Finally, reconstruction and prediction are produced in the form of conditional marginal probability distribution $p(x_\alpha|\mathscr{V}^*)$ of hidden variables in $\mathscr{V}\backslash\mathscr{V}^*$, conditionally to the actual state of the observed variables in the set $\mathscr{V}^*$.

In addition to this microscopic view, it is highly desirable to enrich the description with macroscopic variables, able, in particular, to capture and encode the temporal dynamics of the global system. These can be obtained by some linear analysis, e.g., PCA or with nonlinear methods of clustering providing possibly hierarchical structures. Once some relevant variables are identified, we can expect to have a macroscopic description of the system, which can potentially be easily coupled to the microscopic one, by adding some nodes into the factor graph. These additional degrees of freedom would be possibly interpreted in terms of global traffic indexes, associated to regions or components.

The binary latent states are used to model the interactions in a simplified way enabling for large-scale applications. Trying to model exactly the pairwise dependencies at the observation level is potentially too expensive from the statistical as well as the computational viewpoint. So the pairwise model sketched in Fig. 10.3 corresponds to

$$P(x_i, x_j) = \sum_{\tau,\tau'} \hat{p}_{ij}(\tau,\tau')P(x_i|\tau)P(x_j|\tau'),$$

with $P(x|\tau)$ given in (10.21) and $\hat{p}_{ij}$ to be determined from empirical frequency statistics. Since a probability law of two binary variables requires three independent parameters, two of them are already being given by individual marginal probabilities $\hat{p}_i^1 \stackrel{\text{def}}{=} P(\tau_i = 1)$ according to (10.22). For each pair of variables, one parameter remains therefore to be fixed. By convenience we consider the coefficient

$$p_{ij}^{11} \stackrel{\text{def}}{=} P(\tau_i = 1, \tau_j = 1)$$

and write a moment matching constraint in the traffic index space.[1] We obtain

$$\hat{p}_{ij}^{11} = \hat{p}_i^1 \hat{p}_j^1 + \frac{\widehat{\text{cov}}\big[\Lambda_i(x_i), \Lambda_j(x_j)\big]}{\big(2\hat{\mathbb{E}}[\Lambda_i(x)] - 1\big)\big(2\hat{\mathbb{E}}[\Lambda_j(x)] - 1\big)},$$

involving the empirical expectation of indexes $\hat{\mathbb{E}}[\Lambda_i(x)]$ and empirical covariance between indexes $\widehat{\text{cov}}\big[\Lambda_i(x_i), \Lambda_j(x_j)\big]$ obtained from observation data.

[1]Potentially any arbitrary mapping $\phi(x)$ could be considered to perform the moment matching.

10.4.3 MRF Model and Pseudo Moment Matching Calibration

At the microscopic level, the next step is to define the MRF, i.e., the Ising model, on which to run BP with good inference properties. Recall that we try to answer two related questions:

- Given the set of coefficients $\hat{p}(\tau_{i,t})$ and $\hat{p}(\tau_{i,t},\tau_{j,t})$, considered now as model input, what is the joint law $P(\{\tau_{i,t},(i,t)\in\mathscr{V}\})$?
- Given actual observations $\{x^*_{i,t},(i,t)\in\mathscr{V}^*\}$, how to infer $\{x_{i,t},(i,t)\in\mathscr{V}\backslash\mathscr{V}^*\}$?

The solution that we have been exploring [9] is based on the Bethe approximation described in Sect. 10.4.2. It consists to use the Bethe approximation (10.2) for the encoding and the belief propagation for the decoding, such that the calibration of the model is coherent with the inference algorithm. In particular, when there is no real-time observation, the reference point is given by the set of historical beliefs, so we expect that running BP on our MRF delivers precisely these beliefs. Stated differently, we look for the ϕ and ψ defining the MRF in (10.1) such that the beliefs match the historical marginals:

$$b_i(\tau_i)=\hat{p}_i(\tau_i) \qquad \text{and} \qquad b_{ij}(\tau_i,\tau_j)=\hat{p}_{ij}(\tau_i,\tau_j).$$

As explained in Sect. 10.3 there is an explicit solution to this problem, because BP is coherent with the Bethe approximation [34], and thus any BP fixed point $\mathbf{b}$ has to verify

$$\mathscr{P}(\tau)=\prod_{i\in\mathscr{V}}\phi_i(\tau_i)\prod_{(i,j)\in\mathscr{F}}\psi_{ij}(\tau_i,\tau_j)\propto\prod_{i\in\mathscr{V}}b_i(\tau_i)\prod_{(i,j)\in\mathscr{F}}\frac{b_{ij}(\tau_i,\tau_j)}{b_i(\tau_i)b_i(\tau_j)}. \tag{10.26}$$

As a result, a *canonical choice* for the functions ϕ and ψ is simply

$$\phi_i(\tau_i)=\hat{p}_i(\tau_i), \qquad \psi_{ij}(\tau_i,\tau_j)=\frac{\hat{p}_{ij}(\tau_i,\tau_j)}{\hat{p}_i(\tau_i)\hat{p}_j(\tau_j)}, \tag{10.27}$$

along with $m_{i\to j}(x_j)\equiv 1$ as a particular BP fixed point. In addition, from the re-parametrization property of BP [29], any other choices verifying (10.26) produce the same set of fixed points with the same convergence properties. Note that more advanced methods than the strict use of the Bethe approximation, presented in the preceding section, could be used as well, but as we shall see in the next sections, the hypothesis that traffic data could be well represented by one single BP fixed point might not be fulfilled. In that case the linear response, which takes a BP fixed point as a reference starting point, might be of limited efficiency. Instead of trying to use more accurate version of the linear response, we have followed a different route, by enriching the Bethe approximation with an adjustable parameter, interpreted as an inverse temperature, in order to better calibrate the model in a multiple BP fixed point context. This will be explained in the next section.

Next, for the decoding part, inserting information in real time in the model is done as follows. In practice, observations are in the form of real numbers like speed or travel time. One possibility is to project such an observation onto the binary state $\tau_i = 0$ or $\tau_i = 1$, but this proves to be too crude. As explained in Sect. 10.4.2, since the output of BP is anyway in the form of beliefs, i.e., real numbers in $[0,1]$, the idea is to exploit the full information by defining a correspondence between observations x_i and probabilities $p^*(\tau_i = 1)$. The optimal way of inserting this quantity into the BP equations is obtained variationally by imposing the additional constraint $b_i(\tau_i) = p^*(\tau_i)$, which results in modified messages sent from $i \in \mathscr{V}^*$, now reading [7]

$$n_{i\to j}(x_i) = \frac{p_i^*(\tau_i)}{m_{j\to i}(\tau_i)}.$$

This results in a new version of BP in which convergence properties have been analyzed in [19]. This works well in practice, in particular when compared to some heuristic method consisting in giving a bias to the local field of the observed variables as shown in Fig. 10.7 discussed in the last Sect. 10.6.

10.5 Multiple BP Fixed Points for Multiple Traffic Patterns

Some experiments with a preliminary version of this procedure [9] indicate that many BP fixed point can exist in absence of information, each one corresponding to some congestion pattern, e.g., congestion/free flow. We have analyzed in [8] the presence of multiple fixed points by looking at a study case, and we outline some of the results in this section. In this study, we considered a generative hidden model of traffic in the form of a probabilistic mixture, with each component having a simple product form:

$$P_{\text{hidden}}(\tau) \stackrel{\text{def}}{=} \frac{1}{C} \sum_{c=1}^{C} \prod_{i\in\mathscr{V}} p_i^c(\tau_i). \tag{10.28}$$

C represents the number of mixture components. Although (10.28) is quite general, the tests are conducted with $C \ll N$, with well-separated components of the mixture. The single-site probabilities $p_i^c \stackrel{\text{def}}{=} p_i^c(1)$, corresponding to each component c, are generated randomly as i.i.d. variables

$$p_i^c = \frac{1}{2}(1 + \tanh h_i^c)$$

with h_i^c uniformly distributed in some fixed interval $[-h_{max}, +h_{max}]$. The mean of p_i^c is therefore $1/2$ and its variance reads

$$v \stackrel{\text{def}}{=} \frac{1}{4}\mathbb{E}_h\big(\tanh^2(h)\big) \in [0, 1/4].$$

This parameter ν implicitly fixed by h_{max} fixes the average level of "polarizability" of the variables in each cluster: $\nu = 0$ corresponds to $p_i^c = 1/2$ while $\nu = 1/4$ corresponds to $p_i^c \in \{0,1\}$ with even probabilities. The interpretation of this model is that traffic congestion is organized in various patterns, which can show up at different times. We then studied the behavior of our inference model on the data generated by this hidden probability by adding a single parameter α into its definition (10.27):

$$\phi_i(\tau_i) = \hat{p}_i(\tau_i), \quad \psi_{ij}(\tau_i,\tau_j) = \Big(\frac{\hat{p}_{ij}(\tau_i,\tau_j)}{\hat{p}_i(\tau_i)\hat{p}_j(\tau_j)}\Big)^{\alpha}, \tag{10.29}$$

where $\hat{p}_i$ and $\hat{p}_{ij}$ are again the $1-$ and $2-$ variable frequency statistics that constitute the input of the model, while (10.28) is assumed to be unknown. This parameter α, which can be interpreted as an inverse temperature in the Ising model, is there to compensate for saturation effects ($\alpha < 1$), when the coupling between variables is too large. This is due to some over-counting of the dependencies between variables which may occur in a multiply connected graph. Still, in complement, some sparsity can be imposed to the factor graph with help of some link selection procedure that reduces the mean connectivity to K.

The typical numerical experiment we perform, given a configuration randomly sampled from (10.28), is to reveal gradually the variables $\tau_{\mathscr{V}^*}$ in a random order and compute conditional predictions for the remaining unknown variables. We then compare the beliefs obtained with the true conditional marginal probabilities $P(\tau_i = \tau|\tau_{\mathscr{V}^*})$ computed with (10.28), using an error measure based on the Kullback-Leibler distance:

$$D_{\text{KL}} \stackrel{\text{def}}{=} \Big\langle \sum_{\tau\in\{0,1\}} b_i(\tau)\log\frac{b_i(\tau)}{P(\tau_i = \tau|\tau_{\mathscr{V}^*})}\Big\rangle_{\mathscr{V}^*},$$

where $<>_{\mathscr{V}^*}$ mean an average taken on the set of hidden variables.

A sample test shown in Fig. 10.4b indicates, for example, that, on a system with 10^3 variables, it is possible with our model to infer with good precision a mixture of 20 components by observing 5 % of the variables. To interpret these results, letting $s_i = 2\tau_i - 1$, we first identify the Ising model corresponding to the MRF given by (10.29):

$$\mathscr{P}(\mathbf{s}) = \frac{1}{Z}e^{-\beta H[\mathbf{s}]},$$

with an inverse temperature β and the Hamiltonian

$$H[\mathbf{s}] \stackrel{\text{def}}{=} -\frac{1}{2}\sum_{i,j} J_{ij}s_is_j - \sum_i h_is_i.$$

Fig. 10.4 (**a**) Phase diagram of the Hopfield model and optimal points found experimentally. (**b**) D_{KL} error as a function of observed variables ρ for the single-parameter model with $N = 1000$ and $C = 20$ and various pruning levels and for the multiparameter model $N = 100$ $C = 5$ with various number of calibrated parameters ranging from 1 to 30 (**c**)

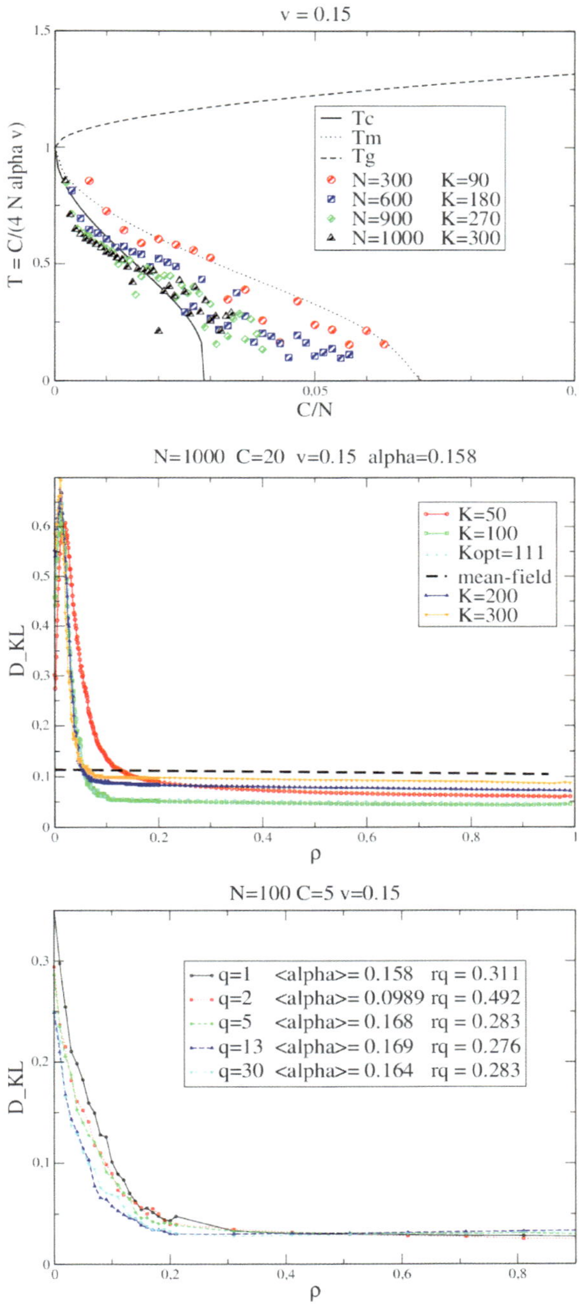

The identification reads:

$$\beta J_{ij} = \frac{\alpha}{4}\log\frac{\hat{p}_{ij}(1,1)\hat{p}_{ij}(0,0)}{\hat{p}_{ij}(0,1)\hat{p}_{ij}(1,0)},$$

$$\beta h_i = \frac{1-\alpha K_i}{2}\log\frac{\hat{p}_i(1)}{\hat{p}_i(0)} + \frac{\alpha}{4}\sum_{j\in i}\log\frac{\hat{p}_{ij}(1,1)\hat{p}_{ij}(1,0)}{\hat{p}_{ij}(0,1)\hat{p}_{ij}(0,0)}.$$

Then, in the limit $C \gg 1$, $N \gg C$ and fixed average connectivity K, we get asymptotically a mapping to the Hopfield model [14]. The relevant parameters in this limit are $\eta = C/N$ and the variance $v \in [0,1/4]$ of the variable bias in the components. In this limit, the Hamiltonian is indeed similar to the one governing the dynamics of the Hopfield neural network model:

$$H[\mathbf{s}] = -\frac{1}{2N}\sum_{i,j,c}\xi_i^c\xi_j^c s_i s_j - \sum_{i,c} h_i^c \xi_i^c s_i,$$

$$\text{with}\quad \xi_i^c \stackrel{\text{def}}{=} \frac{p_i^c(1)-\frac{1}{2}}{\sqrt{v}} \quad\text{and}\quad h_i^c = \frac{C}{2\alpha K\sqrt{v}} - \frac{2C\sqrt{v}}{K}\sum_{j\in i}\text{Cov}(\xi_i^c,\xi_j^c),$$

the inverse temperature given by the mapping reads

$$\beta = \frac{4\alpha v K}{C}.$$

Using mean-field methods, the phase diagram of this model has been established [1]. There are 3 phases, separated by the transition lines T_g, between the paramagnetic phase and the spin-glass phase, and T_c, between the spin-glass phase and the ferromagnetic phase (see Fig. 10.4a). The latter corresponds to the so-called *Mattis states*, i.e., to spin configurations correlated with one of the mixture components of direct relevance w.r.t. inference. Locating the various models obtained in this diagram as in Fig. 10.4a helps to understand whether inference is possible or not with our MRF model.

We have also tested a multiparameter version of the model in [8], where the links sorted according to the mutual information they contribute for and grouped them into a certain number of quartiles: to each quartile q we associated a parameter $\alpha_q < 1$. Using a calibration procedure based on a stochastic optimization algorithm CMAES [12], we can see on some examples a significant improvement of the model, as seen on the example presented in Fig. 10.4c.

To summarize, the main lessons of this theoretical study are the following:

- The various components of a probabilistic mixture with weak internal correlations maybe correctly accounted for by our inference model. It is able to to associate in an unsupervised way one BP fixed point to each component of the mixture.

- The mechanism for that can be understood by some asymptotic analysis which reveals a connection with a Hopfield model, where the main patterns corresponds to the components of the mixture. The phase diagram of the Hopfield model gives then relevant indications on whether inference will be easy, difficult, or impossible depending on the ratio $N_{states}/N_{variables}$ and on the mean internal variance of the variables v within each state.
- The model can be easily generalized to a multiparameter version to improve its accuracy with help of a calibration based on a robust optimization strategy like the CMAES algorithm, for example.

An example of BP fixed points associated to the components mixtures is given in Fig. 10.5. Note that in this figure, the 3-d projection space corresponds to the first principal components of the travel time vectors. The set of beliefs corresponding to each fixed point is converted into travel time through the inverse mapping given in (10.25) and projected on this 3-d space. The reconstruction experiments shown in Fig. 10.5 but explained in the next section show that the model, albeit very economical as compared to the K-nearest neighbor (K-NN) predictor, is able to predict correctly real-valued hidden variables.

10.6 Experiments with Synthetic and Real Data

Using both synthetic and real data, we perform two kinds of numerical tests:

(i) Reconstruction/prediction experiments
(ii) Automatic segmentation and BP fixed point identification

In experiments of the type (i), the dataset is divided into two parts, one corresponding to the learning set, used to build the model and the other part corresponding to the test set, used to perform the tests. In the reconstruction tests traffic configuration corresponding to a single time layer is extracted from the test set; a fraction ρ of variables are chosen at random to be revealed, and while this fraction is progressively increased, travel time is inferred for the hidden one. In prediction experiments, traffic configuration corresponding to successive time layers is selected, with present time t_0 separating the time layers into two equal parts, one corresponding to past and the other to future. Observed variables are necessarily selected in the past window time; variables with time stamp $t = t0$ (present) or $t > t0$ (future) are inferred, i.e., reconstructed or predicted, respectively. In the type (ii) experiments, on one hand an automatic clustering of the data on reduced dimensional space is performed with machine learning techniques[11]. On the other hand the BP fixed points obtained at $\rho = 0$ are listed[7, 8] and compared to segmentation in the reduced dimensional space.

A first set of data has been generated with the traffic simulator "METROPOLIS" [5] for the benchmark network called Siouxfalls shown in Fig. 10.6a. An example of the automatic clustering of spatial configurations with the corresponding

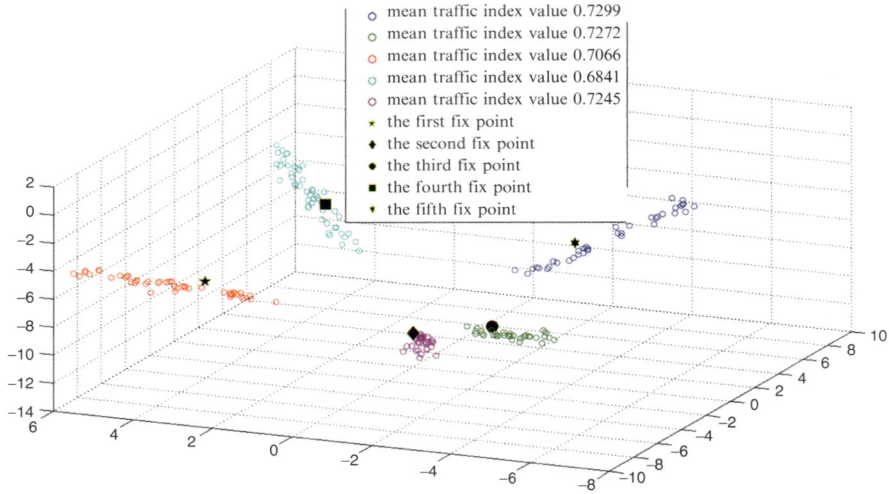

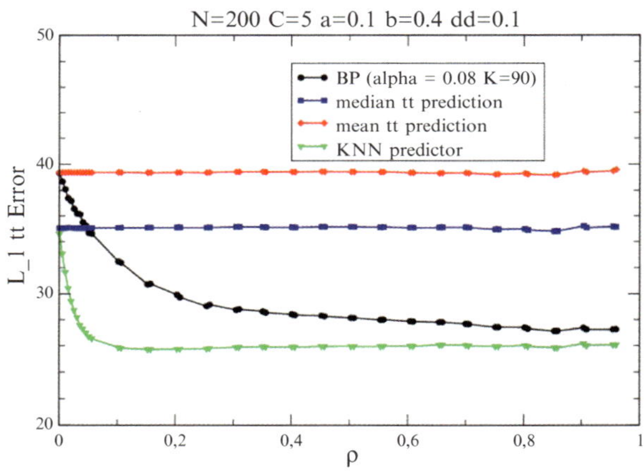

Fig. 10.5 Segmentation and BP fixed point identification for synthetic travel time data corresponding to a mixture with five components with internal correlations. Corresponding reconstruction experiment where the L_1 travel time error is plotted against the fraction ρ of observed variables and compared with a K-NN predictor considered here as ground truth

BP fixed points associated to free flow and congestion is shown in Fig. 10.6b. In Fig. 10.7a, b two ways of inserting information are compared, the variational one mentioned in Sect. 10.4.3 with a heuristic one based on local fields. In both cases the optimal tuning of α yields two BP fixed points, but the variational method yields better results on reconstruction test and is more consistent with clustering results (same optimal value of α). In Fig. 10.7c a reconstruction experiment on simulated Siouxfalls Metropolis data is shown, using the cumulative distribution for both the encoding (10.24) and inverse decoding (10.25) of traffic indexes. The performance

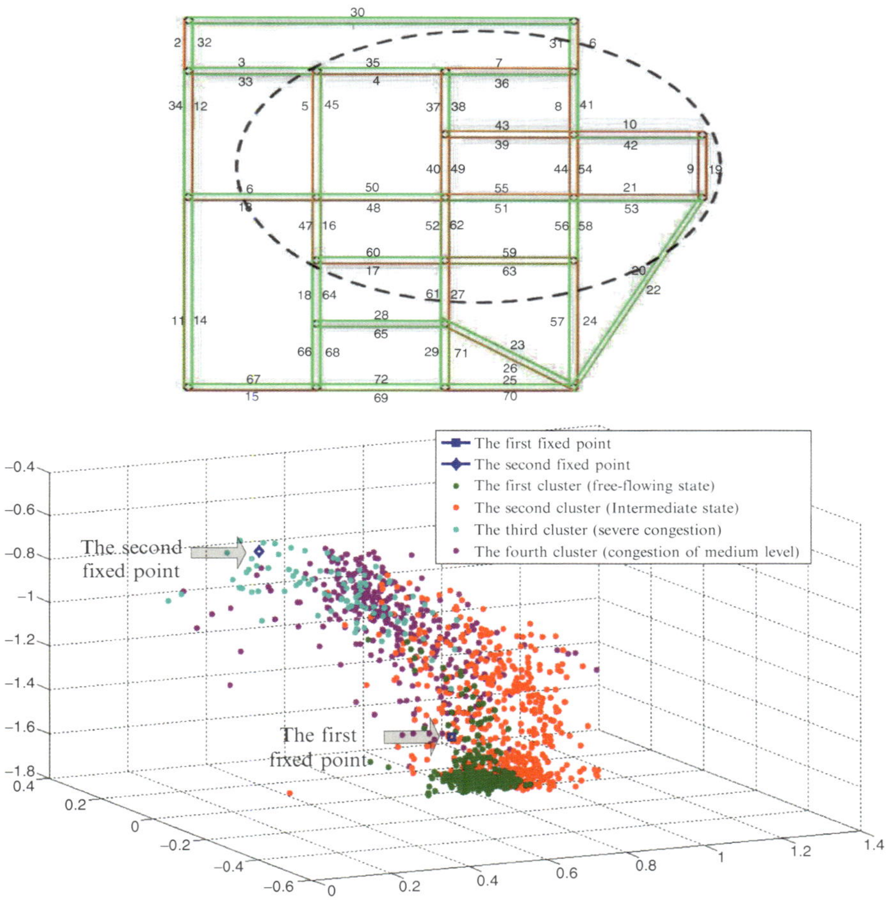

Fig. 10.6 (**a**) Siouxfalls network. (**b**) Automatic segmentation of simulated Siouxfalls data and BP fixed point identification projected in the 3-d main PCA space

is similar to a K-NN predictor, although much more economical. To set a scale of comparison predictors obtained from historical mean and median values are also shown on the same plot.

To perform tests on real data, we have also considered a dataset consisting of travel times measured every 3 min over 2 years of a highway segmented into 62 segments. We use for each segment $i = 1 \dots 62$ a weighted cumulative travel time distribution based on the automatic segmentation for the traffic index encoding. The automatic segmentation using nonnegative matrix factorization techniques [11] is displayed in 3-d Fig. 10.8a. Results of a short-term horizon prediction test are displayed on Fig. 10.8b, showing reasonable performance even though highway data do not correspond to the situation for which the model was designed.

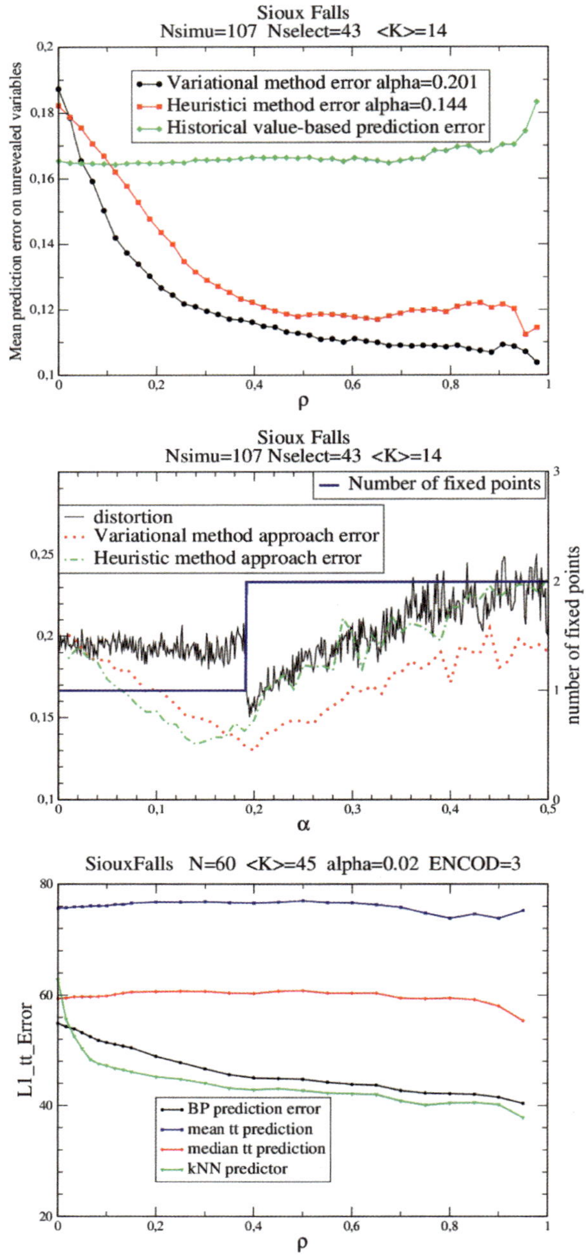

Fig. 10.7 Comparing a heuristic method and a variational one for inserting real-time information (**a**) and (**b**) for the Siouxfalls network data. (**b**) indicates the optimal value of α for traffic reconstruction is coherent with the best clustering value in the variational case. Reconstruction experiment on the same data using the mapping based on the cumulative and compared with a K-NN predictor (**c**)

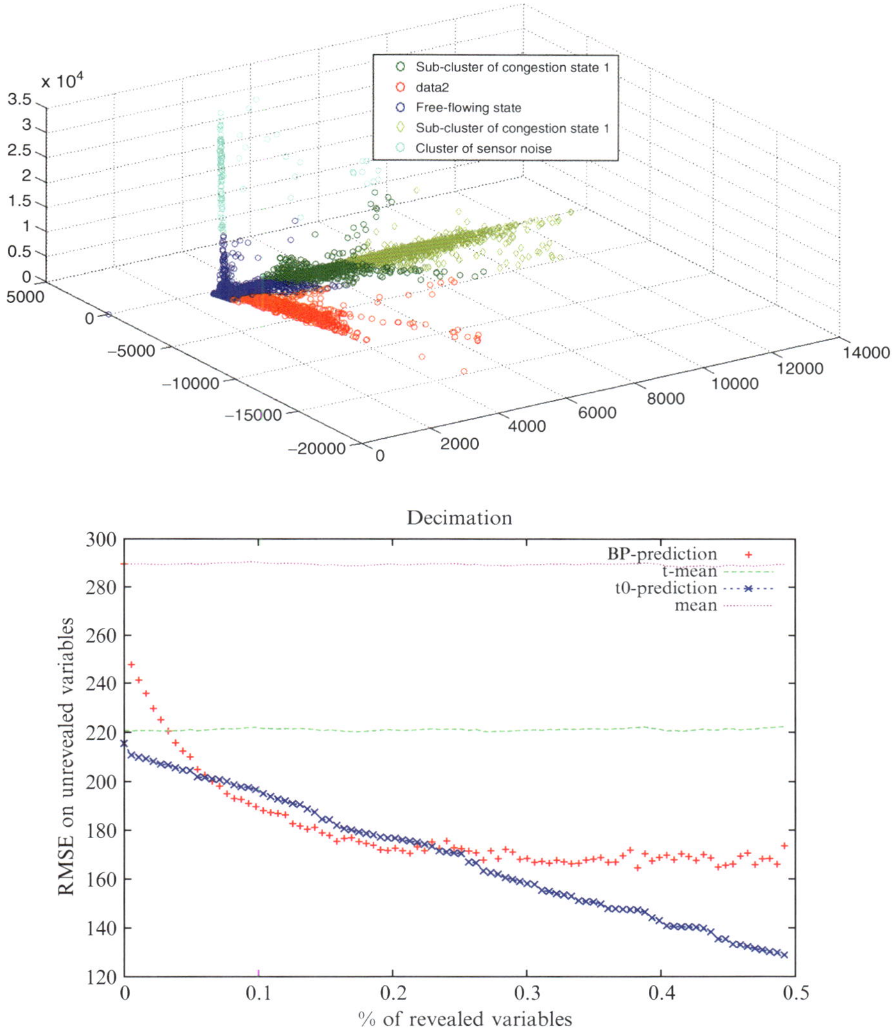

Fig. 10.8 (**a**) Automatic segmentation of highway data projected on 3-d dominant PCA space. (**b**) Error on travel time for a BP prediction of three time layers in future as a function of the fraction of observed variables at t_0. Comparison is made with a predictor combining recent available observations with historical time-dependent mean

10.7 Conclusion

The work concerning the application of belief propagation and related Boltzmann machine to traffic data is related to some ongoing projects [27, 28]. It is based on mean-field concepts in physics and basically related to the linear response theory.

When combined with machine learning techniques like the automatic segmentation methods it can lead to efficient models able to cope with real-time constraints on large-scale networks. We advocate for an Ising model for traffic statistical modelling and propose a proper way for defining traffic indexes which could be also useful for traffic management systems. Still a natural concurrent approach not exposed here can be built analogously using a multivariate model, for which Gaussian belief propagation would apply. More real data will help to decipher from these two possibilities. The main hypothesis underlying our Ising-based approach assumes that traffic congestion is well represented by multiple distant pattern superposition. This needs validation with real data on networks. Our reconstruction schema seems to work already with simple underlying binary indexes, but more work is needed for the dynamical part to be able to perform prediction.

Acknowledgment This gives me the occasion to express my warm thanks to my colleagues Victorin Martin and Jean-Marc Lasgouttes with whom it is a pleasure to collaborate on the main subjects discussed in this review. I am also grateful to Anne Auger, Yufei Han, Fabrice Marchal and Fabien Moutarde, for many aspects mentioned in this work concerning ongoing projects. This work was supported by the grant ANR-08-SYSC-017 from the French National Research Agency.

References

1. D.J. Amit, H. Gutfreund, H. Sompolinsky, Statistical mechanics of neural networks near saturation. Ann. Phys. **173**(1), 30–67 (1987)
2. H. Chau Nguyen, J. Berg, Bethe-peierls approximation and the inverse ising model. ArXiv e-prints, 1112.3501 (2011)
3. S.Cocco, R. Monasson, Adaptive cluster expansion for the inverse Ising problem: convergence, algorithm and tests. arXiv:1110.5416, 2011
4. S. Cocco, R. Monasson, V. Sessak, High-dimensional inference with the generalized hopfield model: Principal component analysis and corrections. Phys. Rev. E **83**, 051123 (2011)
5. A. de Palma, F. Marchal, Real cases applications of the fully dynamic METROPOLIS tool-box: an advocacy for large-scale mesoscopic transportation systems. Networks Spatial Econ. **2**(4), 347–369 (2002)
6. B. Frey, D. Dueck, Clustering by passing messages between data points. Science **315**, 972–976 (2007)
7. C. Furtlehner, Y. Han, J.-M. Lasgouttes, V. Martin, F. Marchal, F. Moutarde, Spatial and temporal analysis of traffic states on large scale networks. In *Intelligent Transportation Systems (ITSC), 2010 13th International IEEE Conference on*, pp. 1215 –1220, 2010
8. C. Furtlehner, J.-M. Lasgouttes, A. Auger, Learning multiple belief propagation fixed points for real time inference. Physica A: Stat. Mech. Appl. **389**(1), 149–163 (2010)
9. C. Furtlehner, J.-M. Lasgouttes, A. de La Fortelle, A belief propagation approach to traffic prediction using probe vehicles. In *Proceedings IEEE 10th Intelligent Conference Intelligent Transport System*, pp. 1022–1027, 2007
10. A. Georges, J. Yedidia, How to expand around mean-field theory using high-temperature expansions. J. Phys. A: Math. Gen. **24**(9), 2173 (1991) .
11. Y. Han, F. Moutarde, Analysis of Network-level Traffic States using Locality Preservative Non-negative Matrix Factorization. In *Proceedings of ITSC*, 2011
12. N. Hansen, A. Ostermeier, Completely derandomized self-adaptation in evolution strategies. Evol. Comput. **9**(2), 159–195 (2001)

13. T. Heskes, On the uniqueness of loopy belief propagation fixed points. Neural Comput. **16**, 2379–2413 (2004)
14. J.J. Hopfield, Neural network and physical systems with emergent collective computational abilities. Proc. Natl. Acad. Sci. USA **79**, 2554–2558 (1982)
15. E.T. Jaynes, *Probability Theory: The Logic of Science (Vol 1)* (Cambridge University Press, Cambridge, 2003)
16. Y. Kabashima, D. Saad, Belief propagation vs. tap for decoding corrupted messages. Europhys. Lett. **44**, 668 (1998)
17. H. Kappen, F. Rodrguez, Efficient learning in boltzmann machines using linear response theory. Neural Comput. **10**(5), 1137–1156 (1998)
18. F.R. Kschischang, B.J. Frey, H.A. Loeliger, Factor graphs and the sum-product algorithm. IEEE Trans. Inf. Th. **47**(2), 498–519 (2001)
19. V. Martin, Modélisation probabiliste et inférence par l'algorithme Belief Propagation, Thèse de doctorat, Ecole des Mines de Paris, 2013
20. M. Mezard, T. Mora, Constraint satisfaction problems and neural networks: A statistical physics perspective. J. Physiology-Paris **103**(1–2), 107–113 (2009)
21. M. Mézard, G. Parisi, M. Virasoro, *Spin Glass Theory and Beyond* (World Scientific, Singapore, 1987)
22. M. Mézard, R. Zecchina, The random K-satisfiability problem: from an analytic solution to an efficient algorithm. Phys. Rev. E **66**, 56126 (2002)
23. T. Minka, Expectation propagation for approximate bayesian inference. In *Proceedings UAI*, pp. 362–369, 2001
24. J.M. Mooij, H.J. Kappen, On the properties of the Bethe approximation and loopy belief propagation on binary network. J. Stat. Mech. P11012 (2005)
25. J. Pearl, *Probabilistic Reasoning in Intelligent Systems: Network of Plausible Inference* (Morgan Kaufmann, San Mateo, 1988)
26. T. Plefka, Convergence condition of the tap equation for the infinite-ranged ising spin glass model. J. Phys. A: Math. Gen. **15**(6), (1971, 1982),
27. PUMAS project, (2010–2013). http://pumas.inria.fr/public/document.
28. TRAVESTI project, (2009–2012). http://travesti.gforge.inria.fr/.
29. M.J. Wainwright, *Stochastic processes on graphs with cycles: geometric and variational approaches*. PhD thesis, MIT, 2002
30. Y. Watanabe, K. Fukumizu, Graph zeta function in the bethe free energy and loopy belief propagation. In *Advances in Neural Information Processing Systems*, vol. 22, pp. 2017–202, 2009
31. Y. Weiss, W.T. Freeman, Correctness of belief propagation in gaussian graphical models of arbitrary topology. Neural Comput. **13**(10), 2173–2200 (2001)
32. M. Welling, Y.W. Teh, Approximate inference in boltzmann machines. Artif. Intell. **143**(1), 19–50 (2003)
33. M. Yasuda, K. Tanaka, Approximate learning algorithm in boltzmann machines. Neural Comput. **21**, 3130–3178 (2009)
34. J.S. Yedidia, W.T. Freeman, Y. Weiss, Generalized belief propagation. Adv. Neural Inform. Process. Syst. **13**, 689–695 (2001)

Index

X. Leoncini and M. Leonetti (eds.), *From Hamiltonian Chaos to Complex Systems: A Nonlinear Physics Approach*, Nonlinear Systems and Complexity 5,
DOI 10.1007/978-1-4614-6962-9,

Printed by Printforce, the Netherlands